BRITAIN'S RARE FLOWERS

The Botanists by Joseph Southall (1928) shows two fashionably dressed ladies, one searching on her knees for a wild flower while the other admires the view of the River Fowey in Cornwall. Their viewpoint is accurately painted and can be identified: and it is, indeed, botanically rich. They may have been looking for scarce clovers such as Fenugreek (*Trifolium ornithopodioides*), which grow there. [Watercolour on silk, Hereford City Museums.]

BRITAIN'S RARE FLOWERS

Peter Marren

T & A D
POYSER
NATURAL
HISTORY

Published in association with
PLANTLIFE AND ENGLISH NATURE

First published in 1999
by T & A D Poyser Ltd
24-28 Oval Road, London NW1 7DX

This book is printed on acid-free paper

Text set in Sabon
Design by Peter Champion
Printed and bound in Spain by Mateu Cromo, S.A. Pinto (Madrid)

A CIP record of this book is available
from the British Library

ISBN 0-85661-114-X

Contents

To Jane
who made me do it

An old woman who lived on a Yorkshire moor told me that one day a car drew up at her cottage and the occupants asked to be directed to a certain spot on the moor. A few days later another car arrived and there was the same request. Car followed car for a month or more. Next summer the same thing happened. 'What did they want?' she asked plaintively. I explained that no doubt a rare plant had been discovered on the moor. I think she scarcely understood, for she shook her head with 'It's very queer'.

Andrew Young, 'Botanists and Botanophils' in
A Prospect of Flowers (1945)

Acknowledgements

It is a great pleasure to thank all the friends and colleagues who have helped me write this book. They are living embodiments of the 400-year-old tradition among British botanists for generosity, hospitality and good companionship – the snippets and anecdotes they shared with me would fill another volume the size of this one. I have pleasant memories of days spent in the field with Andrew Branson, Jo Dunn, Sue Everett, Ro FitzGerald, Marc Hampton, Mark and Clare Kitchen, Camilla Lambrick, Yvonne Leonard, Liz McDonnell, David Pearman, Derek Ratcliffe, Tim Rich, James and Joanna Robertson, Tony Robinson, Francis Rose, Jane Smart, Jonathan Spencer and Phil Wilson. I also thank the many people who have helped with information and in other ways, notably Francis Abraham, John Akeroyd, Steve Alton, Arthur Chater, Clive Chatters, Rob Cooke, Jon Cox, Reg Crossley, Ruth Davis, Dave Green, Lynne Farrell, John Finnie, Michael Greenhill, Barbara Jones, Roger Key, Miles King, Phil Lusby, Richard Mabey, Alison Macdonald, Len Margetts, Roy Maycock, Rose Murphy, David Northcote-White, Philip Oswald, Chris Page, George Peterken, Rosemary Parslow, Ron Porley, Hugh Raven, Steven Rotheroe, Martin Sanford, Michael Scott, Richard Seamons, Alan Showler, Muriel Smith, Christopher Smout, Nick Stewart, Bernard Thompson, Derek Wells and Martin Wigginton. I also wish to thank the Botanical Society of the British Isles as a body, whose published journals and newsletters I have heavily mined.

I thank the librarians and staff of the Druce-Fielding herbarium at the University of Oxford and the herbarium at the University of Reading, and the following institutions and trusts: the British Pteridological Society, the Countryside Council for Wales, English Nature, the Joint Nature Conservation Committee, the National Trust, the National Trust for Scotland, the Royal Botanic Gardens at Kew and Edinburgh, Scottish

Natural Heritage, the Scottish Wildlife Trust and the following county Wildlife Trusts: Berks, Bucks and Oxon, Dorset, Essex, Hampshire, Kent, Northamptonshire, Somerset, Suffolk, Sussex and Wiltshire.

My particular thanks to Lady Rosemary FitzGerald who helped me research the byways and unexpected corners of the rare plant world, and to Ro, Tim Rich, David Pearman, and Simon Leach for kindly reading and commenting on the manuscript. I also thank Jane Smart and the staff of Plantlife, and Andy Clements, Sue Ellis, Roger Mitchell, Ron Porley and Ian Taylor of English Nature for reading the draft and commenting on it from their points of view.

Many of the pictures in this book were lent to us at reduced rates or even free of charge, for which generosity the publishers and myself are extremely grateful. In particular I thank Bob Gibbons for coming to our rescue with some of the more elusive species, and also Sidney Clarke, Sue Everett, Andrew Gagg, Marc Hampton, Hugh Lang, Christopher Page, Plantlife, Derek Ratcliffe, Peter Roworth, Michael Scott and Ray Woods. For permission to reproduce artwork, we thank HarperCollins, Hereford Museum, Nick Stewart, the National Trust for Scotland, Annie Soudain and the Post Offices of Ireland, Jersey and Guernsey. Christina Hart-Davies painted three plant portraits specially for this book. The distribution maps were plotted using the DMAP program written by Dr A.H. Morton, and I thank Chris Preston of ITE for permission to include them.

To Maureen, my faithful maestro of the fast keys, renewed homage. To my publisher Andrew Richford I owe special thanks for his enthusiasm and creative flair.

The research and writing of this book was generously sponsored by English Nature and Plantlife. Roger Mitchell and Jane Smart have been the most understanding of hosts, and I hope the book repays some of their patience and constant support.

That it was possible at all was to a large part due to Jane, who inspired the idea and was its good shepherd from start to finish. To her this book is dedicated, with gratitude and affection.

A note on the text

This book is about the wild flowers and ferns of Great Britain, that is England, Scotland and Wales. I have included a few examples from the Channel Islands, the Isle of Man and from Ireland, but only where they have some relevance to the flora of Britain. Seed-bearing plants and pteridophytes (ferns and their allies, the horsetails and clubmosses) are normally treated together in the floras and that tradition is followed here. Hence 'wild flowers' is really short-hand for wild flowers, ferns, trees, shrubs, sedges and grasses. By 'wild', I mean native or long-established. The botanist Max Walters suggested that a native plant is one that arrived here independently of mankind and has been with us for at least 2000 years. A long-established plant has had an independent existence for at least 400 years. Nearly all the species of 'conservation value' are either native or long-established.

I define the word 'rare' fairly loosely. It includes, of course, all the 300-odd Red-listed species, but I have also included some 'scarce' species which are more widespread but not necessarily less rare. I also include extinct plants. The number of formerly extinct species which have been refound is quite impressive, and in any case you cannot get any rarer than extinct. Hence I deal in all with the 500 rarest wild flowers and ferns – about one third of the British flora.

For names I have used the current standard work, the second (1997) edition of Stace's *New Flora of the British Isles*. In general I give the scientific name in brackets on the first mention of a species (except where it seemed inappropriate), and thereafter use the English names. This is not a perfect arrangement, but the alternatives seemed worse. Although some of the English names are clumsy, artificial and unfamiliar, the scientific names are apt to change rather frequently, especially in groups like ferns and grasses. And a diet of scientific names alone would give the wrong impression. This book is a celebration of our rare flowers, not a science textbook.

Many of the projects mentioned in this book were funded by English

Nature, Scottish Natural Heritage, or Plantlife. For reasons of readability, I have not always said who funds what, but further details may be found in the Appendix 3. Details of the distribution and localities of our rarest native plants will be found in the *Red Data Book of Vascular Plants,* due for publication in 1999.

The text is lightly referenced, with a list of the principal works consulted at the end. All species are indexed by their English and Latin names.

I cannot, I hope for obvious reasons, answer correspondence about the exact localities of rare flowers.

PLANTLIFE

Plantlife – The Wild-Plant Conservation Charity is Britain's only national membership charity dedicated to conserving plants in their natural habitats: the champion of plants in the wild.

Plantlife helps to prevent common wild plants becoming rare, to rescue wild plants on the brink of extinction, and to protect the natural habitats of wild plants. We carry out practical conservation work, we influence relevant policy, and we shape relevant legislation.

Plantlife's membership scheme gives everyone the opportunity to become better informed about wild plants and to become actively involved in their conservation.

Britain's native wild plants are under severe pressure. Almost 300 species are threatened with severe decline, and many others are vanishing from the countryside. Wild camomile is now lost from 23 of the counties in which it once grew wild, for example, and the native fritillary is now found in an area reduced by three-quarters. Vast tracts of habitat have been destroyed since the 1930s: 98% of wild-flower meadows, 75% of open heaths, 96% of open peat bogs, and 190,000 miles of hedgerows.

Plantlife works to stop this destruction and neglect. Plantlife is harnessing the commitment of all sectors of the community in a wild plant conservation action plan, which has four main programmes:

Species conservation is targeted at specific species under threat. Research work directly informs practical action plans. The remaining populations of these species are then safe-guarded and, wherever practical, they are restored to places where they have been lost.

Habitat conservation is carried out by acquiring and managing nature reserves. Plantlife reserves include flower-rich meadows, ancient woodland, limestone pavement and chalk grassland. We also campaign to highlight threats to sites and habitat types.

Community action focuses on volunteers who help to manage habitats, and survey or monitor species in important plant sites.

Campaigning for wild plants centres on increasing awareness of wild plants and their habitats, primarily through the magazine *Plantlife,* and through public relations campaigns. We work with government to improve legal protection for wild plants and their habitats, and to develop policies that will safeguard our plant heritage. Finally, we provide guidance to local authorities, and others, in the use of appropriate native plants in planting schemes.

English Nature is the Government's statutory conservation agency, and aims to sustain and enrich the wildlife and natural features of England.

To achieve this aim, English Nature seeks to influence Government and European policies to prevent or reverse unfavourable trends in the quality of the environment. We provide high-quality advice based on sound science and practical experience, and also work to enhance the public understanding and enjoyment of England's rich and varied natural heritage.

English Nature works in partnership with others to increase the human and financial resources deployed for nature conservation, in particular to deliver the *UK Biodiversity Action Plan* species and habitat targets. We work with over 32,000 owners and land managers to achieve positive management and favourable condition of Natura 2000 sites and Sites of Special Scientific Interest. We are also responsible for the declaration and management of 200 National Nature Reserves, as examples of some of the best places for wildlife in England.

English Nature's association with Plantlife is a good example of a working partnership which enhances the capacity of both organisations to play an effective role in nature conservation. In particular, the *Back from the Brink* project with Plantlife is already implementing action on some 30 plant species of conservation concern, and work on more plants is planned. Since the partnership was formed, 13 plant species have reached initial recovery milestones through habitat management to secure existing populations and, in some cases, to establish new ones.

Both this partnership initiative and this book are funded through English Nature's *Species Recovery Programme*. This Programme began in 1991 with recovery projects on seven species of plants (and seven animals). Now, in 1999, it encompasses about 200 species, half of which are plants.

The inspiring stories of some of these species recovery projects recounted here by Peter Marren will do much to promote a greater understanding and enjoyment or rare wild flowers and so will help to reverse the decline of plant species.

Foreword

Many people find themselves seduced by the idea of rarity. We have been brought up to believe that what is rare is also precious. And precious things must be looked after and cherished. There are many compelling reasons for wishing to safeguard our rarest wild flowers; our concern for their conservation is, in essence, why we felt it so important to support this book.

Peter Marren shows us that rare plant conservation is not always as straightforward as we might think. First of all, of course, we need to understand why a particular plant is rare. Some species have *always* been rare, only ever known from a few sites that have just the right conditions for their survival. But many others have had rareness thrust upon them, like those once-common farmland plants that modern agricultural methods have all but eliminated.

For the first group, our chief task must be to protect the places in which they grow. This, as Peter is quick to point out, does *not* mean erecting fences and 'Keep Out' notices. In many cases, it requires a great deal of human intervention and management. For example, Starfruit – an always scarce plant of muddy pond margins kept open by trampling livestock – has almost disappeared along with the cattle it depended on. In such circumstances positive intervention through schemes such as Plantlife's *Back from the Brink* project and English Nature's *Species Recovery Programme* can make all the difference. Volunteers cut back the overgrowth and, with the aid of shovels and bulldozers, re-create the open muddy fringes it so badly needs. In this way Starfruit is now making a comeback at several of its old haunts. It is extraordinary the degree to which our rare wild flowers actually require *disturbance* to survive.

As for those plants that have had their rarity thrust upon them, the situation is perhaps less bleak now than it was a decade ago. Recent policies and stewardship schemes developed by the Ministry of Agriculture, with active support from the country conservation agencies and Plantlife, are encouraging farmers to change the way they manage the land, to leave hedgerows standing (and plant new ones), to keep field edges unsprayed and meadows and downland unploughed. Such initiatives offer hope to many of our not-so-common wild plants, along with other wildlife that has been devastated as a result of the post-war agricultural revolution.

One of the more encouraging messages in this book is that recovery is possible for plants. They may not move like birds – though orchid seeds can be carried miles on the wind – yet their seed can survive a long time in the soil, awaiting the moment when suitable management permits them to reappear. Thus, for example, many of our lost arable 'weeds' still present on some sites, buried in the soil seed-bank, are ready to show their heads again once agricultural policy has done its bit to create the right habitat conditions.

Peter Marren is a master story-teller and has an almost unique gift for making his subject live. After reading this book one cannot fail to appreciate that our rare wild flowers are a national treasure at least as important as any other aspect of our heritage. This is a timely and passionate book, and one deserving to be widely read.

Baroness Young of Old Scone
Chairman – English Nature

Adrian Darby, OBE
Chairman – Plantlife

Abbreviations

BAP	Biodiversity Action Plan
BEC	Botanical Exchange Club
BSBI	Botanical Society of the British Isles
CAP	Common Agricultural Policy (of the EC)
CCW	Countryside Council for Wales
CPRE	Council for the Preservation of Rural England
CTW	Clapham, Tutin and Warburg, *Flora of the British Isles*
ESA	Environmentally Sensitive Area
ITE	Institute of Terrestrial Ecology
IUCN	International Union for the Conservation of Nature and Natural Resources (now the World Conservation Union)
JNCC	Joint Nature Conservation Committee
NCC	National Conservancy Council
SNH	Scottish Natural Heritage
SPNR	Society for the Promotion of Nature Reserves
SRP	Species Recovery Programme (English Nature)
SSSI	Site of Special Scientific Interest
WWF	World Fund for Wildlife

Chapter 1

Introduction:

The allure of rare flowers

A portrait of Early Spider Orchid (*Ophrys sphegodes*) by Christina Hart-Davies. The orchid is the symbol of Dorset Wildlife Trust, and grows in some plenty above the chalk cliffs of Purbeck in early May. It is the orchid that is early, not the spider.

As far as I remember, I managed to reach the age of 18 without ever setting eyes on a rare wild flower. There was no doubt about which flowers *were* rare, because, in the *Collins Pocket Guide to Wild Flowers*, Messrs David McClintock and R.S.R. Fitter had thoughtfully picked them out for us with stars. The rarest flowers of all had three stars for maximum desirability. They were so rare that in some cases there was no picture, as though even the great McClintock and Fitter had never found them. One could only dream about what a Ghost Orchid or a Thistle Broomrape might look like: barely imaginable flowers, as elusive as the pot of gold beneath the rainbow. For myself these dreams hardened to desires after being given a copy of the Revd Keble Martin's *Concise British Flora* for my fifteenth birthday. Even Keble Martin had evidently been unable to find a Ghost Orchid, but most of the rest of the flora was there in colour, illustrated as they might appear in life, pale flowers against green leaves, one leaf overlapping the next and all crowded together as in the hedgerow. Keble Martin was even more sparing in his remarks than McClintock and Fitter, but, at least for a boy locked up in boarding school in the Midlands, there was a yearning magic in those mysterious 'calcareous pastures', 'subalpine woods' and 'sandy lake margins in the north' where the rare flowers grew. Some rarities seemed extremely choosy. One grew only 'On rocks at 3,000 feet', another 'on a moorland in Perthshire' and a third 'on a heath near Bournemouth'. And the one – not illustrated – that occurred only 'In Esthwaite Water, *without flowers*' seemed to be taking choosiness to a near-suicidal extreme.

It was hard to put into words why one wanted to find flowers (it still is). One could be technological about it: I wanted to take pictures of them with my new camera, but it was the flowers that had brought forth the camera, not the other way round. I was drawn to them in a visceral way that is beyond words. I did not want to pick them or grow them or even to paint their portrait (though I might have done if I was any good at painting). I just wanted to see them in their natural surrounding and learn a bit more about calcareous pastures, for it was the places as well as their plants that were part of the magic. I can see it in my mind's eye now, that first special place. It was on the white cliffs above St Margaret's Bay in Kent, with larks singing in a May sky and the distant thrum of a hovercraft returning from Calais. I had impressed upon my mother, with all the teenage pomposity I could muster, that the flower we were seeking was so very special that if we found it we must stay absolutely calm, otherwise someone would spot us, come over and pick it. The species in question was the Early Spider Orchid, *Ophrys sphegodes*, which McClintock and Fitter award with two stars, and Keble Martin celebrates with two drawings. We did not really expect to find it, though it had been a good morning, with Nottingham Catchfly, Wild Cabbage and other good things on the cliffs above the beach. But without much searching, there all of a sudden it was, a little group of Early Spiders perched near the cliff edge where white chalk had started to show through the turf. Forgetful of any prowling pickers, I roared with delight, waved my arms about (or so it was claimed) and fell to my knees. They were both like and unlike the pictures in the books. The latter were accurate enough in their botanical details, but gave you no real idea of the *presence* of the plant in its setting of turf, rock and sea, nor its soldierly bearing, with military greeny-yellowy-browny colours and lead-grey markings glinting in the sunshine like wet enamel paint on an Airfix model. The flowers bulged in a way that did remind you of a garden spider which had somehow got its head stuck in a flower. In short, it was everything I had hoped for and more. In retrospect, at least, finding it was a transcendent moment in which wild flowers, chalk cliffs and the colours, sounds and scents of nature seemed to imprint themselves in my bones. I have never really got over it (neither, perhaps, have you). Vladimir Nabokov had expressed it with Russian abandon, as: 'ecstasy, and behind the ecstasy is something else, which is hard to explain. It is like a momentary vacuum into which rushes all that I love. A sense of oneness with sun and stone. A thrill of gratitude to whom it may concern – to the contrapuntal genius of human fate or to tender ghosts humouring a lucky mortal' (from *Speak Memory*, the memoirs of Vladimir Nabokov).

Shortly after photographing the Spider Orchids we met up with an elderly friend who was taking a holiday among the May time orchids of Kent. He had seen more Early Spiders than I had seen exam results. I took him to see *our* little group. He looked down at them in silence for a moment, and then looked me in the eye. 'Were you a bit disappointed?' he suggested, gently.

Since that day, rare wild flowers and ferns have filled a great many hours and days of my working and leisure existence – not continually, and rarely nowadays at the white-hot intensity of youth (when, one summer, I spent every weekend travelling around Britain photographing wild orchids. Ridiculous.). But they have always been there in the mindscape, if not the landscape, and there is always the same pleasure at finding a new species, or even an old flower in a new place. The Passion takes you to places where sensible people seldom stray: rock ledges in the Highlands, staring down into the abyss; muddy off-route places in the New Forest or featureless plains in the Breck, bereft of conceivable interest to anyone except a naturalist. And along the way there are often adventures, like the day David Pearman and I searched some dizzying cliff-tops in Dorset for Stinking Goosefoot (*Chenopodium vulvaria*), before finding it in a golf bunker. I have found, and sometimes studied, rare flowers in most parts of Britain from Land's End to Dunnet Head and from the Dover cliffs to the wastes of Inverpolly. And I have helped to conserve some of the places where rare flowers grow in north-east Scotland and the Thames Valley, and have written scientific papers about this plant and that. They form a fairly constant motif in my life, seldom a dominant one, but they never go away. I had to write this book. I was hooked.

In it I have tried to share some of the fascination I find in rare flowers. It is not just about the individual flowers and ferns but about where our flora came from and how it has survived; why some species are rare and why a few have died out; and about the measures we take to protect them and why. We will spend time in some of the special places where large numbers of rarities are found, often side by side. And, inevitably, I find, writing about British flowers means writing about British botanists. Left entirely to themselves, our rarities are mute: they have science and nothing else. It is humankind that places values on them and brings them into our own history. The way we have regarded them has varied from time to time, from the 'herberizing' of apothecaries, the collection of rarities for herbariums and horticulture, to the conservation ethic of today, but always with fascination. Through it, like a continuous vein of lode, runs the long tradition of botanical recording: the perpetual curiosity of the British naturalist. It is the interplay of flowers and botanists, whether scientific, artistic, acquisitive, protective or Utopian, that is the main theme of this book. I have tried to keep the humour and drama in the eternal quest for flowers at the forefront of the story while jargon and technical science, on the other hand, stand well muffled at the rear. I hope it will contain much that is new even to seasoned naturalists and botanists. But I hope it will also be found interesting to people who may not call themselves botanists at all.

What does rare mean?

Rare plants are seldom rare everywhere. Someone living on the Lizard peninsula of Cornwall would regard Cornish Heath (*Erica vagans*) as one of the commonest plants. To the miners of Upper Teesdale, the Spring Gentian (*Gentiana verna*) was the 'spring violet', as familiar there as dog-violets in a lowland lane. Around certain ponds and lakes in the New Forest, the Hampshire Purslane (*Ludwigia palustris*) adds a strange reddish colour to the scenery which impresses people who may never have heard of the plant.

Contrariwise, relatively common plants are usually rare somewhere. In north-east England, Yellow-wort (*Blackstonia perfoliata*), so familiar to walkers on the southern chalk and limestone, is confined to a few coastal cliffs. In Somerset, Mossy Saxifrage (*Saxifraga hypnoides*) is regarded as one of the special flowers of Cheddar Gorge, though in the north you might walk over and ignore it. Even a generally widespread species may be rare in certain counties, like Moonwort (*Botrychium lunaria*) in Devon; or inexplicably rare in a particular district, like Meadow Barley (*Hordeum secalinum*)

Nationally rare plants can be common in places. Dominating the scenery on parts of the Lizard peninsula is Cornish Heath (*Erica vagans*). [Bob Gibbons]

Above: Massed Hampshire Purslane (*Ludwigia palustris*) lends an exotic reddish fringe to this shallow pond in the New Forest. [Peter Marren]

Inset: Closer up, the margins reveal a whole community of rare plants. These few square centimetres contain four rare species: Pennyroyal, Coral Necklace, Wild Chamomile and Hampshire Purslane. [Peter Marren]

on Exmoor. Near my home in Wiltshire, certain banks are scented in late summer by the dense white flower-clusters and wrinkled leaves of Round-leaved Mint (*Mentha suaveolens*), but outside our district you would need to travel to Devon or South Wales to find it in quantity. From an individual point of view, rare and common are relative terms. Even the slugs that eat rare orchids and broomrapes probably take their meals for granted.

Rare, in the everyday botanical sense, means nationally rare. The basic unit of plant records is the ten-kilometre square or hectad. When a plant is found in 15 or fewer ten-kilometre squares throughout Great Britain it is regarded as rare, even when, as in the case of the Spring Gentian, it can be locally abundant on home ground. These plants are included in the British Red Data Book (2nd Edition, 1983), where they are subdivided into 'endangered', 'vulnerable', or simply 'rare' according to the degree in which they seem threatened. About 320 native wild flowers – some 20 per cent of our native flora – are in one of these categories. The forthcoming third edition of the *Red Data Book* will refine these categories further into 'extinct', 'extinct in the wild', 'critically endangered', 'endangered', 'vulnerable', and 'lower risk – near threatened'.

There is a second tier of plants that are found in between 15 and 100 ten-kilometre squares in Britain, and these are regarded as 'lower-risk – nationally scarce'. By coincidence, there are roughly as many 'scarce' plants (260) as there are rare ones, so that some 40 per cent of our flora comes into the 'risk' category. As it happens, some 'scarce' species, like the Deptford Pink (*Dianthus armeria*), are in terms of individual plant numbers probably rarer than some of the 'rare' ones! While the acres of Cornish Heath would satisfy every garden nursery in the land (though God forbid), you would probably have difficulty filling a florist's window with Deptford Pinks. Of course both scarce and rare categories are subject to constant revision as more records come to light or if the plant itself changes status. The Ground Pine (*Ajuga chamaepitys*) passed from scarce to rare recently, partly through genuine decline, partly through a fluke of recording – for some of its sites were on the line, and so a single site had become two full grid squares! On the other hand Green Figwort (*Scrophularia umbrosa*) is no longer considered even scarce. Probably it had been under-recorded, but there is no doubt that it has increased its range, if not its numbers. Unfortunately there are not many species about which you could say as much.

Why are some flowers rare?

The stock answer is that mankind has made them so, through habitat destruction, pollution and development. The reality is more complicated. You could divide our rare flowers into three Shakespearean categories: some that were born rare, some that achieved rarity, and the rest that had rarity thrust upon them – by us. The ones that were 'born rare' have been so since botanical recording began, and in some cases for long before that. These plants are naturally rare for geographical or climatic reasons. Britain is one of the wildlife crossroads of Europe. We are a southern outpost for arctic plants and a northern one for Mediterranean ones. We have species confined to the Atlantic fringe, and others that are most at home on the dry plains of eastern Europe. Species at the edge of their natural range here may find refuges in localized warm spots in the south, or dry places in eastern England or on the highest mountains in Scotland, Wales and northern England. Such plants tend to become specialized; that is they have more exacting requirements in Britain than nearer the centre of their world range. The Alpine Catchfly (*Lychnis alpina*), for example, is confined to a few metal-rich rock outcrops in Scotland and the Lake District. These are toxic to most plants, and hence reduce the competition from more aggressive species. In Scandinavia, on the other hand, this arctic-alpine flower is widespread and nothing like as fussy. Some of these 'born rare' plants were once more widespread, but only during the distant past when the climate and landscape were quite different from today. Pollen from Alpine Catchfly (see picture on p. 8) has been found in peat deposits at scattered places across Scotland, indicating that it was more at home about 12 000 years ago when Britain was a cold, treeless place. But the present-day distribution of such plants has little or nothing to do with farming or development. They could not be anything other than rare. They have piled all their survival chips into a few localities, where, for one reason or another, they are able to hang on.

Our second, smaller, category is of flowers that have 'achieved' rarity, that is they have become rare more recently for reasons at most only indirectly associated with mankind. Hybridization threatens the survival of a few species, like Tuberous Thistle (*Cirsium tuberosum*), where the rarer species has become swamped by a common one, in this case the Dwarf Thistle (*C. acaule*), which is partly interfertile. Climate change threatens others, especially mountain-top plants, which have nowhere else to go. At least a few species suffer from apparent breeding failure, like the Twinflower (*Linnaea borealis*). Others have had their already limited habitats reduced by natural succession, or suffered from marauding rabbits, deer and invertebrates.

Unfortunately a great many flowers have had rarity thrust upon them. Not a few of our Red-listed plants were relatively common a century ago. Until farming became mechanized, much of the natural landscape was shaped by agricultural labourers. Some habitats were maintained by sustainable harvests – of hay, reed-thatch or coppice-wood. There was open-range livestock grazing on downs and commons as yet untouched by chemical fertilizer. Natural hollows were almost invariably damp, and meadows and pastures had their patches of sandy soil or dry ant-hills, and levels of floodland where the river spilled over in winter. Although this landscape was artificial in the sense that mankind had shaped it, it also consisted mainly of wild plants, and was fairly stable from one year to the next. Only the seasons changed. Within this landscape there were many niches for plants that are either ecologically very choosy or are poor competitors. Among them were crop weeds, like Corncockle (*Agrostemma githago*), Cornflower (*Centauria cyanus*) or Thorow-wax (*Bupleurum rotundifolium*), which found space to flower and seed through what is nowadays regarded as inefficient farming. Colourful plants like Fritillary (*Fritillaria meleagris*) grew in riverside meadows where the land flooded in winter (and was encouraged to do so, river silt being a valuable natural fertilizer) but dried out in the spring. Commons on the poorest, sandy soils were a refuge for flowers like Tower Mustard (*Arabis glabra*) and Greater Broomrape (*Orobanche rapum-genistae*). Mud by shallow ponds on commons and greens were the habitat of once familiar flowers like Pennyroyal (*Mentha pulegium*) and Starfruit (*Damasonium alisma*). Virtually none of these habitats are useful or commercially viable today. Modern agriculture is more intensive and reliant on the products of ICI. It works by homogenizing the landscape, on replacing nature with crops. The rural system that supported these species, and many others like them, has virtually disappeared.

Alpine Catchfly (*Lychnis viscaria*) is among the most specialized British plants, being confined to metal-rich rocks on mountains in Cumbria and Angus. [Hugh Lang]

The coarse mesh of the ten-kilometre square distribution maps, like those in the *Atlas of the British Flora*, can distort the true state of affairs and makes things look rosier than they are. On the map, the status of species like Corn Buttercup (*Ranunculus arvensis*) or Fly Orchid (*Ophrys insectifera*) still looks quite healthy. However, each dot means only that the species is present somewhere within 100 square kilometres – just a single colony will do. Moreover the information for the more widespread species may be out of date, and record the situation ten, twenty or even fifty years ago rather than now. Some plants, like the Corn Buttercup are, in fact, very thinly spread and scarce almost everywhere. In his Presidential address to the Botanical Society of the British Isles,[138] David Pearman suggested that some of these maps are really a shroud – a representation of reality for what is really mostly empty space.

This was brought home to me a couple of years ago when, at a conference on recording, Keith Porter of English Nature showed distribution maps of the Pasqueflower (*Pulsatilla vulgaris*) at an increasingly larger scale. At a ten-kilometre scale, things looked quite favourable, with comfortable clusters of dots marking its main strongholds in the Chilterns and the Cotswolds. But at the finer tetrad scale (two-kilometre squares) this solidity dissolved into an open scatter of dots. And when the *actual* colonies were marked – something not yet possible for most species – they were mere flyspecks. Reality for the Pasqueflower is not a solid range but tiny islands measured in metres in an ocean of sterile farmland. The Victorian botanists like C.C. Babington and Druce, who regarded the Pasqueflower as a familiar flower within its limited range, would surely agree that rarity has been well and truly thrust upon it.

The best sites for Pasqueflower were listed as 'worthy of preservation' as early as 1915. Unfortunately they all too seldom were preserved (see map on p. 10). [Peter Wakely/English Nature]

Rarity is not, in general, a healthy state to be in. A small population is not necessarily 'threatened'. Some tiny colonies have lasted a very long time. The colony of Spotted Cat's-ear (*Hypochaeris maculata*) at Humphrey Head in Cumbria seems to have remained about the same, as far as numbers are concerned, for at least 200 years, and no doubt there are other colonies whose survival could be measured in thousands of years. But being rare often means your genetic resources are limited, and that makes you vulnerable to change. Small populations are also liable to be picked off by chance events, like a shrub growing up in front of you, or that new borehole for a new housing estate dug a few metres away. Conservation has recently come to the aid of many threatened plants, drawing many a little further back from the brink, but perhaps at some cost to their inherent wildness (a subject I will return to in Chapter 13). But the wise conservationist will aim at preventing flowers from reaching that state of extreme vulnerability in the first place. A working definition of nature conservation might be to stop widespread species becoming scarce, scarce ones from becoming rare, and rare ones from becoming extinct.

But being rare should not imply that there is anything 'wrong' with the plant. For most of their history, rare flowers have been as good at the survival game as common ones. If there was anything wrong with them ecologically, they would not be here at all. It does less than justice to their qualities of toughness and persistence to regard them merely as victims of our own selfish progress. There are many reasons for rarity – doubtless more than we know of. A plant may require unusual conditions, it may have a specialized life, or it may just be unlucky enough to grow in the wrong place. Each is an individual, each with its own story. In this book, I will try, as the American writer John Burroughs once said of Gilbert White, 'to seize the significant and interesting features and to put the reader into sympathetic communication with them'. Let us therefore begin at the beginning, with those who discovered our rare flora, and see what they made of it.

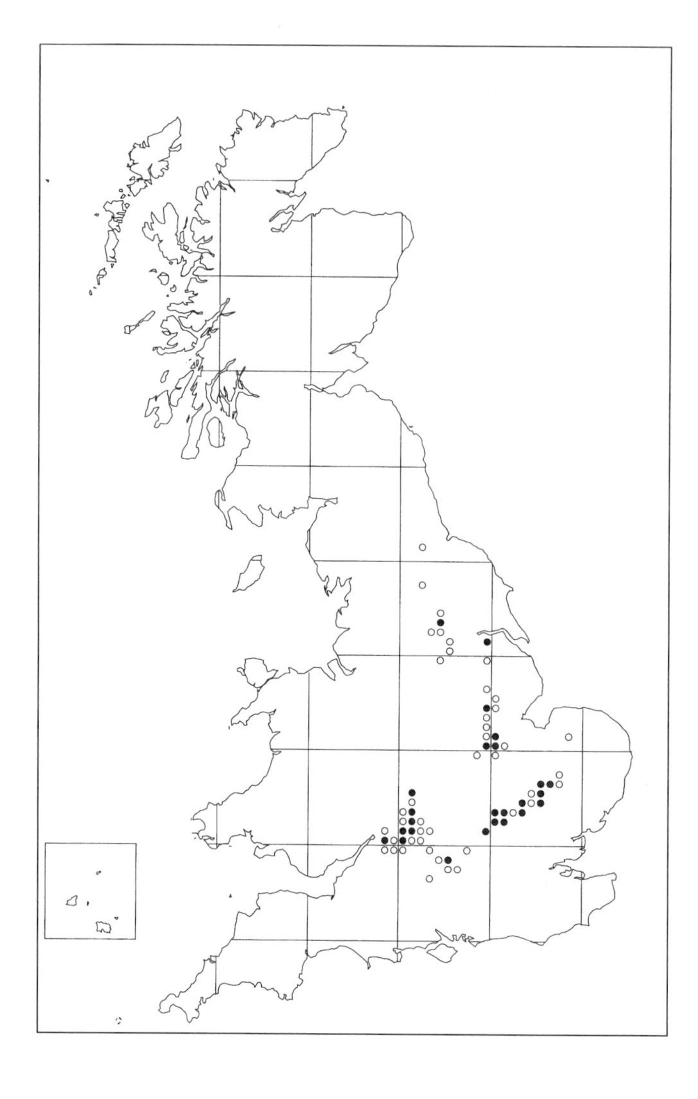

Distribution of Pasque flower by 10-kilometre squares. Solid dots are post-1970 records, open circles pre-1970.

Chapter 2

A root of Honewort:

how our rare flora was discovered

The cliffs below the summit of Ben Lawers form natural hanging gardens of alpine plants.
[Bob Gibbons]

Ben Lawers is the highest hill in Perthshire and the ninth highest in Britain. As big hills go it is also one of the most accessible, especially if you take the 'tourist route' to the summit from the National Trust for Scotland's visitor centre instead of the 'purist route' from the shores of Loch Tay. Ben Lawers is green from top to bottom, with easy tracks and springy turf nearly all the way to the top. Perhaps that is one reason why the botanical wonders of Ben Lawers became such a magnetic draw for naturalists from the 1790s onwards. A large number of our rare mountain plants were first discovered on Ben Lawers, many of them by two nurserymen, James Dickson of Covent Garden and George Don of Dovehillock, Forfar.

Such is the draw of the summit and the rich cliffs nearby that relatively few walkers move on to explore the ridges, gullies and corries beyond. And so, while Ben Lawers itself has become badly worn in places, you usually have its neighbours to the north and east to yourself, except for a lot of sheep, and the occasional Munro bagger. When, therefore, two middle-aged ladies set off to explore the wild hinterland of Ben Lawers, 75 years ago, they knew they were on barely trodden ground as far as botanizing was concerned. They were rewarded with something most of us can only dream about: on that day, on 25 July 1923, they found a plant new to Britain. It is called *Carex microglochin*. Since this is a spare little plant even for a sedge, and moreover bears a passing resemblance to the much commoner Few-flowered Sedge (*Carex pauciflora*) or the Flea Sedge (*Carex pulicaris*), it is to the credit of both ladies that they not only noticed it, but knew what it was. Perhaps they knew that its occurrence in Britain had been predicted by the great amateur botanist, Arthur Bennett, and were looking out for it. How one enjoys such a moment is rarely divulged in print. Discoverers of wild flowers are usually anxious to preserve proper scientific rectitude and maintain a dignified silence on any little jigs they might have performed at the time. In this case, I imagine that one of the ladies, sprightly 49-year-old Gertrude 'Gertie' Bacon (1874–1949), might have leapt into the air with joy. The other, the 58-year-old Lady Davy (1865–1955), whose Christian names, Johanna Charlotte, were not used outside the family, would have straightened her shoulders and removed her hat. Then they would both have taken a powerful nip from their flasks, and (I like to think) sung the national anthem.

They were born to discover a new plant. Lady Davy was one of the respected elders of the Wild Flower Society, famous not only for knowing exactly where all the rare flowers grew but also for her formidably forthright manner. A disciple, Kit Rob, remembered how as a nervous young woman she had accompanied Lady Davy to Upper Teesdale. It had been 'impressed upon me that I must behave well. I literally shook with fear, and the only thing I could think of was to rush off and open gates almost as soon as they were in sight.' Eventually she noticed that her frantic efforts to please were causing vast amusement. 'I saw the twinkle in her eye and knew that all was well. She often teased me about that day.'[204]

Lady Davy's companion and lifelong friend Gertrude Bacon (who later became Mrs Foggitt) was the life and soul of many a Wild Flower Society ramble between the wars. Her approach to flower hunting can be savoured

A portrait of Bristle Sedge (*Carex microglochin*) by the late Olga Stewart.

in the titles of some of the articles she wrote for the *Wild Flower Magazine*, like 'Those Carices!' and 'O Willow, Tit Willow'. She also published comic photographs, like the one illustrating a search for the tiny Dwarf Pansy (*Viola kitaibeliana*) with everyone on all fours, bottoms high, over the caption 'Stern Realities of Botanizing'. Gertie's other passions included flying and hot-air ballooning, and she is said to have been the first woman pilot in the world. She was one of those people who never do things by halves, whether roaring through the clouds in a flying machine, or sitting on the magistrates bench at Thirsk, or amassing one of the largest herbaria in private hands. If you were looking for two people who embodied the spirit of the Wild Flower Society in its heroic years, you would stop at Gertie Bacon and Lady Davy. Long after their deaths, they are still recalled with a mixture of awe and affection.

Their walk that day in 1923 must have taken them along the ridge that horseshoes around the headwaters of the Lawers Burn and encloses

Lochan nan Cat, which from above really does look like a pussy cat, sitting up and purring against the leg of the hill. On either side of the ridge rise a series of mountain rills, seeping up through gravel and moss and rippling down through tussocky pastures to join Glen Lyon or the Lawers Burn. It was at one of these alpine springs, a level of bare wet gravel at the foot of a steep slope where snow often lies late, that they found the *Carex microglochin*. That was what they called it. Only quite recently has this plant acquired an accepted English name, Bristle Sedge (though, as the latest national flora has misprinted it as Brittle Sedge, that may become its new name, and generations yet unborn will ask themselves what is so brittle about *Carex microglochin*). Both the scientific and common names are taken from the distinctive stiff bristle which projects between two short barbs from each utricle of the spike. The person who described and named this species thought its glumes resembled the projecting barb of an arrow, and hence combined the Greek words *mikro* and *glochin* (pron. glok'in) as 'the little arrow barb'. Gertie Bacon was a connoisseur of sedges, and would have been delighted by this one, despite its diminutive size. *Carex microglochin looks* like an alpine, with its spiky head, stiff habit, creeping roots and short thick leaves. Though most pictures show it as greenish, the living plant often has a reddish hue, and it likes to grow in exclusive company, with other rare sedges of alpine springs like the delectable Scorched Alpine-sedge, *Carex atrofusca*. After admiring it at length, Davy and Bacon did what every good botanist would have done in the circumstances: 'Be it ever so abstruse, pack it off to Dr Druce.' A specimen was duly packed off and, according to legend, the very next day the shutters went down on Druce's chemist's shop, and the great man was on his way north to Ben Lawers on the milk train. In boring fact, it was about a month later, but at least he went. And so it was Druce who, in the Botanical Exchange Club's annual report for 1923, revealed a new British plant to the waiting world.

Botanists being botanists, it was to the spot where Davy, Bacon and Druce had found it that people went to see *Carex microglochin*. Only quite recently has the sedge begun to turn up in other places, most notably when my bearded friend Sandy Payne explored the flora of the hills north and west of Ben Lawers while studying eagles in the 1980s. As a result of these new finds, *Carex microglochin* is no longer an extreme rarity so much as a *characteristic* plant of high 'micaceous' springs and bogs in this part of Perthshire. It is more widespread than that single dot in the *Atlas of the British Flora* implies: over an area of roughly ten by ten kilometres you are almost bound to find it in the right habitat. It is one of a small group of alpines, including Alpine Forget-me-not (*Myosotis alpestris*), Alpine Pearlwort (*Sagina nivalis*) and Scorched Alpine-sedge, which are mostly confined to the central Highlands for reasons we can only guess at.

Given its modest appearance and remote habitat, it is not surprising that *Carex microglochin* was missed by earlier botanical explorers. Had it not grown near a famous botanical locality it might have escaped attention even longer, or quite possibly it might not have been noticed at all!

The old masters

Like *Carex microglochin*, most of our rare flowers were discovered by chance. A few were discovered 'indoors' by specialists comparing the characters of closely related plants. But most have been found by amateur naturalists on country walks, often in places already noted for their rich plant life. In a logically ordered world the more showy species would have been discovered first, and the rarest or least distinctive species last. In practice, of course, it was not like that at all. In Tudor and Stuart England, long-distance travel was arduous, expensive and sometimes even dangerous. Those who knew the names of flowers recorded the ones on their doorstep, which more often than not was around London. Hence they knew about now rare plants like Ground Pine, Pennyroyal and Stinking Goosefoot, which grew on commons, village greens or chalk downs not a dozen miles from St Pauls, yet knew little or nothing about northern flowers, and considered Cloudberry (*Rubus chamaemorus*), for example, to be an extreme rarity. Nor did they tend to distinguish between closely related species. To a herbalist like William Turner or John Gerard, a leafless plant with a turnip-like root was simply a broomrape, and a neat yellow flower with divided leaves a crowfoot. The finer distinctions of broomrapes and buttercups came later (though they could distinguish three poppies and several speedwells). Nor did the writers of English herbals always distinguish clearly between wild flowers and garden flowers. Gerard knew and figured the Lady's Slipper, but apparently only as a garden plant. They were less interested in the distribution or habitat of a plant than in its properties. Where a wild locality is mentioned, it was to inform their fellow apothecaries and herbalists where to gather it.

The first Englishman to write about the wild flowers of his native country was William Turner (1508–68). Hence he has been called the Father of English Botany and the first English naturalist. Because he came first, all the identifiable flowers described by Turner are considered to be first records. Some of them were no doubt familiar to Turner's learned contemporaries, like the Doctors Clement, Wendy, Owen and Wotton, who, he admitted, possessed 'as much knowledge in herbes' as he. It does seem, though, that the study of wild plants in Britain had not advanced very far since Roman times. Turner himself was largely self-taught, and complained that at Cambridge he learned nothing 'of any herbe or tree, even amongst the physicians, such was the ignorance at that time'. As things stood, said Turner in his blunt north-country way (for he was born in Northumberland), physicians relied on apothecaries for their herbs and potions, and apothecaries relied in turn on the 'old wives' who did the actual gathering, and that meant, in Turner's opinion, that patients were in danger of being poisoned in their beds. That was why he decided to write his great work, the *Herbal*, in plain English instead of in Latin, so that anyone could use it for the 'sekyng of herbes and markynge in what places they do grow'. The same reforming spirit that drove men to risk their necks by translating the Bible into English lay behind Turner's *Herbal.* Out went the old wive's tales and in came observation and inquiry. 'Let every man follow that which he findeth to be most true, both by reason and by experience,' proclaimed William Turner, and his sermon holds as good for natural history as it did for his religion.

And so we come to what I think is the most seminal sentence in the whole of British field botany, the one in which Turner tells us he found a 'root of Honewort at Saynt Vincentis rock, a little from Bristow'. Undramatic as it sounds, it marks the very beginning of Britain's long tradition of recording the sites of rare flowers. How appropriate then, that Turner found his Honewort (*Trinia glauca*) at one of the classic botanical sites: at St Vincent's Rock in the Avon Gorge, where the Clifton Suspension Bridge now stands. Honewort is neither a striking plant, nor one known to the Latin and Greek authors. It was probably a personal discovery. Turner must have come across it during his years of exile in central Europe as the controversial author of 'The Hunting of the Romish Wolf' – the Pope, that is. Honewort is common in parts of central Europe, where it is known as a 'steppe-runner' from its habit of uprooting in the wind and rolling along like tumbleweed. In Britain, however, it is rare, and confined to the Avon Gorge, the Mendips and limestone headlands around Torbay, all, as it happens, windy places. Honewort is an umbellifer (it is our only dioecious one, with separate male and female plants), and apparently has a hot-tasting root, like a little parsnip (I have nibbled its seeds, which taste a little like coffee). Gerard noted that 'the roote is thicker than the smallness of the herbe will well beare' and that 'among the people about Bristowe . . . this hath been thought to be good to eate'. That is, of course, the reason why Turner 'marked the place' where it grew, and why he mentions the root and not the flower. Among the many botanists that have since paid court to the Honeworts of the Avon Gorge are Thomas Johnson, who called it 'Dwarf Hogges-fennell or Rock-parsley', and Sir Joseph Banks, who came across it during his evening walk 'in full bloom plentifull, just above the Rock House'. Three and a half centuries after Turner, the place to see Honewort had moved half a mile downstream to Black Rock Gully, and the Honewort's virtuous root had been forgotten. It still occurs there today, in company with another of the Gorge's special plants,

Dwarf Sedge (*Carex humilis*). If ever a plant deserved a blue plaque, surely it is the Honewort at Bristol.

In his various works, Turner mentions many good places 'for the seeking of herbs'. It may well be, as legend insists, that he got to know the cliffs of Dover waiting for the cross-channel boat while fleeing from religious

Honewort (*Trinia glauca*), one of the first rare plants to be recorded in Britain. This is a male plant growing in Avon Gorge not far from the original site. [Andrew Gagg's Photoflora]

Sea Pea (*Lathyrus japonicus* ssp. *maritimus*), a beautiful shingle plant, found mainly on unspoiled beaches in south-east England. [Peter Wakely/English Nature]

oppression. Thus he was the first to write about the Wild Cabbage or 'Colewort' (*Brassica oleracea*) – 'in Dover cliffes, where as I have onely seene it in al my lyfe' – a record which helps to establish Wild Cabbage as an ancient, if not a native, plant. It still grows there. Turner also mentions Samphire (*Smyrnium olusatrum*) on the steep white cliffs, which may be the source of its fleeting appearance in the Dover scene in *King Lear*, where '*halfway down hangs one that gathers Samphire, dreadful trade! Methinks he is no bigger than his head*.' At the opposite end of England, in his native Northumberland, Turner identified some 144 plants, including the strange 'Libbardsbane' or Herb Paris (*Paris quadrifolia*) of Cotting Wood. Cotting Wood still exists. As it happens, back in the 1970s, I did my best to have it scheduled as an SSSI on the grounds of its association with the young Turner. But, I was informed, 450-year-old records of plants did not come within the definition of 'special scientific interest'.

One of the most amusing 'first records' of a rare plant is that of Turner's contemporary, John Key, or Caius, royal physician to King Edward VI and Queens Mary and Elizabeth. He tells of the famine year of 1555, when the poor people of the Suffolk coast were reduced to scavenging pea pods they found above the shoreline. They probably belonged to the Sea Pea (*Lathyrus japonicus*), which still grows abundantly on the shingles of Orfordness. Woven into the yarn later was the tale of a vessel loaded with peas which broke up offshore, sending its leguminous cargo rolling on to the shore in the tide, and into the baskets of the famished peasantry. Most plants included in the earliest herbals had some similar useful quality, whether for eating, like the Sea Pea, or for medicine like Henbane, or at least for their attractiveness or oddity, like Pasqueflower or Bee Orchid.

Turner gave us the first records of some 300 identifiable plants. Not all of them were personal finds; for example, his Pasqueflower 'greweth about Oxford, as my frende Falconer told me'. Within fifty years of his death, the British flora had doubled in size. The travels of the Flemish herbalist Matthias de l'Obel (Lobelius) added Grass of Parnassus (*Parnassia palustris*), Flowering Rush (*Butomus umbellatus*) and Yellow Star-of-Bethlehem (*Gagea lutea*) to our flora, as well as that well-known physician's plant, Deadly Nightshade (*Atropa belladonna*), which Turner had surprisingly omitted. The Revd Dr Thomas Penny provided a few northern species, among them Bird's-eye Primrose (*Primula farinosa*) and Melancholy Thistle (*Cirsium helenium*), but the Scottish flora at that time was almost completely unknown to the savants of London. The first record of a plant in Scotland is said to be the Chickweed Wintergreen (*Trientalis europaea*), illustrated in Bauhin's *Prodomus* of 1620, though John Parkinson had a rough idea that Scots Pine grew there: 'in Scotland *I am assured*', he wrote, in his famous *Theatre of Plants*. But no one from England got to see the special flowers of the Scottish pine forests until the year 1775.

The herbal that eclipses all the others – for most people it is *the* Herbal – is Gerard's *Historie of Plants*, first published in 1597. The reputation of John Gerard (1545–1612) has suffered at the hands of scholars, who never tire of pointing out that Gerard's original version of the herbal rambled on a great deal, and that he borrowed most of his material and passed it off as his own. The tauter, more readable version of 'Gerard's Herbal', available in paperback today, was thoroughly revised after his death by Thomas Johnson. Far from being a mere disciple, Johnson was a sometimes astringent critic of 'our author', drawing attention to Gerard's mistakes and want of classical learning. Yet, despite all, the obstinate fact remains that Gerard is the most quoted and quotable English botanist who ever lived. Surely it is more sensible to accept that people did things a little differently back in the 1590s; Gerard's contemporary, Shakespeare, for example, borrowed most of his plots, yet we still think that *King Lear* is pretty good. What we appreciate in Gerard is his skill at portraying wild plants, with a sharpness of observation combined with vivid, often charming, use of words. Gerard clearly *enjoyed* flowers, for their own sake. It was he that invented the name Traveller's Joy, and there is something slightly heartcatching in his account of '*Pulegium regium*' which 'is so exceedingly well knowne to all our English nation, that it needeth no description, being our common Pennie Royall'. In those days Pennyroyal grew 'on the common near London called Mile End . . . whence poor women bring plenty to sell in London markets'. (Alas it is common no longer except as a garden herb, and anyone caught selling it now would be poorer still after a hefty fine.) At other moments, you find yourself smiling – where, for example, Gerard puts out cat poison to revenge himself on the stray felines which scraped in his garden, 'spoiling both the herbes and seedes new sowen', or in his occasional stab at verse, like this ode to the stinging nettle: 'Neither without desert his name he seems to git/As that which quickly burns the fingers touching it.'

Gerard added more species to the English flora than anyone except John Ray: among his 182 'first records' were some now rare and elusive species like Tower Mustard (*Arabis glabra*), Hog's-fennel (*Peucedanum officinale*), Smooth Rupturewort (*Herniaria glabra*) and Baneberry or Herb Christopher (*Actaea spicata*). Of course, since so much in Gerard is second-hand information, we cannot be sure whether Gerard was himself the finder. One plant he certainly knew, and which is very rare today, is his *Atriplex olida* or 'Stinking Orache', now better known as Stinking Goosefoot. Stinking Orache then grew 'upon dung hills' and other 'filthy places' on waste ground and on old walls in and around London. The crushed leaves smelled so awful that Gerard portrays it almost as a kind of joke flower, noting that those growing by brick kilns smelled 'like tosted cheese' while those 'which groweth in his natural place smels like stinking salt-fish ... more stinking than the rammish male Goat' – which certainly sounds like personal experience.

Stinking Goosefoot (*Chenopodium vulvaria*) – as I say, it is not much to look at, but once sniffed, never forgotten. Here it is growing in a golf bunker in Dorset. [Peter Marren]

Inset: 'Our Pennyroyal' (*Mentha pulegium*), once familiar in late summer on village greens but now an endangered species. [Bob Gibbons]

The Elizabethans were the first to notice the Stinking Goosefoot for: 'There hath bin nothing set downe by the Antients, either of [its] nature or vertues.' Today, far from the dung hills and brick kilns of London, it ekes out an uncertain Red Data Book existence on open ground near the sea.

'For the sake of the plants'

Gerard's Herbal is full of references to places around London where he and his fellow gardeners and apothecaries combed the commons, heaths and waysides in search of herbs. Hampstead Heath was a popular destination, and there was a well-trodden route across the North Downs as far as Canterbury or Dover, and another along both sides of the Thames estuary. At that time Middlesex and the hinterland of London may have been as rich in plants as any part of Britain, with its tapestry of heaths, ancient woods, chalk downs and marshes by the river. Very likely apothecaries had been gathering herbs in such places for centuries, but by the time of Thomas Johnson (*c.* 1605–1644) there was a new spirit about – not simply of collecting but of careful recording and the systematic listing of plants in a locality. The document known as the *Iter Plantarum Investigationis*, which appeared in 1629, is the earliest printed account of an organized botanical excursion in Britain. It is not unlike an account of a modern excursion, with the same companionship and jokes and misadventures that have always characterized botanizing parties. Like Gerard, Thomas Johnson was an apothecary, with premises at Snow Hill in the City of London. His main stamping grounds were Kent and Hampstead Heath, but occasionally he travelled more widely, to the West Country and twice to Wales. In all, he and his helpers added some 170 new wild flowers, including many that are now rarities, to the British list. The first journey he records in detail was an excursion into Kent with nine companions, all medical men, in July 1629 – a journey made expressly 'for the purpose of discovering plants'. In those days, even a modest tour of North Kent needed a lot of planning. The party decided to make their way to Gravesend by river in two cargo wherries moored at Blackfriars. But river traffic was at the mercy of the weather, and no sooner had they set off than a storm blew up and one of the boats was forced to dock at Greenwich. By the time it was ready to resume the journey, the tide had turned, and so, rather than delay any longer, the stranded

Wild Chamomile (*Chamaemelium nobile*), the 'herb of humility', so familiar as to be proverbial, yet now restricted to scattered localities in southern England, mainly the New Forest and the south-west.
[Peter Marren]

botanists decided to abandon the river and hired horses to catch up with the others. The next day the reunited excursionists moved on to Chatham where, as tourists do, they sought out and inspected the main attraction of the town, the new 55-gun warship, Royal Prince, and also recorded plants growing around the docks. (I once did the same at Chatham while waiting for a train to arrive – if you are ever stuck there in July, walk up to the open space called The Lines, where you should soon find the rare Red Star-thistle (*Centaurea calcitrapa*), known there since the eighteenth century.) To reach Sheppey, they took the ferry and lodged for the night in two inns at Queenborough. Their departure the next day was delayed while they were summoned by the mayor who wanted to know why so large a party of strangers was travelling through – evidently no common sight at that time. Johnson explained politely that 'they were addicted to the study of medicine and had come to those parts to investigate what rare plants grew there'. One of his companions added, no doubt in a helpful spirit, that an additional reason was to meet a man of such surpassing merit as the mayor of Queenborough. The mayor took that compliment as his due, and the interview ended amicably with a loyal toast with tankards of ale. In what remained of the day, the party botanized along the shore, and then crossed on a barge to the Isle of Grain. After a hot and thirsty walk followed by yet another hearty dinner, all but Johnson and a man named Styles suddenly felt themselves overcome by lassitude. 'And so we parted from our comrades', wrote Johnson, who were left in a brewer's cart, 'lolling among the barrells and entrusted to the care of the drivers.' Johnson himself continued on foot as far as Cliffe before eventually returning by river to London. At least one commentator has objected to Johnson's references to ale and barrels, and the meals he enjoyed *en route*, as vulgar and unbotanical. But most of us find his description of travelling life nearly 400 years ago just as fascinating as what he found.

It is quite possible to retrace Johnson's steps today and in 1975 the Kent Field Club actually did so.[18] What impressed them most was the durability of our wild flora – to their surprise, they refound many of the plants recorded by Johnson, and in much the same places despite all the modern industrial and housing developments in this part of Kent. The rare White Mullein (*Verbascum lychnitis*) still grew on the surviving scrap of Dartford Heath, as did most of Johnson's seaside plants on the Isle of Grain. The explorers even traced the chalk-pit, mentioned by Johnson as a good place for plants, and refound eight of his records there. What has changed beyond recognition is not so much the wild places, much reduced as they are, but the farmed landscape. On the road from Gravesend to Rochester, Johnson's party walked past crop fields full of flourishing weeds like Corncockle, Corn Marigold, Thorow-wax, Cornflower, Shepherd's Needle and Corn Parsley. By the wayside, they found Turner's 'great Broom Rape', *Orobanche rapum-*

genistae. All these plants have gone, and even the crops themselves have changed. In Johnson's day the fields near Chatham and Cliffe grew hemp (*Cannabis*) for the Royal Navy and rape for cattle feed – though the latter now has a counterpart in Kent's yellow vistas of oil-seed rape.

In 1634, Johnson made a more ambitious journey westwards, again in pleasant company and in a growing spirit of independent exploration and discovery, to roam 'around the meadows, stony places, broken ground, and the woods at hand, for the sake of the plants'. They were particularly eager to examine localities around Bath and Bristol, already noted for rare plants and other natural wonders. At Bath, Johnson found the now extinct Hairy Spurge (*Euphorbia villosa*), and was the first to record the town's other special plant, Spiked Star-of-Bethlehem or 'Bath Asparagus' (*Ornithogalum pyrenaicum*). At St Vincent's Rocks, he admired Western Spiked Speedwell (*Veronica spicata* spp. *hybrida*), discovered by John Goodyer a few years before, and in the woods opposite he was the first to marvel at the bursting fishbone pods of Narrow-leaved Bitter-cress (*Cardamine impatiens*).

Johnson was the first Englishman to explore the mountain flora. He set out in July 1639 on a journey to North Wales, and on 3 August ascended one of the highest hills. There has been some argument about which hill it was, but the consensus is that it was Snowdon. For a man more used to the low, rolling hills around Bath and Chatham, it must have been a horrendous, if romantic, experience.

> Here (he wrote), among the other high mountains, the peak raised its head as does the Cypress among the Viburnums. Desirous of seeing the peak we hurried on and obtained the services of a country boy as guide, because not only the summit but the whole mountain was wrapped in cloud. Then, leaving our horses and outer garments, we began to climb the mountain. The ascent is at first difficult, but after a time the way broadens out – though still equally steep, with vast precipices on the left and a difficult climb on the right. After negotiating three miles, we at last reached the crest of the mountain, shrouded in thick cloud. Here the path is very narrow and, as we climbed it, we were appalled at the rugged crags and precipices on both sides, and also the Stygian marshes here and there, the largest of which is called by the natives the Devil's Home. At the summit we sat down in the clouds. First we arranged in order the plants collected at our peril among the rocks and precipices, then ate the food we had brought with us.[158]

'The Stygian marshes' is good – evidently he had Dante's Inferno in mind. Snowdon has lost some of its fearfulness since then, but few of us are likely to find as many alpine flowers as Johnson did, especially so late in the season. The party gathered at least 20 species, of which four, Alpine Saw-wort (*Saussurea alpina*), Mountain Sorrel (*Oxyria digyna*), Moss Campion (*Silene acaulis*) and Alpine Saxifrage (*Saxifraga nivalis*), were new. It must have been a moment like Balboa's first glimpse of the Pacific, or Armstrong's setting foot on the moon, instant, unique, never to be repeated. What intrigued Johnson most, however, was to find seaside plants like Thrift (*Armeria maritima*)and Sea Campion (*Silene uniflora*) so high and so far from the coast. This ascent of Snowdon capped the first of an intended series of journeys 'to travel over most parts of this Kingdom' in search of plants. But alas, it took place on the eve of the Civil War. Johnson volunteered to serve as an officer in the King's party, and, as a Londoner, he ended up in the besieged garrison at Basing House, where he was fatally wounded in 1644, aged only 40. An unknown person paid tribute to him as 'no less eminent in the Garrison for his valour and conduct as a soldier than famous through the kingdom for his excellency as a physician'. The envisaged nationwide botanical survey had to wait for a full generation, and the man who finally undertook it was the greatest English naturalist, John Ray.

A Ray of light

How wonderful it must have been to be a naturalist in seventeenth-century England. The land was teeming with unexplored treasures, when any country walk might produce an unknown bird, butterfly, moss or flower. However, the problems in the way of any would-be explorer were intractable. There were no journals to publish discoveries, no natural history societies – and hardly any real naturalists. Thomas Johnson's work survives because he published it himself. That there were others like him is evident from the use that was made of their records later. George Bowles, for example, travelled much of England recording plants, chalking up an impressive list of new species, including Grass-poly (*Lythrum hyssopifolia*) near the Thames at Dorchester, Sea Stock (*Matthiola sinuata*) at Aberdovey, and Touch-me-Not Balsam (*Impatiens noli-tangere*) by the River Kemlet in Shropshire. Another was the London apothecary Leonard Buckner, who first found the 'Wilde Stynking Horehound', better known as Downy Woundwort (*Stachys germanica*), on the bank of Witney Park, Oxfordshire. The most active of all was probably John Goodyer (1592–1664), a manorial steward at Petersfield, Hampshire, who has been called the first amateur naturalist – that is, one who studies nature not for personal gain nor as an adjunct to a medical practice, but for its own sake. It was Goodyer who first found some of the special plants of his native county like Hampshire Purslane, Pillwort (*Pilularia globulifera*), and Narrow-leaved Lungwort (*Pulmonaria longifolia*), the first two in his home parish. Conservationists busy digging ponds for Starfruit might recall that Goodyer was the first to find it, on Hounslow Heath. But Goodyer had nowhere to publish his discoveries. Instead, his work became part of a botanical bran tub, dipped into by the compilers of the first national floras, How, Merret and Ray. Most of what is known about Goodyer was brilliantly pieced together by the Oxford historian R.T. Gunther from the chance survival of some of his botanical notes, scribbled on the back of bills and odd scraps of paper.

Thomas Johnson had proposed a great national survey of wild plants – a kind of wild life inventory. He saw himself in that role, as the 'Botanical Mercury', winging hither and thither recording plants and bringing the news to fellow botanophiles. He might have succeeded, for he was still comparatively young when struck down by a Roundhead musket ball. Instead, it became the life work of the great John Ray (1628–1705). Ray's contribution to the study of the British flora is enormous. In his various works written between 1660 and 1704 he added some 190 new flowering plants to the British list. What was even more important, however, was the spirit in which he studied nature. No longer was Nature a mere collection of animate objects to be plundered at will, but the natural *world*, with its own hidden laws and patterns. Ray's natural world was still firmly God-centred. He had himself taken holy orders, and saw nature as a piece of divine 'craftsmanship', full of beauty and ingenuity (his contemporary, Newton, saw God in the same inventive role as the universal Clockmaker). Ray's doctrine lay in uncovering God's designs in nature and so revealing the workings of creation. Since he was fundamentally interested in the mind of God, some of his writing sounds to a modern ear more like sermon than science. But if you substitute 'nature' for 'God', he suddenly flips into focus, as the first naturalist driven by *curiosity*.

Ray's first love was wild flowers. Some time in his late twenties, he was forced by illness to rest from his official duties as a Fellow of Trinity College, Cambridge and spent much time walking and riding in the then fertile countryside about the town. In words that we can warm to so easily, because they often remind us of our own feelings, he explained how 'the flame of my enthusiasm' was ignited.

> I had leisure in the course of my journeys to contemplate the varied beauty of plants and the cunning craftsmanship of Nature that was constantly before my eyes, and that so often been thoughtlessly trodden underfoot. Once I had become aware of these wonders, I ceased to pass them by and treat them as matters unworthy of my attention . . . First I was fascinated and absorbed by the rich spectacle of meadows in springtime; then I was filled with wonder and delight by the marvellous shape, colour and structure of the individual plants. While

> my eye feasted on these sights, my mind too was stimulated. I became inspired with a passion for Botany...[53]

Eager to learn their names and more about them, Ray 'searched through the University, looking everywhere for someone to act as my teacher and guide'. Finding no one at all, he decided therefore to 'advance the study of Phytology' himself. He created his own botanical garden, he compared the plants he found with the crude pictures in his books, and learned the shared characters of related plants so that he could assign unknown flowers to the right grouping. Gradually his painstaking study and exploration was turned into the first county flora, or, as Ray himself called it, 'a Catalogue of all the plants I found in the Cambridgeshire countryside'. It was also the first pocket guide, designed for use in the field. Ray was writing to arouse fellow spirits 'whose concern is not so much to know what authors think as to gaze with their own eyes on the nature of things and to listen with their own ears to her voice . . . '

The Cambridge Flora was a remarkable piece of work. Ray found some 700 species of plants near the City, including some like Fen Orchid (*Liparis loeselii*), Fen Ragwort (*Senecio paludosus*) and Crested Cow-wheat (*Melampyrum cristatum*) that were new to Britain, if not to science. As well as listing them, he also described where they grew, sometimes with such precision that we can visit those places today and, where they survive, see the same flowers as Ray did, like Spiked Speedwell (*Veronica spicata* ssp. *spicata*) on its knoll by the road to Newmarket, or Bloody Cranesbill (*Geranium sanguineum*) on Devil's Dyke, or the Perennial Flax (*Linum perenne* ssp. *anglicum*) on the Gogmagog hills. Some of these places, like Cherry Hinton chalk-pit and Newmarket Heath have been visited regularly by botanists ever since, providing a 300-year-old sequence of records.

Having completed his first 'Catalogue' and also, as a non-conformist, having lost his living at Cambridge Ray became, in effect, a professional naturalist. He had little money of his own (Ray was a blacksmith's son), but his companion and former pupil Francis Willughby defrayed his travel expenses and later left Ray an annual legacy to act as tutor to his children and continue his work. The 1660s were an auspicious time for a naturalist. The troubles of the previous twenty years were at last over, and long-distance travel was made easier by the first good road maps – the famous pictorial 'Britannia' series. From 1660 onwards, Ray and Willughby travelled the roads of England, Wales and Scotland, and later much of Europe, collecting specimens for a projected Synopsis, a methodical attempt to show the natural relationships of all the plants, animals and insects in God's creation. Fortunately Ray kept a travel diary or Itinerary, and so we not only know where they went but also many of the things they saw on the way. They include new plants. On the first of these journeys, in 1660, he managed to cross to the Isle of Man, and on his way from the landing place at Ramsey discovered the yellow mustard-like plant, known today as the Isle of Man Cabbage (*Coincya monensis* ssp. *monensis*) – and, though Ray had no way of knowing it, one of our few endemic plants. The jetty is in the same place today, and so is the plant. Perhaps the most rewarding journey was their fourth, made in 1662, when Ray and Willughby travelled to Snowdonia, in the footsteps of Thomas Johnson, and thence toured the coast as far as Lands End. Among the botanical treasures they uncovered was Blue Gromwell (*Lithospermum purpurocaeruleum*) which Ray describes as 'an elegant plant on a bushy hill near Denbigh'. A boat trip to Bardsey yielded Spring Squill (*Scilla verna*) and Sea Spleenwort (*Asplenium marinum*) and near St Ives, Ray found a dainty, trailing plant he dubbed the 'Bastard Chickweed', though later botanists gave it a more kindly name, Cornish Moneywort (*Sibthorpia europaea*). On the beach of Penzance, perhaps while eating the seventeenth-century equivalent of an ice-cream, he casually recorded two of our rarest wild flowers, Cottonweed (*Otanthus maritimus*) and Purple Spurge (*Euphorbia peplis*). Two July days by the Tamar produced an array of exciting plants, beginning with the Bastard Balm (*Melittis melissophyllum*) 'on a woody bank by a comb to the south of Saltash, growing in great plenty', and, after taking the ferry crossing to Plymouth, the Field Eryngo (*Eryngium campestre*) 'on a rock which you descend to the ferry'. Many of these rare flowers discovered in the reign of Charles II can still be seen in much the same places as Ray found them – you can follow in his steps to admire the Blue Gromwell on its limestone knoll, or the Field Eryngo on rocks near Saltash. But reading Ray, you are occasionally brought back to his time with a bump, when, for example, he and Willughby noticed the head of Scotland's former governor stuck on a tollbooth spike, or when he remarked, in passing, that witches 'to the number of about 120' had been burned *that month*.[182]

Though Ray travelled extensively in the mountainous north of England, he seems to have followed a trail blazed recently by others, notably by Thomas Willisell (d. 1675), an old soldier, who, like the Scottish naturalists a century later, 'could endure hardship and live as well upon oatcake and whig [buttermilk] as another man upon flesh and wine, and ramble over hills and mountains and woods and plains'. Ray called him Tom. Willisell was a paid plant collector – possibly the first – and his travels took him to Edinburgh, where he discovered the Sticky Catchfly (*Lychnis viscaria*), to Nottingham where he found a related plant, which ever since has borne that city's name – the Nottingham Catchfly (*Silene nutans*) – and to Thetford where his hawk eyes spotted the Fingered Speedwell (*Veronica triphyllos*). In 1671, Willisell took Ray on an extended foray through the Yorkshire Dales, Teesdale and as far as Holy Isle, during which time they found Shrubby Cinquefoil (*Potentilla fruticosa*), Dark-red Helleborine (*Epipactis

Field Eryngo (*Eryngium campestre*) still grows near the sea at Plymouth where it was first found, 350 years ago. [Peter Wakely/English Nature]

atrorubens), Angular Solomon's-seal (*Polygonatum odoratum*), Hairy Stonecrop (*Sedum villosum*), Dwarf Cornel (*Cornus suecica*) and Water Lobelia (*Lobelia dortmanna*) to add to the list of plants in *Catalogus Plantarum*. A find which soon took the fancy of gardeners was the Purple Saxifrage (*Saxifraga oppositifolia*), which, Ray noted with characteristic precision, grew 'on the North side of Ingleborough Hill, near a bog by the side of an underground river called Cromock'. Since Purple Saxifrages from Yorkshire were put on sale at Covent Garden for upwards of a shilling a clump, it was just as well that it grew on other hills. Quite a number of alpine flowers were first thought to be confined to Ingleborough, apparently because that was the one hill people seemed prepared to climb! Other exciting finds of that tour were the first record of Alpine Bartsia (*Bartsia alpina*) 'near Orton by a stream running across the road to Crosby' (and still grows there) and 'the Greek Valerian, called by the vulgar Jacob's Ladder' (*Polemonium caeruleum*), which flowered against the splendid backdrop of Malham Cove (and still does, though the spot where Ray saw it has since been grazed flat by sheep). Another hitherto unknown species was the dainty Scottish Asphodel (*Tofieldia pusilla*) which Ray discovered in a bog near Berwick. It was later named after Thomas Tofield (1730–1779), a choleric Yorkshireman and 'patron of the sciences', who was himself the first to find Bithynian Vetch (*Vicia bithynica*) in Britain.

Ray inspired a golden age of botanical discovery. Important among a younger generation of plant-searchers was Edward Lhwyd (1660–1709) who found about 50 species new to Wales, and several new to Britain, like Alpine Mouse-ear Chickweed (*Cerastium alpinum*), Rock Cinquefoil (*Potentilla rupestris*), and Alpine Meadow-rue (*Thalictrum alpinum*). Lhwyd was perhaps the first botanist to specialize in the mountain flora. His greatest discovery, and the one that bears his name, was the Snowdon Lily (*Lloydia serotina*). It says a lot for Lhwyd's powers of observation that this plant was not in flower when he first spotted it; perhaps its pea-sized bulbs appeared on his trowel as he was filling his basket. Lhwyd's explorations also took him to Ireland as far as the hills of Sligo, Kerry and Galway, where he chalked up St Patrick's Cabbage (*Saxifraga spathularis*), and St Dabeoc's Heath (*Daboecia cantabrica*), and also found Shrubby Cinquefoil in much greater quantity than Willisell had by the Tees. Many other naturalists – for we can surely use that word by now – helped Ray in his later years, when his own travelling had had to be curtailed by family commitments. There was Thomas Lawson, a schoolmaster at Penrith and the first of a long line of Quaker botanists in the north of England; and at the opposite end of the country, Isaac Rand, a Chelsea apothecary, and Samuel Doody, who discovered Nit-grass (*Gastridium ventricosum*) at Tunbridge Wells in 1688. Another was Jacob Bobart the younger (1641–1719), keeper of Oxford Botanic Garden who was the first to find Perfoliate or Cotswold Penny-cress (*Thlaspi perfoliatum*) 'among the stone pits between Witney and Burford in Oxfordshire' – and apparently not seen again until the Revd John Lightfoot rode by 80 years later. A close friend and neighbour from Ray's own birthplace at Braintree was Samuel Dale (1659–1739), a physician, who travelled through much of East Anglia in search of plants. It was probably Dale that discovered two rare flowers that share the same species name: the Narrow-leaved Cudweed (*Filago gallica*) and the Small-flowered Catchfly (*Silene gallica*) – an unlucky name, one might think, for the one is now nationally extinct and the other is fast declining.

Ray's celebrated *Synopsis Stirpium Britannicarum* (1690) became the backbone of British botany for almost a century, and many new records were added to the 1724 edition, revised by Jacob Dillenius, like Marsh Saxifrage (*Saxifraga hirculus*) on Knutsford Heath in Cheshire or Field Cow-wheat (*Melampyrum arvense*) in a Norfolk cornfield. In Ray's own lifetime, botanists also used the rival standard floras from Oxford, by William How and Christopher Merret. Unsatisfactory though they were in some respects – 'the world is glutted with Merret's bungling *Pinax*', complained Ray – these works also record many wild flowers for the first time. How's *Phytologia Britannica* owed much to Thomas Johnson, Thomas Penny, and other early botanists, for How himself was too busy to waste time searching for wild plants. Merret's *Pinax* contains rather more original work. Two new species are of particular interest. The London Rocket (*Sisymbrium irio*), attracted attention when its scruffy yellow flowers spread over burnt-out areas after the Great Fire of London in 1666 (see Chapter 8) – an occasion Merret had cause to remember, since it consumed most of the first edition of his *Pinax*! The other was the Hairy Vetchling (*Lathyrus hirsutus*) which grew on the earthen ramparts of Hadleigh Castle in Essex for over 300 years until over-zealous tidying-up removed it in the 1950s. Many early botanists, including Ray, Dale and John Blackstone, paid their respects to it: it seems odd now that much was made of a now little-known plant whilst a large proportion of our rare flora was still unknown.

The flora unfolds

The discovery dates of rare wild flowers provide a rough idea of levels of botanical activity during the past 400 years. The springtime of discovery lay roughly between the publication of Gerard's Herbal in 1597 and Thomas Johnson's expedition to Snowdon in 1639. Those were the days when someone like George Bowles could discover two new species (Touch-me-not Balsam and Ivy-leaved Bellflower, *Wahlenbergia hederacea*) on a single outing. One senses an awakening of enthusiasm for flowers during this period. To Thomas Parkinson, for example, a garden was literally heaven on earth, and people like Bowles, Goodyer and Johnson seem to have been working within an already well-established tradition of botanical recording. After the grim interlude of the Civil War and the Commonwealth came a brilliant period inspired by John Ray, and what amounted to a systematic 'stocktaking' of Britain's wild life. Wonderful localities were discovered along the way, like Tor Bay, Breidden Hill, the Lizard, Teesdale, Cheddar Gorge, Malham Cove and Cader Idris. Like-minded people began to communicate their thoughts and finds, mainly by correspondence. New species were now described scientifically using the principles established by John Ray and later developed by Linnaeus and others, and their relationships established. There was the beginnings of co-operation between the local naturalist and the expert, a sharing of skills.

Take the discovery of the Cornish Bladderseed (*Physospermum cornubiense*), an umbellifer with large inflated seeds, found mainly in Cornwall. It first came to notice after the Revd Lewis Stephens of Menheniot collected specimens from a local wood, and, not finding it in Ray's *Synopsis* or any other work in the vicarage library, sent them off to Adam Buddle (1660–1715). Buddle is forgotten now, but he wrote a manuscript English Flora and was so highly regarded in his day that Linnaeus named a colourful bush in his memory – the Buddleia. Buddle was unable to name the plant either, and concluded, correctly, that it had not yet been described. He sent it on to James Petiver in London, who ran a kind of production line for new species, writing formal Latin descriptions amid a delightful squalor of mouldering plants, boxed insects and dusty manuscripts. On the curling herbarium sheet Petiver wrote, in his crabby hand, '*Tragoselinum maximum cornubiense umbella candida. A. Buddle a D. Stevens e Cornubia missum*' – a big white umbellifer found in Cornwall sent to me by Buddle and Stevens from Cornwall. When he came to the plant, Linnaeus shortened the name, but he decided to retain *cornubiense*, the Roman name for Cornwall, for its species name. And that is why a plant found in many countries still bears the name of Cornwall, the English county in which it was first discovered. The original well-travelled specimens passed in due course from Buddle's private herbarium into the vast collections of Sir Hans Sloane, and today are in the Natural History Museum, as the type material of the Cornish Bladderseed.

The discovery of Cornish Bladderseed came towards the end of field botany's golden age. Much of the impulse to record plants had seemed to fade during the middle of the eighteenth century. Until the publication of William Hudson's *Flora Anglica* in 1762, naturalists had to make do with their old copy of Ray's *Synopsis*. Ray's genius had departed abroad to Sweden, where Linnaeus was busy adapting many of Ray's ideas to bring order out of chaos and lay the foundations for a scientific system of naming species based on their sexual characteristics – in our case, the flowers. What sold the system to British naturalists, however, was the handy way in which Linnaeus had condensed 'scientific' names into just two words – the name of the genus followed by the species. Hudson used the new system in his popular flora, and by the end of the century Linnaeus was all the rage. It was through his leading advocates in Britain that the modern era of scientific botany was forged.

The greatest single advance of the late eighteenth century was the lasting establishment of learned societies, together with journals in which work could be published without delay, and in detail. An early symptom of the botanical revival was the publication of a guide that was not only accessible but – at last – written in English: William Withering's *Botanical Arrangement*. It went through several editions from the 1770s, and announced the discovery of several new wild flowers, among them the Strapwort, complete with its new Linnaean name *Corrigiola litoralis*, meaning 'little string on the sand', which is apter than the English name. In 1788, James Edward Smith (1759–1828) and six others founded the Linnaean Society, which published its first *Transactions* three years later – our oldest periodical devoted to natural history. Most plant discoveries, however, appeared in another landmark publication, *English Botany*, often called Smith's English Botany from its principal compiler, or Sowerby's English Botany (to the great displeasure of Smith) from its brilliant illustrator, James

Cornish Bladderseed (*Physospermum cornubiense*), discovered by a Cornish vicar, is so-named from its distinctive inflated seeds.
[Bob Gibbons]

Sowerby. *English Botany* came out in 36 volumes between 1790 and 1814, and, among other things, reported the fruits of the great renaissance of field botany in Scotland, following the publication of Lightfoot's *Flora Scotica* in 1777 (see Chapter 3).

One of the most beautiful and original works of the century was the *Flora Londinensis* by William Curtis (1746–99), which appeared in folio parts from 1775 to 1798 with 435 sumptuous hand-coloured plates. Its scope is not confined to the London area, and the work was on hand to record and illustrate an interesting new discovery from a heath near Axminster in

Heath Lobelia (*Lobelia urens*), with flowers similar to garden Lobelias, contains a caustic juice that half-blinded its first illustrator.
[Bob Gibbons]

Devon. This was the Heath Lobelia (*Lobelia urens*), first found in 1768 by Lord Webb Seymour who described the site with the precision of a more innocent age. It was Curtis himself who learned what was meant by its specific name *urens*: 'Mr Sydenham Edwards, my draughtsman . . . having handled a branch of this plant broken off from the main stem, and afterwards rubbed his eyes slightly, had a violent pain and temporary inflammation excited in them thereby, which however soon went off, on washing them with cold water.'[36] It also had, they found, 'a hot and burning taste'. The Heath Lobelia is still found on its home ground near Axminster, though most of its heath is now a wood. Nearby is a house called Lobelia Cottage. Geoffrey Grigson went there to find the flower blooming 'with a pleasant incongruity among hen-coops and the rusty ends of old bedsteads and old tyres sliced in half for chicken troughs.'[75] The plants are now looked after by the Devon Wildlife Trust, who harrow the ground from time to time to create space for the seedlings.

By the first decades of the nineteenth century we are already moving into the sunshine of the Victorian age, with the famous *British Flora* (1830) by Sir William Hooker, Director of Kew Gardens, and a spate of new species reported in the journal of the Botanical Society of London. Between 1828 and 1848 hardly a year went by without at least one new species to add to the flora. This was the high summer of British field botany when the great names of the early Victorian age were in the field – H.C. Watson, Charles Cardale Babington, J. Hutton Balfour, Edward Newman, the Backhouses, William Borrer and George Bentham among others. New discoveries were made in old localities, like the trio of rare clovers found at the Lizard between 1838 and 1847, or the Backhouses' discovery of Teesdale Violet and Teesdale Sandwort (see Chapter 3). Other rich localities were being explored almost for the first time, like the Isles of Scilly. Unlike the discoveries of earlier generations, a large proportion of nineteenth-century first records are of rare or critical species. Many are plants that only a botanist would notice, like the Dwarf Spike-rush (*Eleocharis parvula*) or Red-tipped Cudweed (*Filago lutescens*). But some remarkably attractive flowers had slipped through the net. Perhaps one reason why no one had recorded the Wild Gladiolus (*Gladiolus illyricus*) lurking among the bracken in New Forest glades until 1856 was that it was assumed to be a garden escape (see picture on p. 36). Certainly H.C. Watson thought so, when its discovery by W.H. Lucas was made known, on the grounds that it grew suspiciously near some planted Rhododendrons. It was only when the Gladiolus began to turn up in remote places throughout the Forest that people acknowledged that it was beyond doubt native. It is even possible that our plants, so much daintier and particular about where they grow than those in mainland Europe, are a distinct subspecies.

One family that had confused the earlier field botanists was the sedges. Quite a crop of them had been left to the Victorians to sort out, including such distinctive ones as Bird's-foot Sedge (*Carex ornithopoda*), not noticed until 1874 – though John Ray had probably trodden on it. The so-called Rare Spring-sedge (*Carex ericetorum*) is a good example of Caricetal obscurity. The first specimens had been gathered near Cambridge in 1833 – by C.C. Babington no less, though they were not correctly identified until 1861. Then older specimens of this plant began to turn up in herbaria, like the one mislabelled as Pill Sedge collected from Newmarket Heath in 1829. Very probably Sir John Cullum's 'Carex montana', collected from Newmarket

Heath in the previous century was also *Carex ericetorum*, for the real *C. montana* does not grow there. Rare Spring-sedge, it seemed, was an East Anglian plant, and no one thought of looking for it anywhere else. Then, in 1944, E.C. Wallace found it in Yorkshire. Since then it has turned up in a scatter of places on dry soils between Hertfordshire and Cumbria. It is probably not rare at all but merely overlooked. If so, it needs a new name – perhaps the 'Not-All-That-Rare Spring-Sedge'?

The Victorians made a cult of rare flowers and ferns, which led to competitive herbarium-stuffing as botanists attempted to obtain specimens of every species of wild flower and fern. This sense of chase must have spurred on discovery. The epitome and champion of the 'twitchers' was surely G.C. Druce, whose tireless pursuit of rare flowers was proverbial. Oddly enough, and not for want of trying, Druce discovered few new species himself – or rather, he thought he had, but few of them have stood the test of time and have been seamlessly reabsorbed.[4] An undoubted 'supertwitch' was his discovery of Loddon Pondweed in 1893. It was thought to be new to science, and Alfred Fryer named it *Potamogeton drucei* in Druce's honour. Unfortunately later workers were able to show that *P. drucei* was in fact the same plant as the Continental *Potamogeton nodosus*, and *drucei* had to go into the bin. Another species described but not found by Druce was the Thistle Broomrape (*Orobanche reticulata*), a plant which now has its own newsletter. But his finest hour was perhaps the rediscovery of a much dimmer plant, the Somerset or Dillenius' Hair-grass (*Koeleria vallesiana*). What made this find so unusual is that the specimen had been dead for 200 years. One day, in 1904, Druce was going through some unlabelled sheets of plants, collected by Jacob Dillenius in 1726, when he found three specimens of a grass unknown to Britain. They turned out to be *Koeleria vallesiana*, a southern European species whose nearest station lies south of the Loire. Ingenious detective work led to the discovery of a loose label and notes that indicated that the plants had been collected at a place called Uphill, near Weston-super-Mare. Druce was off there like a shot and, as he puts it, 'within a quarter of an hour I succeeded in finding some flowerless specimens growing in the turf . . . and eventually saw it in plenty on the steep limestone terraces near the great quarry which faces the sea'.[4] It still grows there in quantity, easily noticed by its tight clumps of blue-green, Thrift-like leaves. Strictly speaking, the grass was discovered by Dillenius, not Druce, but the chase must have given the great man a rare satisfaction.

The noonday flood of new plants has long since ebbed into a trickle, but discoveries continue to be made. The last years of the Victorian age produced yet more sedges: Club Sedge (*Carex buxbaumii*) and String Sedge (*Carex chordorhiza*), in different but equally God-forsaken bogs in northern Scotland, and Limestone Woundwort (*Stachys alpina*) in a coppiced glade near Wootton-under-Edge in the Cotswolds. The twentieth century flora

Half-hidden under a canopy of bracken, the Wild Gladiolus (*Gladiolus illyricus*) needs searching for on hands-and-knees despite its beauty. Perhaps that is why no one apparently noticed it until 1854. [Bob Gibbons]

began well with the Welsh Mudwort (*Limosella australis*) in 1901, and may end, as far as new discoveries are concerned, with the Proliferous Pink (*Petrorhagia prolifera*) in 1992, during which time we have added Scottish Dock (*Rumex aquaticus*), Breckland Speedwell (*Veronica praecox*), Lundy Cabbage (*Coincya wrightii*) and Shetland Pondweed (*Potamogeton rutilus*) among many other new native plants (not to mention the scores of new 'aliens' and micro-species). The odd thing about the most recently discovered flowers is that some of them are quite glamorous. We somehow managed to miss a native tree (*Sorbus domestica*), a gentian with a blue flower more than an inch across (*Gentianella ciliata*) and a Star-of-Bethlehem as bright as a gold bar (*Gagea bohemica*). With such blind-spots, we have surely overlooked a few more Mudworts and Hair-grasses. From its springhead in the Tudor herbals our rare flora continues to swell as someone, somewhere, examines a plant more closely than anyone else, or in a different way, or simply finds a new one by accident.

Chapter 3

Botanical stations:
finding rare flowers

A love of field botany and a love of maps go well together. There cannot be many wild flower lovers who have not spent hours poring over the distribution maps in the *Atlas of the British Flora* or their incredibly detailed equivalents in the county floras, where you can trace the occurrence of a plant along the river valleys or observe its confinement to certain geological formations. At a more personal level are the sketch maps we send to one another with directions on how to find such-and-such a plant. I have known people to keep a journal with beautifully coloured maps on where a rare species is to be found in which 'X' marked the spot, as if it were buried treasure. They come with all sorts of field signs and directions to follow – I seem to remember three tortoiseshells set in a stone wall in Galway pointing towards a patch of Giant Butterwort. My all-time favourite is the directions to finding *Ajuga genevensis* on some Cornish dunes, culminating in the remark that 'it is at the top of the letter 'L' in "EXPLOSIVES" '.

The most interesting maps in the *Atlas of the British Flora* are those of relatively widespread species, which you can play with using the transparent overlays of rivers, counties, rainfall, etc. which the publishers have thoughtfully provided. The rarest flowers, on the other hand, have only a few dots, and taken individually the maps are not usually very informative. It was not until Ron Porley decided to combine together *all* the recent records of the 300-odd 'Red List and Near Threatened' species using English Nature's sophisticated computer-mapping system, that a truly astonishing overall picture emerged of where rare plants grow. The map looks like the distribution of a single widespread but extremely local species. You can pick out the Chilterns, the North and South Downs and the Mendips by their troops of dots, and the even denser 'dot-cities' marking the Breck, the New Forest and Poole Harbour, the coastlines of Cornwall, Dorset and North Devon, the Yorkshire

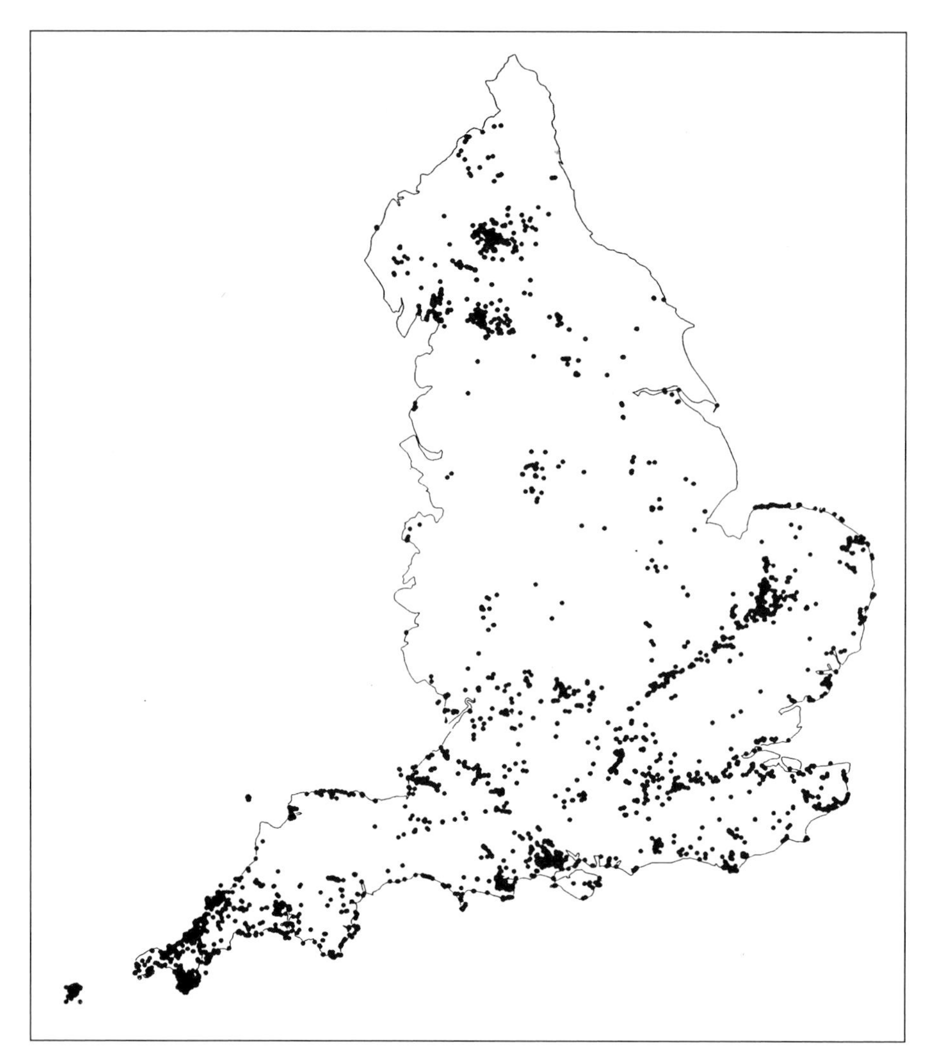

This map, produced by English Nature, plots the distribution of all 200 or so of the rarest plants in England. In effect, it records the richest places for wild flowers: see how they hug the coast and the chalk and limestone and with isolated clusters in places like Breckland, Teesdale and the New Forest. [English Nature]

Dales, Upper Teesdale and the Isles of Scilly. There are smaller clusters – the county towns as it were – in places like the Norfolk Broads, Dover, the White Peak, Silverdale and the Tamar valley. Here, if it is needed, is the proof that rare species live in exceptional places. Put together, the areas of high dot-density would cover an area not much larger than Kent. By contrast, much of England has few or no rare plants. English Nature has emphasized this by superimposing on the dot map their 'natural areas' map which highlights those big bare spaces in the Midlands, eastern England between Teesmouth and the Fens, and the East Anglian interior. The polarity between rich areas and barren ones must have been widened by housing and modern agriculture, and some of the best agricultural land now has the fewest rare plants. But much of this pattern is natural. Most rare flowers are very choosy about where they grow. In a sense, our map shows the *peculiar* places of England, the wildest nooks and the islands of strange geography, like inland sand, toxic rocks and frost-free coastlines. It may also reveal the areas that remained relatively open when most of England was dense wild woodland (I will go into this a bit more in Chapter 4).

A map produced of the species in *Scarce Plants in Britain* (1994)[184] – the ones found in between 15 and 100 ten-kilometre grid squares – also shows an overall clumped pattern, but these species are much more widespread. There are few parts of England where there is not at least one scarce species. On the other hand, the chances of having a Red-listed plant in your parish are only about one in ten.

This chapter is about localities that are unusually rich in rare flowers. The forthcoming Red Data Book of wild flowers will describe the ecology of these places, and what I prefer to do instead is to maintain the historical approach of the previous chapter, describe the circumstances in which these famous communities of flowers were revealed, and look in more detail at some of their botanical highlights, and what they tell us about the locality. Because of their special allure, I also include a short history of our orchid flora, the one family where every species has rarity status, if not in fact then at least by honorary association.

The Avon Gorge

The Avon Gorge, near Bristol, is the original rare plant 'locality'. William Turner had gone there to pluck his seminal root of Honewort 450 years ago and in his footsteps came many of the great names of British botany, some of whom discovered, and plucked, rarities of their own. It adds a certain *frisson* to know that you are treading where Johnson, Gerard, Ray, Curtis and Sir Joseph Banks have trod, and might even be admiring the same plants as they did. It has to be said that you need all the incentives you can find to stay long on the Bristol side today, as you are forced to cross and recross a busy main road, sneak round various fences and gates, and put up with the roar of traffic, and the sight of litter piled up against the rocks. But the flowers do not seem to mind, and most of them can be seen without climbing. The odd thing about the flora of Avon Gorge is that it took botanists about 400 years to get to the bottom of it. Either Turner, Gerard, etc. were a bit short-sighted, or else they saw only what they wanted to see – useful plants. The season here is an unusually long one. Hutchinsia or Rock Pepperwort (*Hornungia patraea*) flowers not so much on the first day of spring as on the first *hint* of spring; it can flower happily in melting snow. On the other hand, to see Autumn Squill (*Scilla autumnalis*) and some of the local garlics and whitebeams at their best, you need to return in August or September, while a midsummer visit would miss early flowering specialities like Bristol Rock-cress (*Arabis scabra*) or Dwarf Mouse-ear (*Cerastium pumilum*).

Turner's famous Honewort was the object of a celebrated botanical

St Vincent's Rock today, somewhat compromised by the bridge and covered road. This site has the longest sequence of rare plant records in Britain. [Peter Marren]

muddle. A report had been made of the presence near Bristol of another umbellifer, *Meum mutellina*, a continental plant useful as a condiment or spice. No less a personage than John Gerard was sent there to gather it, and, in his words, 'upon Saint Vincents rocke by Bristowe, I spent two daies to seeke it, but it was not my hap to find it, therefore I make some doubt of the truth thereof'. The *Meum* (now called *Ligusticum mutellina*) has never been found wild in Britain, and there seems little doubt that Gerard had been sent on a wild-goose chase. His near-contemporary John Parkinson, suspected that someone had mistaken Honewort for the missing *Meum*.

We do not know whether Gerard found anything else to compensate for his two days on St Vincent's Rock. One other local plant the herbalists went there to collect was 'wild Asparagus': Parkinson comments on its abundance near Bristol; Johnson found it in the river marshes below St Vincent's Rock in 1631, and it was still there in 1767, when Sir Joseph Banks gathered it. This was unlikely to be the true Wild Asparagus (*Asparagus officinalis* ssp. *prostratus*), however, which is a plant of coastal cliffs, dunes and ungrazed islands. As the herbalists looked for plants like these, the real native flora of the Gorge began to be unveiled, bit by bit. Thomas Johnson came here with a band of pleasant-living medical colleagues and a local guide, John Price, 'a not less learned Apothecary than a jovial companion'. Unfortunately Johnson's journal for the trip is even more concerned than usual with the fine meal he enjoyed on 'laden tables crowned with vines', and he seems to have been taken more with rock crystals and an interesting spring of tepid water than with the plants they found. Goodyer came this way too, to rather greater purpose, for he found the first British Rock Stonecrop (*Sedum forsteranum*) and Western Spiked Speedwell, both new to Britain at that time. John Ray passed through the Gorge in 1662, and again in 1667, and added the Autumn Squill, Hutchinsia, and the rayless variety of Sea Aster (*Aster tripolium*), still common on the tidal mud by the river. A much rarer plant, the Bristol Rock-cress, was reported to Ray in 1686 by his helper, Dr James Newton. This became another famous flower every naturalist wanted to see; in 1773, for example, John Lightfoot and Sir Joseph Banks sought it out on a rock ledge 'a little above high-water mark beyond the new well-house', on the same visit that they added Compact or Madrid Brome (*Bromus madritensis*) to the flora. Of this botanic journey, which also took in the Wye Valley and parts of Wales, Lightfoot recalled that 'I can truly say I never became a Party in any Scheme which afforded me more Satisfaction or sincere Delight . . . I believe it may be said without vanity that few, if any, Botanical excursions in Great Britain have exceeded our Collection, either in Number or Rarity of Plants or Places.'[203]

Still they came, on foot and by coach: curates, physicians and philosophers. William Curtis, author of *Flora Londinensis*, added Stiff Saltmarsh-grass (*Puccinellia rupestris*), and Dwarf Mouse-ear, then new to science, in 1793. A Dr Broughton discovered Slender Hare's-ear (*Bupleurum tenuissimum*), in what one could be tempted to call the usual place, below St Vincent's Rock. The Bath-based surgeon William Sole trumped that with the first British record of Dwarf Sedge (*Carex humilis*) – a remarkable one, since the Sedge is quite hard to find on its rock ledges at Bristol, and yet it carpets acres and acres of downland in Wiltshire and Dorset, and once even grew inside the circle at Stonehenge. Clearly more botanists went to Avon Gorge than to Salisbury Plain!

The Avon Gorge is unique in that its special plants were discovered so early, and by such eminent people. One species, however, went unrecorded until July 1847, when Dr H.O. Stevens found a strange leafless *Allium* 'on

Below left: The genus *Sorbus* includes many closely related small trees which reproduce without sex and grow mainly on isolated rock outcrops, especially hard limestones. This is Wilmott's Whitebeam (*Sorbus wilmottiana*), found wild nowhere else in the world outside Avon Gorge. [Peter Wakely/English Nature]

Below right: Round-headed Leek (*Allium sphaerocephalon*) has its only mainland British site at St Vincent's Rock, where it may be a nineteenth-century introduction. [Andrew Gagg's Photoflora]

steep declivities of the cliffs' at St Vincent's Rocks, by now crowned by the Clifton Suspension Bridge. This was the Round-headed Leek, also known as Round-headed Garlic (*Allium sphaerocephalon*). As anyone that has seen it will know, this is a conspicuous plant, which projects from crevices and ledges on the bare, white limestone slabs like long purple lollipops. How could it possibly have been missed? – especially as a member of the onion family would certainly have interested the apothecaries. It could surely not have been, and there are some clues to its possible origin. The Garlic has increased since J.W. White counted 20 plants in 1882, despite having suffered 'from the pranks of scrambling boys who, in attempting to gather the flowering heads, pull up the root and all from the thin loose soil'.[203] It has also turned up in another part of the Gorge, on Durdham Down. It seems that certain Bristol flower-lovers used to scatter the seed of garden plants, a foolish practice compared by White to 'painting the lily and gilding refined gold'. They seem to have been particularly fond of the onion family for there is quite a collection of wild Alliums at Bristol – Rosy Garlic (*Allium roseum*), Honey Garlic (*Nectaroscordum siculum*) and Field Garlic (*A. carinatum*) are locally common on roadbanks; Keeled Garlic (*A. carinatum*) has spread into grassy glades in nearby Leigh Woods, and now looks quite wild there; Few-flowered Garlic (*A. paradoxum*) is well-established and increasing in the Gorge.[64] Once established, wild leeks, garlics and onions can be extremely persistent, and may come to dominate verges, banks and cliff-tops at the expense of the native vegetation. It seems very odd to me that Round-headed Garlic is an accepted native species at Bristol, whilst most of these other Alliums are rightly regarded as introductions. Perhaps one explanation is that the Round-headed one behaves itself, while some of the others are highly invasive and a great nuisance. My theory is that someone introduced it by dropping bulbils from Brunel's newly built suspension bridge, for it was first seen some years after the latter was opened. It grows wild nowhere else in the British Isles outside the Channel Islands, and probably established at Bristol thanks to an unusually warm, mild local climate.

The great botanical discoveries of the twentieth century lie not so much in the wild flowers of the Gorge but its trees. In the 1930s, E.F. Warburg and A.J. Wilmott began to sort out its whitebeam trees. They distinguished several different kinds on the basis of leaf-shape and other characters, some of which had been named before, but others were new to science, and two of them, *Sorbus bristoliensis* and *Sorbus wilmottiana*, have not been found anywhere else in the world; they almost certainly evolved here. These interesting trees occur in some of the richest woodland found in Britain, and containing other scarce natives like Wild Service-tree (*Sorbus torminalis*) and Large-leaved Lime (*Tilia platyphyllos*). It seems probable that, despite the proximity of a large city, with its bridges, street lights and busy roads, the slopes of Avon Gorge are an ancient landscape which has preserved plants that have long since died out elsewhere. I recommend whitebeam-hunting at Bristol. Once you learn to recognize the different species, with the help of the arboretum in Nightingale Valley, you will soon start to find them on the slopes and quarries on the Somerset bank. *Sorbus bristoliensis* is a particularly fine tree, with a graceful shape and glorious orange berries. The elders of Bristol should have its portrait in the City Hall. From certain viewpoints you are looking at most of the world's population.

Concern for the survival of this special flora is not new. In the first volume of *The Phytologist*, in 1841, T.B. Flowers claimed that one individual had offered rewards for Bristol Rock-cress 'in order to render it scarce'. Later in the century, White and others complained bitterly about the 'huge and hideous quarries' now marring the Somerset bank, accusing their fellow townsmen of 'selling the sublime and beautiful by the boatload'. On the Bristol bank, meanwhile, the City authorities were unobtrusively, but with far more damaging effect, planting exotic and invasive trees and replacing natural grass with lawn grass, thus restricting the rare plants to steep rocky slopes 'for which the golfer and footballer have no use'. Fortunately the now disused quarries and their railway have become good refuges for wildlife. The main threat today comes from another quarter – sycamore. It has invaded so much of the formerly open gorge, and none of its numerous conservation designations have made any difference. Indeed, many people seem to regard the wooded slopes as an improvement. But there we must take our leave of this resonant and paradoxical place.

Upper Teesdale

Long ago, and for only a brief summer, my workplace was the treeless fells and windy summits of Upper Teesdale. Except at weekends you hardly met anybody, and my usual companions were the melancholic pipe of the Golden Plover, and the laryngeal squawk of angry Lapwings, sometimes stooping so low in their rage that you felt the breath of their wings. Once, though, I fell into conversation with a walker, who had been watching me push cocktail sticks into the turf with a look of generous amusement (I was recording gentians). After a moment or two, he decided to remind me of the beauties of the scene. 'What a magnificent wilderness,' he exclaimed. I could see what he meant. 'You know,' he went on, confidingly, 'I doubt this view has changed much since the beginning of time.' 'Apart from your cocktail sticks,' he might have added. If I seemed a little reluctant to agree with him, it was because his view consisted of a reservoir, some disused lead mines, a grouse moor and a wireless mast on a distant hill. And he was standing on a tarmac road as he said it. But, botanically speaking, he wasn't far wrong. Although, on the map, the famous names of Widdybank and Cronkley Fell

An isolated farm set among hay meadows and stone walls in the heart of Upper Teesdale. Virtually every uncultivated verge, bank and meadow here has a rich flora.
[Bob Gibbons]

look like any other square mile of northern Pennines, they are on the ground a botanical oasis where rare alpines, sedges and lady's-mantles crowd together leaf by leaf in sheep-cropped turf scarcely more than an inch high. This is one of the coldest parts of England, more like Iceland than Cornwall, and that, coupled with the thin, open limestone soils, has maintained these few square miles of turf in a kind of time-warp. If you want to see what an Ice Age field looked like, go to Upper Teesdale. Here are flowers whose heyday was 12 000 years ago, before woods filled the dales, and long before the high meadows turned into peat bogs.

Few people did pass this way until the end of the eighteenth century. The first known botanist to visit the dale was old Tom Willisel, Ray's faithful collector and guide. He evidently missed the high fells, but crossed the Tees somewhere near Middleton and saw Shrubby Cinquefoil on the river shingles. In his steps came Thomas Lawson the Quaker, who also filled his trug with the cinquefoil after admiring the waterfall at High Force. But the first person to unveil the real treasures of the dale was a resident lead miner, John Binks, who lived at Middleton-in-Teesdale. Perhaps he was an official rather than a labourer – James Backhouse described him as resembling 'a smart little French doctor'. In April 1797, Binks took an unnamed wild flower to a local physician, and also to the Revd John Harriman, curate of Eggleston, who identified it and sent specimens to Sir J.E. Smith in London. This resulted next year in a note in *English Botany* announcing the discovery of the Spring Gentian (*Gentiana verna*) in England. Harriman received all the credit, but it seems that the locals had long known about the pretty blue flowers that appeared every spring, for Smith was told that 'the inhabitants of the forest know it well by the name of Spring Violet'.[72]

Such discoveries attracted botanical tourists to the dale, among them

Nathaniel Winch, author of an early *Botanist's Guide to Northumberland and Durham*, and Edward Robson, who produced one of the earliest printed lists of rare northern plants. Between them, they put on record many more special plants, including the first British record of Tufted Sedge (*Kobresia simpliciuscula*). Another regular visitor was James Backhouse senior, who befriended John Binks and went on excursions with him. Backhouse and his son James Backhouse junior were the first to record two of the most celebrated Teesdale plants. One, the Teesdale Sandwort (*Minuartia stricta*), is a maddeningly elusive flower, more like a pearlwort than a sandwort, and hard to find even when you know where to look. In a hot summer the tiny white florets are over in days, and in a cool one they hardly bother to open. It was discovered 'by an alpine rill on Widdybank Fell' in June 1844, and the younger Backhouse recalled the moment in his diary: 'To my late friend, Mr G.S. Gibson of Saffron Walden, belongs the credit of this discovery. We

Teesdale Sandwort (*Minuartia stricta*) is an arctic flower with a single outpost in Britain at Upper Teesdale. [Andrew Gagg's Photoflora]

Teesdale Violet (*Viola rupestris*). Is there a special magic to flowers with a place in their name? [Peter Marren]

were together, and I believe saw the plant almost at the same moment; but that he first said, "what is that?" fixing my attention specially upon it.' Backhouse Senior wrote to Sir William Hooker, Director of Kew to tell him that: 'In the course of our Teesdale visit this season, we have met with the accompanying little plant, which we suppose to be *Spergula saginoides* [Mountain Pearlwort]. What is thy opinion of it?' (Backhouse was a Quaker, as well as a Yorkshireman, and stuck to his thee, thy and thous.) Hooker identified it as a rare sandwort hitherto known only from Lapland, and the specimen, accompanied by a note by Backhouse, is now in the Yorkshire Museum at Bowes. Though no one but a botanist would look twice at it, the Teesdale Sandwort is a plant of great biogeographical interest as a high arctic plant with a single outpost in Britain.

The other flower named after the dale is the Teesdale Violet (*Viola rupestris*). It too came under the discriminating eye of the Backhouses in 1862, 'growing upon the sugar limestone at the upper end of Teesdale on the north side of the river'. Unfortunately, the Backhouses' journal for the period has been lost, and so this time we are unable to savour the moment with them. By 1862, a great many botanical feet had trodden the turf of Widdybank Fell, and probably on the violet. But it too has a short season, and can be confused with the upland form of Common Dog-violet, (*Viola riviniana*), which grows in similar places. Once noticed, the Teesdale Violet is a fairly distinctive little plant, with its neat, rounded, hairy leaves, and pubescent flower stalks. Even so, it is probably no accident that it was the Backhouses who found it, for they were hawkweed specialists, and so used to finding minute differences in closely related plants.

In just a few decades from 1797, Upper Teesdale had become renowned for its rare and exciting alpine flowers. Reaching the hallowed ground became much easier when the railway from Darlington was built with a halt at Middleton. From the 1840s, the Backhouses made regular, almost annual, visits, and for every Backhouse or Robson there were dozens of other plant lovers who followed in their steps to see the latest Teesdale plant. Unfortunately the fame of Teesdale distracted attention from other parts of the North Pennines, so that while the dale itself was intensively botanized, more distant hills and valleys were ignored. For a long time the Teesdale Violet was known only from Widdybank, yet in fact it occurs in limestone grassland across northern England. Another unfortunate aspect of fame is that it tends to attract the wrong sort of visitor. Whether or not the sales of potted Spring Gentians and Bird's-eye Primroses did any permanent harm is uncertain, but collecting by botanists and nurserymen certainly eradicated the Oblong Woodsia (*Woodsia ilvensis*), and nearly wiped out the Holly Fern (*Polystichum lonchitis*) and various mountain hawkweeds, which are especially vulnerable to over-collecting as their populations are often small and circumscribed.[72]

Although wild-flower lovers continued to visit Upper Teesdale, botanical discovery lapsed for half a century until the 1940s, since when many new and surprising discoveries have been made. Among the most interesting are the lady's-mantles, first studied here by Max Walters, and later by Margaret Bradshaw.[69] Possibly because most of them grow on road verges and in meadows further down the dale, the lady's-mantles seem to have been overlooked by the hill-orientated Victorians. Upper Teesdale and nearby Weardale have in fact more species of lady's-mantle than anywhere in Britain. Other overlooked plants included Rare Spring-sedge, first found in 1949 by T.G. Tutin, Alpine Foxtail (*Alopecurus borealis*), in 1945 by J.K. Morton, and, most surprisingly, Dwarf Birch (*Betula nana*), discovered in a boggy part of Widdybank as late as 1974. Each new discovery adds a little to the dale's significance as an outpost and meeting place of northern and arctic-alpine plants: the 'Teesdale Assemblage'. Better known to the wider public was the outcry over the construction of the reservoir at Cow Green in the 1960s, built to regulate the water supply to ICI's chemical factories in Teesmouth. The rising waters drowned the lower slopes of Widdybank Fell, including some of the precious sugar limestone and its associated alpine rills. By a terrible irony the sedge *Carex aquatilis* was discovered only months before its site was drowned by the rising waters. Several rare flowers, including Teesdale Violet, were significantly reduced. On the other hand, the case was a milestone along the way to a more sympathetic treatment of landscape and wildlife. And the 'blood money' offered by the developers in compensation gave the botanical world an unparalleled opportunity to study the Teesdale flora. The consequent work on vegetation history and mapping, and the ecological studies of the individual species, have contributed significantly to the study of the British flora.[29]

The Lizard

One good way to get the most out of a natural desire to travel to the extremities of Britain is to look for wild flowers. Go north to Dunnet Head in Caithness and you should soon find *Primula scotica* along the cliff path. At the opposite end of Britain, Wild Cabbage and Early Spider Orchid grow near the walls of Dover Castle. But the most rewarding extremity of all is the Lizard peninsula, the 'second toe' of Cornwall, simultaneously the southernmost and almost the westernmost part of England (especially as Land's End itself is relatively dull, give or take the odd sea-lavender). You can find more rare flowers in half an hour at the Lizard than anywhere in Britain.[22]

The Lizard is a remote place, and most of its flora was discovered in the railway age. However John Ray came this way in 1667, during one of his perambulations of Britain, and he was the first to record the Cornish Heath

The Lizard coast near Mullion.
[Bob Gibbons]

(*Erica vagans*), or, as he called it, 'Juniper or Firre-leaved Heath with many flowers' which grew 'by the way-side going from Helston to Lizard-point in Cornwall, plentifully'. Having reached Lizard Point itself, and passing lilac-trimmed acres of 'Firre-leaved Heath' all the way, he added two much less conspicuous species, Fringed Rupturewort (*Herniaria ciliolata*) and Wild Asparagus.[182] But his were the first and last botanical records for nearly a century.

Two rare clovers, discovered within a few years of each other, and confined mainly to the Lizard: left Upright clover (*Trifolium strictum*) and right Long-stalked or Lizard Clover (*Trifolium incarnatum* ssp. *molinerii*). [Bob Gibbons]

The Lizard was really put on the naturalist's map by the Helston schoolmaster, the Revd C.A. Johns, who revealed some of its natural glories in *A Week at the Lizard* (1848). Having heard a great deal about the scenic beauty and rock plants of Kynance Cove, he 'determined to see for myself whether accounts I had heard were true'. They were, and over the years many more rare plants were found, some of them new to the British flora. The most celebrated are the three 'Lizard clovers'. The most conspicuous and first found was the Long-headed Clover (*Trifolium incarnatum* ssp. *molinerii*), discovered by the Revd William Strong Hore in 1838, close to the Lizard lighthouse (where I saw it exactly 135 years later). Twin-headed Clover (*Trifolium bocconei*) turned up a year later, when Charles Babington and William Borrer found it on a wall top near Cadgwith, where it persisted until the 1960s. And, finally, Johns himself came up trumps in 1847 when he found Upright Clover (*Trifolium strictum*) at Caerthillian Cove, where all three rare clovers grew side by side. This was the occasion of Johns' celebrated 'hat-trick', when, as he noted, 'So abundant are the Leguminosae at this spot that I covered with my hat *Trifolium Bocconi* [*sic*], *T. strictum*, *T. molinerii*, *T. striatum*, *T. arvense*, *Lotus hispidus* and *Anthyllis vulneraria*

Pygmy Rush (*Juncus pygmaeus*) has been unkindly called our ugliest rare plant. It lives in cart-ruts on Cornish heaths, and has become endangered mainly through lack of carts. [Peter Wakely/English Nature]

var. *Dillenii*. Had the brim been a little wider,' he added, 'I might have included *Genista tinctoria* and *Lotus corniculatus*'.[92] Although the hat in question was a straw one with a broad brim, this is a pretty impressive square foot or so of turf! And although one would probably need to wear a sombrero to repeat his feat today, all these species and others besides still grow on this south-facing bank near the sea, though they are very much 'hands and knees jobs'. I have taken close-up photographs of Lizard turf, and, years later, I am still noticing tiny clovers and other plants that I didn't know were there!

In the 1870s another Helston-based naturalist, James Cunnack, added Wavy-leaved St John's-wort (*Hypericum undulatum*), Shore Dock (*Rumex rupestris*), Four-leaved All-seed (*Polycarpon tetraphyllum*) and the prostrate form of Juniper (*Juniperus communis* ssp. *hemisphaerica*), to the list. William Beeby discovered the tiny Dwarf Rush (*Juncus capitatus*), otherwise known only from Land's End, in 1872, and followed that up a few days later by finding the even tinier Pigmy Rush (*Juncus pygmaeus*) – in a cart rut! Perhaps the most remarkable discovery of all was the Land Quillwort (*Isoetes histrix*). One day, in June 1919, while emptying his vasculum, Fred Robinson found a strange bulb with withered 'leaves' among the loot he had been gathering at Caerthillian Cove. It was identified as Land Quillwort, hitherto known only from the Channel Islands, and so 'The first for England, I believe'. But George Claridge Druce, the arbiter in these matters, pronounced it an error. It was not until 1937 that the Quillwort was refound, again by accident. It was assumed to be very rare, but a survey in 1982 led by the late Lewis Frost from the University of Bristol suggested a local population of some 100 000 plants![21] The reason why no one had spotted it before is that Land Quillwort is a winter plant, dying back in late spring, but in any case its squill-like leaves are hard to spot, especially when, as they often are, they are growing in a patch of squill! For the same reason, another Lizard rarity, the Early Meadow-grass (*Poa infirma*) eluded everyone until 1950, when John Raven went there early in the year and found it in some plenty on well-trodden paths. On the warm, dry soils of the Lizard, annual plants tend to flower early, as they would in the Mediterranean, and many have frizzled up by the time the holidaymakers are jamming the road to Kynance Cove.

Yet another rare – and early – clover, Western Clover (*Trifolium occidentale*), was not recognized until 1957. In describing it as a species new to science, Dr David Coombe found a reference to what was almost certainly this plant by none other than C.A. Johns. 'Common White Clover', he pronounced in *A Week at the Lizard*, 'is one of the earliest flowers and has its stems, leaves and flowers closely pressed to the ground.' No it isn't; no it doesn't. It seems that, had he known it, he had found a fourth new clover at the Lizard.

Breckland

The Breck, the sandy district on the Norfolk-Suffolk border with Thetford at its heart, is the British counterpart to the steppes of eastern Europe – and a world apart from the rest of East Anglia (see picture on p. 52). The typical Breckland landscape is conjured up for us on the jacket of *The Ecological Flora of the Breckland* [90] – umbrella-shaped pines on a rabbit-cropped plain (the paired ears sticking up in the grass), with one of the Breck's characteristic circular pools and a Stone Curlew calling somewhere. It is the least *lush* part of the English lowlands, with comparatively warm, thirsty summers, and cold, windswept winters. Before 1920, a lot of the Breck was like that. Today, much of the area has been afforested or agriculturally 'improved'. The rare plants hang on where they can, on arable headlands, on road verges, around concrete bunkers and airfields, on open spaces in housing estates, and even in gardens. What remains of the old Breck landscape survives mainly as nature reserves and on one or two large private estates.

Although the Breck is perhaps the most accessible of all the rare plant 'cities', it is also the one where the stranger is most in need of a guide. The area is large, and most of the rare flowers are very small. Breckland botanists spend a lot of their time on hands and knees, with eyes on stalks, searching for that rare speedwell or rupturewort. Fortunately you can now 'get your eye in' by visiting West Stow country park, which has a small garden of Breck specialities, or the rather larger one at the University Botanic Garden in Cambridge. Most Breckland flowers grow in open sandy places. Many are annuals, which germinate in autumn and flower early in the year. In the past the shifting cultivation of the Breck, with crops of oats or rye followed by periods of fallow, provided a niche for such plants, as did the many sand-pits and cart tracks. Very likely, flowers like the Grape Hyacinth (*Muscari neglectum*) or the Fingered Speedwell (*Veronica triphyllos*) have been exploiting the activities of poor farmers and carters since the time when some Neolithic entrepreneur ran a flint factory at Grimes Graves. Today, however, they have been squeezed to the agricultural margins, and the survival of all too many Breck species now depends on the charity of nature lovers. The annual 'weeds' of sandy places are the most endangered group of plants in Britain; two Breckland specialities, Swine's Succory (*Arnoseris minima*) and Jagged Chickweed (*Holosteum umbellatum*), are nationally extinct.

The first printed records of this flora appear in the seventeenth century. During 'long walks of exploration' for his Cambridge Flora, John Ray was 'led on by an impatient longing for novelty, eagerly searching for strange plants in foreign fields'. These explorations took him to Newmarket, where he found a marvellous collecting ground at Newmarket Heath. Tom Willisel, his collector, evidently explored more of the Breck, penetrating as far as

The Breck is a land apart. It feels like a foreign country with its stony plains, windbreaks of pines and blown sand far from the sea. [Peter Marren]

Thetford. By the time Ray wrote his *Synopsis*, several of the famous rarities of the Breck had been found, including Field Wormwood (*Artemisia campestris*), Fingered Speedwell, Spanish Catchfly (*Silene otites*), Hoary Mullein (*Verbascum pulverulentum*) and Perennial Knawel (*Scleranthus perennis* ssp. *prostratus*). Although Ray's *Synopsis* listed species by county, not precise locality, he notes that the Perennial Knawel was gathered from the Elveden estate, which still harbours most of the British population. Willisel must have had exceptionally sharp eyes to spot the Fingered Speedwell and see that it was different. As early as 1802, a resident, Sir Thomas Cullum, thanked a correspondent for a specimen of this plant, which, he wrote, 'had been so extremely scarce that I had given it up in my mind'.

Ray's *Synopsis* alerted all educated naturalists to the exceptional nature of the Breck. It attracted visitors like Adam Buddle of Henley, then working on a new British flora, who was the first to recognize that varied and curiously coloured 'vetch', Sickle Medick (*Medicago sativa* ssp. *falcata*). The first botanist to record Breckland plants on a more systematic basis was Sir John Cullum (1732–85), who lived at Bury St Edmunds and compiled a Naturalist's Journal between 1776 and 1780. In it he includes 500 local plants, including, for the first time, Grape Hyacinth, Hairy Greenweed (*Genista pilosa*) and Spring Speedwell (*Veronica verna*), among many other Breckland specialities. He even knew the elusive Jagged Chickweed, which he had found on old wall-tops in Norwich and Bury St Edmunds. After his death, his equally botanical brother, Sir Thomas, added 40 more new plants to Suffolk, including two more Breckland plants, Smooth Rupturewort (*Herniaria glabra*) and Suffocated Clover (*Trifolium suffocatum*), in 1804–5.

A still more comprehensive recording of Breckland plants was undertaken by W.H. Hind, and published as part of his *Flora of Suffolk* (1889).[84] During this time, Sir James Edward Smith, author of *English Botany*, had discovered two new grasses, Grey Hair-grass (*Corynephorus canescens*) and Blue Fescue (*Festuca longifolia*), the latter on the same road verge at Foxhole Heath where it grows today. More first records were added by the most eagle-eyed: Rare Spring-sedge, gathered at Mildenhall by Sir W.C. Trevelyan in 1829; Bearded Fescue (*Vulpia ciliata*) by J. Townsend in 1846; Purple-stem Cat's-tail (*Phleum phleoides*) by Sir C.J.F. Bunbury in 1850; and Drooping Brome (*Anisantha tectorum*) by the Revd E.F. Linton in 1885.

One might have thought that these men were unlikely to miss a plant. But most botanists look for plants they know, not those they do not know and therefore do not suspect. Perhaps the skills of men like Ray and Willisel, exploring the *unknown*, have been replaced by a zest for recording the known. However that may be, the Victorians did miss Breckland plants, and more than one. One day in April 1933, J.E. Lousley and A.W. Graveson found an unknown speedwell near a track at Tuddenham Gallops, and, soon afterwards, more of it on a headland at Cherry Hill.[105] It was similar to the common Wall Speedwell, but with distinctively shaped fruits and crisply toothed leaves. It was a new British species, *Veronica praecox*, only recently given an English name, the Breckland Speedwell – an unfortunate one, for it promptly turned up in Oxfordshire. This plant was probably there all the time, but overlooked. But could men like Cullum and Hind have missed the Military Orchid (*Orchis militaris*), blooming bright as guardsmen in the chalk pit near Mildenhall, where it was found by Mrs M. Southwell in 1955? That, at least, must surely be a new arrival, probably from seed blown over from Belgium. The single flowering specimen of Large Pink (*Dianthus superbus*) found in tall grass near Grimes Graves in 1993 is even more puzzling. At first, some botanists were convinced that it was a new

Right: The dark indigo-blue wild Grape-Hyacinth (*Muscari neglectum*) is probably native on sandy headlands in the Breck. The often naturalized garden Grape-Hyacinths are a much brighter blue. [Peter Marren]

Far right: Proliferous Pink (*Petrorhagia prolifera*) may turn out to be the last native wild flower discovered in the twentieth century. It was only distinguished from the closely related Childing Pink (*P. nanteulii*) in 1993. [Bob Gibbons]

British plant, either a recent arrival or possibly even an overlooked native. But could anyone possibly overlook so festive a flower, with broad pink petals splitting into narrow tassels, like some barbaric banner? Its habitat resembled the dry grassland in which it grows on the continent, and there were no signs of planting. But the Pink has not reappeared since then, and the feeling now is that it was probably sown. It is commonly grown in gardens. Another new Pink, however, is probably a genuine native plant. A tiny annual Pink, with a compact head of miniature flowers which open one at a time, has long been known from dry banks around Mundford, on the Norfolk side of the Breck. Until very recently it was believed to be the Childing Pink (*Petrorhagia nanteulii*), better known from shingle shores on the south coast. It was only in 1992, when John Akeroyd examined its seeds and found they had a reticulate pattern, not the expected tuberculate one, that its true identity was revealed. It is Proliferous Pink (*Petrorhagia prolifera*) – confusingly an old name for the Childing Pink.[1] As a 'steppe' plant of central and south-eastern Europe, it matches the continental distribution of most Breckland flowers. Interestingly it grows well away from the 'classic' sites visited by generations of botanists. It takes its place in the latest floras as a new native flower.

The list of Breckland rarities is growing in quite another, more sinister, way. A species like Corn Chamomile (*Anthemis arvensis*) would not have turned the heads of Ray or the Cullum brothers, but there are remarkably few recent records outside the Breck, where for the moment it remains fairly common. This is also increasingly true of plants like Tower Mustard and Fine-leaved Sandwort (*Minuartia hybrida*), which are losing their sandy habitats to changing land-use, and today are only really frequent in this part of England. Today's familiars may be tomorrow's rarities.

The Scottish Highlands

Until recently, most people believed that the flora of Scotland was more or less discovered by an Englishman, the Revd John Lightfoot (1735–88), curate of Uxbridge and author of *Flora Scotica* (1777) – the first comprehensive account of wild Scottish plants. However, Lightfoot himself acknowledged that his Flora relied heavily on the work of Scottish botanists. His own fieldwork was limited to three months in 1772, when, with his friend, publisher and eventual biographer, the naturalist Thomas Pennant, and accompanied by the Gaelic-speaking John Stuart of Killin, they toured the Highlands. The resulting Flora was impressive in its scope: it described some 1300 plants and included ferns, mosses, fungi and even algae, as well as flowers. Among them are the first printed records of some 30 species. Lightfoot's was not, however, the first exploration of Scotland's wonderful mountain flora. The credit for that lies with a less celebrated, but much more adventurous botanist, James Robertson (*c.* 1745–96).

Robertson's travels in the Highlands took place between 1766 and 1771, and so preceded Lightfoot's expedition by up to five years. Even then, the Highlands and Islands were not wholly virgin territory for the botanist. Though John Ray had penetrated no further than Stirling, Martin Martin had travelled to the furthest Hebrides, and Robert Sibbald had included mountain plants in his natural history, *Scotia Illustrata*, among them, his eponymous *Sibbaldia*. Moreover, the Highland clans themselves seem to have had an excellent working knowledge of local plants, to judge from the rich array of surviving Gaelic names and folklore. Local people undoubtedly knew all about the virtues of plants like Dwarf Cornel, Bog Whortleberry and Scottish Primrose, long before they were collected and described by botanists. Robertson's explorations were undertaken with the express purpose of providing plants for Professor John Hope (1725–96) of Edinburgh University, whose ambitions to produce a Scottish flora were eventually to be frustrated by Lightfoot. A modern edition of Robertson's journals, recording his wide-ranging and perceptive observations, has only recently become available,[83] and it presents the discovery of the Highland flora in a completely new light. Intriguingly, it becomes clear that Hope did not pass everything on to Lightfoot, but kept some of the plums for himself. For example, the flower Robertson had found high up on Ben Avon in the Cairngorms in 1771, and recorded only by the initials S... C..., can only be the very rare *Saxifraga cespitosa* or Tufted Saxifrage, then unknown to Britain. The relevant passage runs as follows: 'On the top of the mountains there were few vegetables; I observed however the following viz. Thrift, Moss Campion, S... C..., Procumbent Azalea, Cudweed, and the tribe of Lichens.' The Saxifrage was rediscovered on Ben Avon in 1831, and is still there, though a 20-mile-round walk is required to see it.

This element of secrecy, which confuses the discovery of mountain plants in particular, was widespread in the eighteenth and nineteenth century. In Hope's case, it was probably because he wished to receive the proper credit for Robertson's discoveries at a time when record poaching was rife. For one of the leading contemporary botanists in Wales, John Wayne Griffith (1763–1834) it seems to have been a disinclination to publish anything. Although he helped William Withering compile a catalogue of mosses and lichens, he kept some of his most important finds to himself, among them the first record of Wild Cotoneaster (*Cotoneaster cambricus*) from Great Orme in 1783, and of the Tufted Saxifrage from Cwm Idwal in 1778.[93] Another reason for due secrecy was, of course, to preserve rare plants from the collector. Owen Rowlands of Blaen y nant is said to have kept his knowledge of the whereabouts of *Lloydia* a close secret, finally divulged only on his deathbed.

One of Robertson's intriguing coded discoveries is a plant he refers to only as 'A... C...', presumably the initials of its Linnaean name. If so, the only name that fits is *Andromeda caerulea*, now called *Phyllodoce caerulea* or Blue Heath (a rotten name, incidentally; there are no blue heaths anywhere in the world; this one is about the colour of Bell-heather). As it happens, Robertson wrote up the occasion in entertaining detail in his journal, since it proved a memorable one. He had set out early one morning in July 1771 to reach the remote misty hills above the headwaters of the River Findhorn: 'These were very high and had much snow on several parts' he wrote, and it was 'here I found the A... C... not mentioned either by Linnaeus or any other Botanist, as far as I know. I found also Dwarf Honey-suckle, Bastard Cinquefoil, Cloudberries, Mountain Willow herb, Great Billberry bush, Hairy Kidneywort, Mountain Saw-wort and Mountain Hawkweed.' Quite a good haul then. The God of botany might have resented Robertson's good luck that day. Not long afterwards, the heavens opened, and giant hail-stones fell on the unfortunate naturalist as he retreated downhill. Some of the stones, he said, were the size of marbles, and

> fell so thick and violently on my head that I hardly knew whither I went. The little rills were instantly swelled to impassable torrents, down which the masses of rock rolling along, filled with dreadful noise every pause between the different bursts of thunder. These were inconceivably awful, and not without danger, as I was among the clouds from which they issued. The Thunder seemed indeed to burst over my head whence it diffused itself with terrible claps, while the flashes darting in rapid succession thro' the dark clouds, added new horrors to the gloom with which I was incompassed. This dreadful scene lasted two hours . . .[83]

The Blue Heath has not been seen there since, but this is a remote area and it could still be lurking there somewhere. The extreme secrecy attached to this first finding of *Phyllodoce caerulea* seems also to have been extend-ed to the second. In 1812, a Perthshire nurseryman, James Brown gathered specimens from what he said was 'a dry moor . . . near Aviemore'. His site was almost certainly its best known present-day one at the Sow of Atholl, but it is a long way from Aviemore, nor could it be described as a dry moor. *Phyllodoce* is an attractive little heath, and nurseryman Brown had every motive for putting others off the scent.

Many of the first records usually attributed to Lightfoot were in fact made by the young James Robertson. Among them were Creeping Ladies-tresses (*Goodyera repens*) 'in a wood called Cregenon', where he also found Toothed Wintergreen (*Orthilia secunda*) for the first time; Curved Sedge (*Carex maritima*) on the dunes at Invernaver on the north coast; Mountain

Azalea (*Loiseleuria procumbens*), 'never before found in Britain' on the summit of Ben Valich; and most remarkably, Scottish Primrose (*Primula scotica*) by the Kyle of Tongue (though at the time it was considered to be a dwarf variety of Bird's-eye Primrose). It is also possible that he found Brown Bog-rush (*Schoenus ferrugineus*), not confirmed as a native plant until 1884. However his reference to the 'S... Rush which I never met with before and which no Botanist has numbered among the British plants' leaves the matter open; he was near Loch Rannoch at the time, and he might have been looking at the Rannoch Rush, *Scheuchzeria palustris*. But whatever it was, the 'S... Rush' is yet another of the special plants that Hope and Robertson decided to keep to themselves.

Like Johnson's, James Robertson's life is another one of promise deferred. In 1772, the year Lightfoot toured Scotland, he embarked as surgeon's mate on a merchant ship and sailed away to India. When he eventually returned to his native land his health was wrecked and he was no longer able to go botanizing in the Highlands. Instead, he 'took to opium with ardent spirits' and died in 1796, aged about 50. Only now has his full contribution to British field botany become widely known.

The years between James Robertson's excursions in the 1760s and the death of George Don in 1814 were the great years of botanical exploration in the Highlands, during which the majority of rare alpine flowers were discovered. One of the first was the beautiful Purple Oxytropis (*Oxytropis halleri*), found near Loch Leven in 1761, by the Revd John Walker, Professor

Right: Blue Heath (*Phyllodoce caerulea*). Its discovery in Britain remained a secret for many years. [Michael Scott]

Far right: Alpine Fleabane (*Erigeron borealis*) is one of several pretty alpine plants first noticed by professional nurserymen. [Peter Marren]

of Natural History at Edinburgh, and illustrated by James Robertson himself. John Lightfoot seldom wandered sufficiently far from civilization to find many really rare plants, but, thanks to the guidance of his companion, John Stuart of Killin, he was the first to reveal the astonishingly rich flora of the mica-schist hills north of Loch Tay. The title page of *Flora Scotica* contains an illustration of Creeping Spearwort or *Ranunculus reptans* (copied from a drawing in a contemporary Danish flora), one of the special species that Professor Hope *did* pass on to him, having found it at Loch Leven in 1764.

Surprisingly, neither Robertson nor Lightfoot seem to have climbed Ben Lawers, the richest of all the mica-schist hills. Its first recorded botanist was a nurseryman from Covent Garden, James Dickson, who helped himself to some of its more attractive plants between 1789 and 1792. They included Rock Speedwell (*Veronica fruticans*), Alpine Fleabane (*Erigeron borealis*), Alpine Gentian (*Gentiana nivalis*) and Drooping Saxifrage (*Saxifraga cernua*) for the first time in Britain. It was another nurseryman, George Don (1764–1814), who became the greatest of all the pioneer botanists of Scotland. Don's botanical conquests are legendary. On Ben Lawers, which he first visited in 1793, a year after Dickson's second raid, he added Mountain Sandwort (*Sagina saginoides*), Alpine Forget-me-not and Scorched Alpine-sedge to the British flora (the last one not seen again for another 90 years). But his favourite hunting grounds were the eastern Highlands, from the glens of Angus, across the rolling plateau of the Mounth and over the Dee to Ben MacDui and the Cairngorms. Because some of the rare flowers he found still grow in the same very restricted localities, it is possible to retrace Don's steps and gaze down on what is probably the same ledge with its Alpine Sow-thistles (*Cicerbita alpina*) on Lochnagar (1801), recline on the serpentine knoll at Meikle Kilrannoch where the Alpine Catchfly (*Lychnis alpina*) still grows (1795), or gingerly traverse the crumbling band of pale rock in Glen Clova where Yellow Oxytropis (*Oxytropis campestris*) puts forth its creamy, pea-like flowers. Don managed to reach vertiginous places where no botanist had set foot before. A colleague, Patrick Neill recalled how he 'astonished the Highlanders wherever he went by his strange occupation of climbing rocks and hooking down plants, which they regarded as weeds, with a fifteen-foot staff crowned by an iron spaddle'.[170] Don's enthusiasm for alpine plants may be judged from the extent of his garden nursery at Forfar, where, according to Neill, he succeeded in cultivating upwards of 60 species of *Carex* (a decidedly non-commercial venture, one would have thought), 46 saxifrages and 55 hardy speedwells, as well as some of the rarest plants in the British flora – and all without a single greenhouse. A few of his records, like *Ranunculus alpestris* 'about 2 or 3 rocks on the mountains of Clova', have never since been refound, but whether he muddled his notes or indulged in secret transplanting experiments, most of his discoveries are genuine – and there were many of them (see Raven & Walters[158] for a fuller account). Before Don, the Highland flora had only been glimpsed. After him, only a few, mostly critical or inconspicuous flowers remained undiscovered.

They were picked up one by one as botanists visited and revisited the classic places, and prepared their local floras and vast Victorian herbaria. Rather surprisingly, Don missed the beautiful Alpine Milk-vetch (*Astragalus alpinus*) in his regular stamping grounds of Glen Doll; that was picked up later in 1831. The equally attractive Mountain Bladder-fern (*Cystopteris montana*) was a late arrival to the flora probably because botanists went up the wrong mountains. W. Wilson found it on Ben Lawers, where it is not very common, in 1836. But most of the mountain flowers discovered by the Victorians were of the sort that you distinguish back at base rather than recognize in the field. One of the exceptions was Alpine Rock-cress (*Arabis alpina*), one of the remotest of all British plants, discovered in 1887 in the Black Cuillins of Skye by the great Irish botanist H.C. Hart, while on honeymoon. Like Robertson and Don, Hart was a man of prodigious energy. One feels for his poor bride on that rock-cress outing, when we read how he liked to 'stride along the mountain-ridges in all weathers at five or six miles an hour, making frequent forays to explore the cliffs below and stuffing plants in his pockets as he went'.[200] Alpine Rock-cress was an appropriate plant for him to find. The locality has been described as 'an apparently endless scree' topped by a cliff with 'a long perpendicular cleft rather like a rough staircase'. Only blue riband botanists, dwellers of the Wild Flower Society's 'Valhalla' ever reach it. Another successful alpine-hunter was the Revd E.S. Marshall, who described a species of scurvy-grass new to science (*Cochlearia micacea*), in 1894, and followed that up three years later by the discovery of String Sedge in a remote bog near Altnaharra in Sutherland. The subsequent history of the String Sedge reminds us that, in cases like this, it is not just plants that are rare, but also people like Marshall and Hart. Until 1978, this sedge was thought to be confined to Altnaharra. Then someone noticed the plant in a very well-known locality, Insh Marshes in Speyside. And plenty of it too – one surveyor likened the stands of String Sedge to a rice paddy. My own excuse for not noticing it is the fine drake Goldeneye that was calling at the time. You can't do everything.

With the passing of the Victorian age, it seemed for a long time that there were no species left unfound in the Highlands, apart from hawkweeds and eyebrights. Then, in the space of little more than a year – the *annus mirabilis* of modern plant discovery – three new and unquestionably native plants were found. First was the tiny Iceland Purslane, *Koenigia islandica,* in 1950, on the plateau gravels and fine screes of the Storr, on Skye. It had, in fact, been collected there sixteen years earlier, by a palaeobotanist taking peat samples, but misidentified as the common Water Purslane. It is well known that palaeo-botanists cannot identify living

Iceland Purslane (*Koenigia islandica*), a tiny annual, often no bigger than a thumbnail, is confined to Skye and Mull today, but its pollen is widespread in peat deposits from late Ice Age Britain. [Andrew Gagg's Photoflora]

plants. Pollen analysis has shown that there was a time, late in the Ice Age, when *Koenigia* grew like duckweed on the damp gravel moraines left by the retreating ice cap. Now it is confined to a few cold, unstable hillsides in western Scotland. *Koenigia* is one of our smallest flowers, and such attractions as it can offer look best under a X10 lens. There could hardly be a greater contrast with the next new alpine, *Diapensia lapponica* (like Rhododendron, it is still known only by its generic name, Diapensia). It was discovered in July 1951 by C.F. Tebbutt, while birdwatching, and in Britain is still known only from its original barren ridge near Glenfinnan. This lovely cushion plant with its neat cream flowers and glossy green leaves has a brief flowering period, which varies from season to season, and when not in flower could easily be passed over as just another patch of Mountain Azalea. Diapensia is designed to survive some of the bleakest conditions on earth. Its barren Scottish locality might seem comparatively welcoming compared with that on Mt Washington in Alaska, which has the world's highest windspeed – some 234 miles an hour![111]

The last of the miraculous trio was also discovered by a birdwatcher. Norwegian Mugwort, *Artemisia norvegica*, was the unexpected prize of a walk on to the stony ridge of Cul Mor, in Wester Ross, by Sir Christopher Cox. He found it in August 1952, and it was identified the following year. Although not a 'designer-alpine' like *Diapensia*, an *Artemisia norvegica* in fresh flower is at least a pearl among mugworts, with its bee-like ball of bright yellow stamens, held aloft on stout, woolly stems. In some ways, this is the most remarkable of the trio, since its world range is limited to widely separated mountain ranges in Scotland, Norway and the Urals, and in each it is slightly different, our one being the woolliest and most aesthetically pleasing; it has recently been named var. *scotica*. Since 1952, it has been found on high ridges in three separate parts of Ross, each of them in majestic mountain scenery.

There are surely at least a few more mountain plants out there somewhere. In a recent paper[114] Hugh McAllister mentions a number of local forms of mountain grasses which are certainly separate genetic entities, if not full species, like the two distinctive varieties of Wavy Meadow-grass (*Poa flexuosa*). But it remains the ambition of every botanist to find something obviously new, and preferably spectacular – another Diapensia!

Wild orchids

At least one species of wild orchid was well enough known in Tudor England to merit a walk-on part in Hamlet. Unfortunately it is not clear exactly

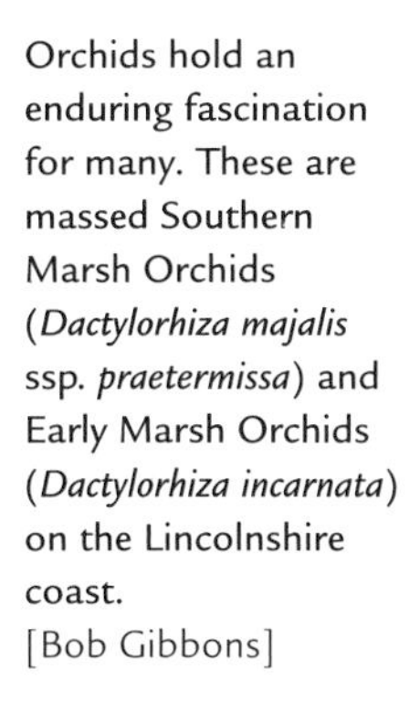

Orchids hold an enduring fascination for many. These are massed Southern Marsh Orchids (*Dactylorhiza majalis* ssp. *praetermissa*) and Early Marsh Orchids (*Dactylorhiza incarnata*) on the Lincolnshire coast. [Bob Gibbons]

which one Shakespeare meant by his reference to the 'long purple' gathered by Ophelia for her sorrowful garland (the artist Millais even assumed he meant Purple Loosestrife). We are only told that liberal shepherds gave it a much ruder name, 'but our cold maids do dead men's fingers call them'. Most of Ophelia's flowers were chosen for their symbolism – a kind of language of flowers – nettles for their sting, crowflowers for wastefulness. The orchid had a venereal meaning. Orchid tubers resembled testicles ('stones'), and, by a sympathetic magic, contained the power to stimulate lust. Ophelia was hinting that Hamlet was much in need of a love potion, or aphrodisiac. But Shakespeare was no botanist, and he seems to have confused or conflated two different orchids: the Early Purple Orchid (*Orchis mascula*), with its long purple spikes, and one of the spotted or marsh orchids with their finger-like tubers. Perhaps the latter was more likely to present itself to Ophelia as she wandered crazily along the bank of the fateful brook.

The herbalists of Shakespeare's day knew at least sixteen kinds of wild orchid. They were not always species in the modern scientific sense, but arranged according to their 'signatures'. Every herbalist knew the Early Purple Orchid, for it was illustrated in their copy of Dioscorides. One of its names, most of which come under Shakespeare's description of 'gross', was 'dog stone', after its pair of round tubers. It seems that they distinguished the daintier Green-winged Orchid (*Orchis morio*), then common in wet meadows, as the 'fool stones' or *morio*, which is still the plant's scientific name. Another kind of orchid could be distinguished by its roots, which were like pudgy fingers instead of paired stones. William Turner called it the 'Hand Satyrion' and Gerard the 'Palma Christi', which had two forms, male (*mas*) and female (*foemina*). They were not of course true gender differences, but recognizably different plants, one bolder and more assertive – one of the marsh orchids perhaps – the other more graceful and delicate - possibly the Common Spotted Orchid (*Dactylorhiza fuchsii*). One of the country names, 'Adam and Eve', seems to reflect this classically-derived preoccupation with dividing the natural world into male and female. Outside the apothecary's shop, lay people also knew wild orchids – they were common enough in meadows and woods nearly everywhere. But, to judge from the surviving folk-names, pretty purplish orchids were all one, dog stone or palmate, spotted or unspotted; they were cuckoo flower and crow flower, bull's bags, beldairy and many other names, varying less from species to species as from county to county.

Turner distinguished several other plants which we now know are orchids. There was one with just two big round paired leaves which was widespread 'in watery middowes and in woddes'. He called it 'Martagon', but to other herbalists it was Twayblade – 'two leaves' – (*Listera ovata*), a name which, fortunately, no one has thought to improve on. There was also a much rarer one Turner knew as the 'Satyrion . . . having a white flower like a lily'. He had seen it only twice, in Suffolk and in Hertfordshire. What it was is uncertain: most commentators have assumed it was Broad-leaved Helleborine (*Epipactis helleborine*). 'Satyrion' was a common name for woodland orchids. Classically speaking, woods were the abode of satyrs, half-man, half-goat, noted for their lust, presumably as a result of gorging on orchid tubers. An even scarcer Satyrion was 'Lady's traces' which 'bringeth forth whyte floures in the ende of harveste'. Turner knew it from only one place, near Syon House in Middlesex; he had recognized it as a flower he had often seen in Germany. It is still called Autumn Lady's-tresses (*Spiranthes spiralis*) to this day, after the spiral arrangement of the flowers, like the ringlets of fashionable ladies.

Gerard introduces us to a group of orchids that were useless in physic, but attracted attention to themselves by their amazing resemblance to various insects. 'Some', he wrote, were 'the shape and proportion of flies, in other gnats, some humble bees, others like unto honey Bees; some like Butterflies, and others like Waspes that be dead.' In his descriptions we can clearly recognize the Bee, Fly and Butterfly Orchids, while his 'wasp orchid' with flowers 'the colour of a dry oken leafe' is probably the Early Spider Orchid. The Butterfly Orchid (*Platanthera chlorantha*) was rare, so Gerard gave details where it grew 'upon the declining of the hill at the end of Hampsted heath, neere to a small cottage there in the way side' and also 'in the wood belonging to a worshipfull gentleman of Kent named Master Sedley of Southfleete'.[67] To Gerard, as for every flower-lover since, these plants were simply delightful in themselves, 'wherewith Nature hath seemed to play and disport her selfe'. In a shady wood near Gravesend, Gerard found an even stranger orchid, all 'parched brown' in colour, with a spike of honey-coloured flowers and 'many tangling roots platted or crossed one over another very intricately, which resembleth a Crows nest made of sticks'. Like Hand Satyrion and Dog Stones, this plant was named after its most interesting part – the roots. This one, though, did not look healthy: it was a 'degenerat kind of orchis', lacking any life-giving chlorophyll. Hence it was a 'Satyrium abortivum', a 'bastard or unkindly Satyrion'. Today, of course, we know it as the Bird's-nest Orchid (*Neottia nidus-avis*).

Did Gerard know the Military Orchid? He certainly wrote a vivid description of what seems to be this plant, though he might have based it on a foreign herbal. He called it Souldier's Satyrion, or Soldier's Cullions (literally 'balls'), for it was a deeply masculine orchid. In Gerard's day, soldiers did not have red uniforms, and the military connection seems to have been found in the resemblance of the flower sepals to the coal-scuttle helmets worn by an earlier generation of military men. Gerard says nothing about where it grows, which he normally did when describing an unusual plant, and the earliest localized date for this rare orchid is in Merret's *Pinax*, written some 70 years later. Still, the description is Gerard at his liveliest, and here it is again.

> Souldiers Satyrion bringeth forth many broad large and ribbed leaves, spread upon the ground like unto those of the great Plantaine: among the which riseth up a fat stalke full of sap or juice, clothed or wrapped in the like leaves even to the tuft of flowers, wherupon doe grow little flowers resembling a little man, having a helmet upon his head, his hands, and legs cut off; white upon the inside, spotted with many purple spots, and the backe part of the flower of a deeper colour tending to rednes. The rootes be greater stones than any of the kinds of Satyrions.[67]

An even rarer orchid was the Lady's Slipper, the Calceolus Maria or Shoe of Our Lady. (Though it looks more like a clog than a slipper, Lady's Clog does not have the same ring, does it?) Gerard grew it in his London garden, 'which I received from Mr Garret, Apothecary, my very good friend'. So did John Tradescant, the leading gardener of the age. But where had Garret got it from? The plant is difficult to cultivate, and a foreign source seems unlikely. More probably he had obtained plants from a supplier in 'the North parts of this kingdom' (the kingdom of England, that is). Thomas Johnson knew it grew wild up there somewhere, but it was John Parkinson who gave the game away and told his readers exactly where: in Helks Wood, Lancashire, supplier: Mrs Tunstall. The rest, as they say, is history.

Our remaining wild orchids were discovered by amateur naturalists over the next 300 years, but not in any order that could possibly have been predicted. John Goodyer, 'the first naturalist' found a new one in a wet meadow called 'Wood-mead' not far from his home in Petersfield, Hants. It was another of the Palma Christi type, but in this case the roots had horizontal runners, '*radice repente*'. Today we forget about roots and call his plant the Marsh Helleborine (*Epipactis palustris*). Possibly this attractive and distinctive flower was already known, for Lobelius provides a wood-cut of it in his *Stirpium Adversaria* of 1601.

Thomas Johnson added a trio of new orchids in his 'Botanical Mercury'. He was the first to distinguish the Fragrant Orchid (*Gymnadenia conopsea*), so common 'in montosis' – for the old botanists called chalk downs 'mountains' for lack of anything taller. With it 'in montosis pratis' – in mountain meadows – grew the Burnt-tip Orchid (*Orchis ustulata*) of the 'dog stones' group. Evidently it was much rarer. The first named locality is 'On Scosby-lease', near Doncaster, where it was found by a Mr Stonehouse. More unexpected is the early first record of Orchis Saurodes or Lizard Orchid (*Himantoglossum hircinum*) 'nigh the highway between Crayford and Dartford in Kent', that well-trodden path of London apothecaries. An early name for it was Great Goat-stones, for it had mighty swollen tubers and stank like a billy-goat. Johnson also found 'Orchis Batrachites' or Frog Orchid (*Coeloglossum viride*), assuming that he is the same Dr Johnson that provided its locality to William How: 'by Barkway' in Herts. Another surprising early discovery was the tiny Bog Orchid (*Hammarbya paludosa*), probably much commoner in the days when most parishes had a bog or two. This one was on 'low wet grounds betweene Hatfield and St Albones' and was recorded in Parkinson's *Theatre* of 1640. It also grew 'in divers places of Romney Marsh'.[47]

The next generation of botanists found an assortment of new orchids, published for the first time in the works of How, Merret and Ray. Christopher Merret's *Pinax* of 1666 produced Lesser Twayblade (*Listera cordata*), found 'near the Beacon on Pendle Hill in Lancashire' and Sword-leaved Helleborine in none other than Helk Wood, where the Lady's Slipper

The apparently wriggling motion of Lizard Orchid flowers is reminiscent of live lizards. The scientific name, *hircinum*, however, refers to their scent – like billy-goats. [Peter Marren]

The Late Spider Orchid (*Ophrys fuciflora*) has always been confined to East Kent. For such a rare plant, it is remarkably variable with dark-lipped forms like this one and others with more extensive markings that resemble the continental Woodcock Orchid (*Orchis scolopax*). [Peter Marren]

grew. No doubt both orchids were sold together on the market stalls of Settle. Merret's biggest catch, however, was his double record of Monkey and Military Orchid on the Berkshire Downs between Wallingford and Reading, supplied to him by William Brown of Magdalen College, Oxford. But, since he mentions this astonishing feat without further comment, one does wonder whether these exotic plants were not known about already under some name or other.

It took the critical faculties of a John Ray to sort out two of the awkward orchid pairings. He distinguished Pyramidal from Fragrant Orchid by comparing the two in the still surviving chalk-pit at Cherry Hinton near Cambridge, where he also found Musk Orchid (*Herminium monorchis*) for the first time. He was also the first to realize that we had two Butterfly Orchids, a Greater and a Lesser (*Platanthera bifolia*). The former grew mainly in woods, the latter 'in pascuis', probably in the wet meads and fens that then surrounded the ancient city of Cambridge. Ray's 'exploratory walks' also turned up the rare Fen Orchid (*Liparis loeselii*), 'in the watery places of Hinton and Teversham Moors'. His later excursions to the further reaches of England and Wales added White Helleborine (*Cephalanthera damasonium*), once known as the Egg Orchid 'in the woods near Stockenchurch, Oxfordshire'; Small White Orchid (*Pseudorchis albida*) on the slopes of Snowdon (1670) and Dark-red Helleborine (*Epipactis atrorubens*) 'On the sides of the mountains near Malham . . . in great plenty' (1677). Also recorded for the first time in Ray's *Synopsis* are Man Orchid (*Aceras anthropophorum*) 'found by Mr Dale in an old Gravel-pit' near Sudbury, Essex in 1690, and in the revision by Dillenius in 1724, the Lady Orchid (*Orchis purpurea*), 'at Northfleet near Gravesend', found by John Sherard. It is surprising that Gerard and Johnson missed the Lady Orchid during their perambulations in north Kent, but perhaps they went too late in the year after the Ladies had gone to seed. It is just possible that the Hampshire naturalist John Goodyer saw Lady Orchid long before, although Merrett's description in his *Pinax* is impossibly ambiguous.

By the year 1700, therefore, our orchid flora stood at 32 species, a few long familiar, others discovered during 'simpling' expeditions around London or the more wide-ranging botanizing of Johnson or Ray. Most of the missing species were either northern, or rare, or taxonomically difficult. The remaining northern ones fell at one swoop to John Lightfoot in his tour with Thomas Pennant, written up in his *Flora Scotica* (1777). These were Coralroot (*Corallorhiza trifida*), 'Ophrys Corallorhiza . . . in a moist hanging wood near the head of Little Loch Broom' in Wester Ross, and in the same neighbourhood, if not the same wood, Creeping Lady's-tresses, 'Satyrium repens . . . in an old shady hanging birch wood'. As we have seen, the latter had already been discovered by James Robertson a few years previously, but had not been published.

Lightfoot's brief foray to the far north effectively drew the botanist's net across most of Britain; but the net was still of a fairly coarse mesh, and a great many elusive plants remained for the field botanists of the nineteenth century. Among the first was the rare and beautiful Red Helleborine (*Cephalanthera rubra*), discovered in 1797 on 'a steep stoney bank' not far from her home by a Mrs Elizabeth Smith of Minchinhampton. Mrs Smith sent a specimen to James Edward Smith, who included its portrait by Sowerby in that year's edition of English Botany. However the Revd W. Lloyd Baker, a botanist and Fellow of the Linnaean Society, told Sowerby that *he* already knew about the plant, having found it in the same place some years earlier, but had kept quiet about it. There is a portrait of his son holding a drawing of the Red Helleborine in his hand, just as heroes of old are depicted holding a baton or a telescope.

The next new orchid was 'Neottia gemmifera', now known as the Irish Lady's-tresses (*Spiranthes romanzoffiana*). It has a strange history. In about 1810, Mr J. Drummond, Curator of the Botanic Gardens in County Cork, found the plant in a salt-marsh near Castletown on the shore of Bantry Bay. Apparently the two specimens he carried away constituted the whole colony, for no one else succeeded in finding it there for another thirty years. Indeed the species remained little known until the 1890s, and all that the authorities had to go on in the meantime were a few dried herbarium scraps. When Druce and A.J. Wilmott toured Ireland in 1926 and compared fresh material from north and south, they thought they had not one Irish Lady's-tresses but two. But, like most of Druce's separations, they were subsequently reunited. The distribution of Irish Lady's-tresses on both sides of the Atlantic Ocean suggests either that it is a very old species, pre-dating the separation of Europe and North America, or that its seeds have the power to cross oceans. Irish Lady's-tresses is no longer a good name, for it has since turned up in Scotland and Devon – Atlantic Lady's-tresses would be better. Another new orchid was discovered in Ireland in May 1864 by a Miss More at Castle Taylor in County Galway. This is the Dense-flowered Orchid (*Neotinia maculata*), a speciality of the Western Irish limestone and nearly as interesting biogeographically as *Spiranthes romanzoffiana* for it is otherwise confined mainly to the Mediterranean and the Iberian peninsula. Its late discovery is not surprising, for the flowers of Dense-flowered Orchid are usually over by the time botanists visit the wonders of the Burren limestone.

Meanwhile, hundreds of miles away, the Revd Gerard E. Smith had been busy botanizing in East Kent and in 1828 had discovered a second Spider Orchid 'on the southern declivities of chalky downs near Folkestone', close to where the Channel Tunnel trains now roar past. It was called the Late Spider Orchid (*Ophrys fuciflora*), not because we were late in finding it, but because it flowers later than the Early Spider Orchid (see picture on p. 65)! On the same excursion Smith found Clove-scented Broomrape – as John

Gilmour once sighed, 'those were the days!' A portrait of the new orchid was duly published, and shortly afterwards botanists began to find it in other parts of Kent. A much less rare plant, the Violet Helleborine (*Epipactis purpurata*), was also illustrated for the first time that year in *Smith's English Botany*. These were plants that had been gathered a full twenty years earlier and, oddly enough, they were not typical Violet Helleborines at all, but strange magenta-hued plants, gathered at Leigh in Worcestershire in 1807 by the Revd Dr Abbot. Abbot considered them to be 'parasitical on the stump of a maple or hazel'.[187] In fact they were just an extreme form of the species which just happened to be growing under maple or hazel, and parasitical only in the sense that it feeds on fungi within its own roots. Like most woodland helleborines, *Epipactis purpurata* remained a rather doubtful species for many years.

When Professor C.C. Babington first visited Jersey in 1837, two new orchids were waiting for him in the same part of the island. The first, and, as it happened, much the less common, was the Summer Lady's-tresses (*Spiranthes aestivalis*), growing in small quantity 'in a wet sandy spot upon the banks of St Ouen's Pond' on 24 July 1837. The other was the Loose-flowered or Jersey Orchid (*Orchis laxiflora*) not known from mainland Britain. Strictly speaking it should be called the Guernsey Orchid. The plant had been gathered long before by that island's first botanist, Joshua Gosselin, in 1788, but he was unable to name it from his only textbook – a battered copy of Parkinson's *Theatre* of 1640![116]

Epipogium aphyllum is elusive for a different reason. Like the Bird's-nest Orchid, it has no green leaves, and furthermore appears above ground to flower only occasionally. Hence it is usually found serendipitiously, like buried treasure, and so seldom that this orchid did not even acquire a generally recognized English name until the 1960s. To Jocelyn Brooke it was 'Epipogon', to Keble Martin 'Leafless Epipogium' and to Victor Summerhayes 'Spurred Coralroot'. But 'Ghost Orchid' provides the best sense of its unpredictable appearances, pale hue and dark gloomy habitat. It was first glimpsed in 1854 by a Mrs W. Anderton Smith near a brook at Tedstone Delamere, on the border of Hereford and Worcestershire. She removed it to the safety of her garden, where it naturally died. Understandably, the ghost was affronted and delayed its return until 1876, when it bloomed for a few years at Ringwood Chase, near Ludlow, before disappearing again, seemingly for good. Between 1923 and 1986, it haunted the beech woods near Henley slightly more regularly, its appearances marked by wooden posts and little tents of twigs.

What are the orchids, so elusive, so hard to distinguish, that they escaped the notice of the giants of the nineteenth century – the Smiths, Symes, Babingtons, Backhouses, Lintons and Marshalls? Surprisingly common and attractive ones! We are back to Shakespeare's 'long purples'. It took a Druce, and after him a Pugsley and a Heslop-Harrison, to sort out the 'Dactylorchids', the orchids with tubers like a dead hand, into two spotted orchids and five or six marsh orchids. And even today, though Floras provide descriptions of the swarm of varieties and subspecies and hybrids in this troublesome group, there is still plenty of room for disagreement about what constitutes a good species. There has, for example, been a different concept of the 'Irish Marsh Orchid' roughly once every decade since its first description in 1937, under the successive names of *Dactylorchis occidentalis*, *Dactylorchis kerryensis*, *Dactylorhiza occidentalis*, *D. majalis* and now *D. comosa*! What we were told was a new species first found in 1986, *Dactylorhiza lapponica*, has been demoted by Peter Sell[177] to a subspecies of Narrow-leaved Marsh-orchid, *Dactylorhiza traunsteineri*, itself segregated only in 1946 – though Stace[180] and the Red Data Book are hanging on to it,

Young's Helleborine (*Epipactis youngiana*) is probably a recently evolved species (if it is a species) of hybrid origin. It lives on mining spoil. [Sidney Clarke]

at least for the moment. With marsh orchids, the idea of a species, with a consistent genetic constitution and a set of distinguishing characters, seems to break down.

The same is true of that other difficult group, the self-pollinated helleborines. Until 1917 they were more or less ignored. Then Mr Hunnybun discovered near Ventnor an odd helleborine orchid whose flowers barely opened. This became the 'Isle of Wight Helleborine', and, once recognized, similar plants began to turn up in a wide scatter of woodland sites, usually in deep shade. In 1919, Colonel M.J. Godfery distinguished another helleborine, which he named *Epipactis leptochila*, the Narrow-lipped Helleborine. This plant was new to science, and for many years was thought to be endemic. Other 'Isle of Wight helleborines' with half-opened drooping flowers were distinguished, named and described, but in 1952 D.P. Young lumped most of them into a single variable species, *Epipactis phyllanthes*, the Green-flowered Helleborine, and there they have stayed. Meanwhile what many of us were brought up to call the Dune Helleborine had been swallowed up by the Narrow-lipped Helleborine, which in turn has now been merged with a European orchid, *Epipactis muelleri*! However, one helleborine bucked the 'lumping' trend. This is *Epipactis youngiana*, 'Young's Helleborine', a strange plant of northern coal bings and mining spoil first described in 1976 by A.J. Richards and A.F. Porter.[165] It may have originated as a cross between Broad-leaved Helleborine and either Narrow-lipped or Green-flowered Helleborine which now breeds true. The distinguishing features are minor but consistent. Should it be considered a separate species? Opinions differ. As with the marsh orchids, it might be better to shed our preconceptions about species altogether, and see these unusual orchids as a case study in the dynamics of evolution with various local forms, half-species and species in the making. What you define as a species depends on where you decide to draw the line.

As for the species of tongue-orchids (*Serapias*) that have turned up in recent years, possibly as heralds of global warming, I must refer you to Chapter 8 on Origins.

Chapter 4

On the trail of the ice

What is the oldest British plant? Confining ourselves to vascular plants (flowers, trees and ferns), it is probably the humble horsetail. The remains of horsetails have been found next to the bones of Iguanodons in the Weald of Sussex, and although these were tree-sized species now extinct, they were of the same genus, *Equisetum*, as the present-day ones of hillside and hedgerow. The horsetail design is clearly an evolutionary winner. Some of the descendants of those hot Mesozoic swamp forests survive on shores beyond the Arctic Circle.

Flowering plants began to evolve in the age of the dinosaurs, probably hand-in-glove with their partners and pollinators, the insects. For a long time the British flora, or such glimpses of it that the fossil record allows, was wholly different to that of today. But over the past 40 million years or so, the parade of plants gradually introduces species which are not dissimilar to the ones we know. For example, in the London Basin of the late Eocene (about 40 million years ago) there was reed-swamp, with bulrush, sedges, water-lilies, pondweeds and even Water Soldier – probably not the same species (it was a lot warmer then), but at least recognizable vegetation, possibly not unlike aquatic habitats today (the animals lurking there would have been much less familiar; that rat-like creature was the ancestor of a lion). By the late Pliocene, some 3 million years ago, about a fifth of the flora consisted of species occurring in Britain today, including willows, elm, holly, willowherbs, daisy and bog myrtle.[88] A million or so years later we start finding limestone flowers like rock-rose and dropwort, as well as familiar things like stinging nettle, sorrel and dog-rose. But at the beginning of the Pleistocene, this Arcadian world is suddenly interrupted by the first of a succession of Ice Ages. Throughout this time Britain was part of the continental mainland – a mere bump on the western edge of Eurasia. As the ice advanced and then retreated, so plant life followed its trail northwards and flourished during the long, warm interglacial periods. During the first interglacial, there were still many strange plants in the British landscape, like the Umbrella Pine (*Sciadopitys*), now confined to Japan, or Wingnut (*Pterocarya*), whose nearest present-day site is in the Caucasus mountains. Overall, though, the scene shifts towards species familiar today – though not always in the same environment. Sea-buckthorn (*Hippophae rhamnoides*), for example, made its first appearance as a hardy colonizer of glacial moraines. Chickweeds and plantains did well where hippos wallowed in the Thames valley. Hawkweeds and dandelions yellowed the well-grazed plains on the migration routes of bison and elephants. But the main difference we would notice if we were transported back half a million years would be the much greater extent of native conifers. Silver Fir and Norway Spruce formed extensive forests, as had Serbian and Hemlock Spruce in an earlier interglacial. They would do so still had not Britain become an island before they had had time to spread back. By the Hoxnian interglacial, about 370 000 years ago, a distinctive flora was forming on the Atlantic fringe of what is now Ireland, with the appearance of species like Pipewort (*Eriocaulon aquaticum*), St Daboec's Heath and Mackay's Heath (*Erica mackaiana*). These are among the oldest known *rare* plants, and some of them may be survivors from the earlier, ice-free Tertiary period.

By the last interglacial, probably most of our present-day flora was in existence. All, or nearly all of it, is popularly supposed to have been wiped away by the last, and probably the coldest, Ice Age, the Devensian. The climate had started to cool down again about 100 000 years ago. Tundra and wet open grassland began to replace the warm temperate forests and hippo-wallows, and eventually ice-sheets spread south from the arctic. During the coldest period of all, between 40 and 20 000 years ago, most of Britain was a polar desert, like Greenland today. The big, still unanswered question is: did this age of extreme cold wipe out every last living flower, fern and tree? If it did, then our entire flora is no more than 15 000 years old at the most – effectively a pioneering flora. But even polar deserts have warm pockets and shorelines that are free of ice. Today some British wild flowers, like Field Horsetail (*Equisetum arvense*), Cuckooflower (*Cardamine pratensis*) and some arctic-alpine plants, grow on the northernmost land in the world. The south-western tip of Ireland, in Kerry and Cork, almost certainly retained vegetation throughout the last Ice Age, probably including open woodland of Scots Pine – for recent genetic research indicates that Scots Pine in the west of Scotland arrived from this direction. Southern Britain, too, was free of ice. Though very cold, it was also very dry. A cold desert of wind-blown sandy deposits, like the Breckland today, could easily have supported an arctic flora of sedges, dwarf willows and dwarf birches and lichen-rich heath (such vegetation can support a surprising variety of animals too – not only voles, lemmings and owls, but reindeer, bovines, mammoths and even horses, at least in summer). Pollen of a few alpine plants has been found in what palaeo-botanists call the full glacial in south-east England. Purple Saxifrage, for example, grew in present-day Cambridgeshire – perhaps not surprisingly, since it is one of the most frost-hardy plants in the world (I thought of that while admiring it on Derek Ratcliffe's Cambridge rockery). Equally hardy but non-competitive plants like Tufted Saxifrage may well be relicts from the coldest period.

There are other plants whose biology and distribution suggests that they might have survived through successive ice ages from a much earlier period. One of these is *Lloydia* or the Snowdon Lily. In the absence of any fossil record for this now very rare flower, one is thrown back on deduction. *Lloydia* is in several respects unique among our mountain flowers. It is our only alpine bulb plant; indeed, as Neville Woodhead pointed out,[210] it is the only monocotyledon with petals in the whole northern hemisphere that lives entirely in Arctic regions or on high mountains. It seems to be a relict species worldwide, dispersed in widely separated mountain ranges from Colorado to the Himalayas. Within Britain, it is confined to the highest mountains of North Wales, including, of course, Snowdon. That, too, is unusual, for the greatest diversity of mountain flowers is found not in Wales but in the central Highlands of Scotland. *Lloydia* survives in Wales probably because it was less severely glaciated than Scotland. Exceptionally among our alpines, it was unable to follow the ice northwards. Recent work by Barbara Jones indicates that it rarely ripens seed and has no obvious way of dispersing them when it does. It relies instead on its bulbs, which multiply vegetatively by budding from creeping rhizomes. Its competitiveness being close to zero, *Lloydia* is confined to rock crevices with little or no vegetation. Moreover it seems to make life difficult for itself by the complicated arrangements for fertilization, some flowers being wholly male, others with long or short styles, as well as stigmas. All in all, it is an intriguing species, destined to be forever rare, though at some point in the distant past it must have been a widespread rock plant.

Snowdon Lily (*Lloydia serotina*) is our only mountain bulb-plant, confined to the highest hills of North Wales.
[Derek Ratcliffe]

There is a different group of plants whose pollen is found almost continuously through deposits laid down during the past million or so years. These are freshwater species, which must have survived the ice ages in the many cold lakes on the tundra near the coast. Among them are floating or submerged species like Yellow Water-lily (*Nuphar lutea*), Water Milfoils (*Myriophyllum*) and some of the pondweeds (*Potamogeton*), as well as emergent plants like Water-plantain (*Alisma*), Spike-rush (*Eleocharis*), Common Club-rush (*Schoenoplectus lacustris*), Branched Bur-reed (*Sparganium erectum*) and Bottle Sedge (*Carex rostrata*). Pollen of Mare's-tail (*Hippuris vulgaris*) and Bogbean (*Menyanthes trifoliata*) has been found in what seem to have been pools in the lee of glaciers.[148] A surprise species of the Late-glacial period is Starfruit, today an endangered flower of pond-margins in the warm south. Judging from their pollen flora, these Ice Age lakes might well have been similar to some northern lakes and lochs of the present day, though in a bleaker setting, without any softening fringe of trees.

Freshwater plants appear well in the record because their remains are readily preserved in the lake sediments and their peaty surroundings. However, freshwater is also less affected by outside temperatures than other habitats, and this is the main reason why many freshwater plants have very large world ranges – for example, some of our pondweeds are found in cool regions throughout the northern hemisphere.[148] It must be significant that, compared with other habitats, Britain has a notably rich freshwater flora. All but one of the European species of pondweeds (*Potamogeton*), for example, are native to Britain. In the lowlands we have recently lost much of it through pollution and drainage, and you have to snorkel dive in some of the still-clear lochs at the edge of the Highlands to see how rich these Late-glacial lakes probably were, with their quillworts and stoneworts, pondweeds and water lobelias, and great rafts of yellow and white water-lilies. The essential point about Britain at the end of the Ice Age is that there was an awful lot of water – we were a lakeland, like present day Finland but minus the forests. Plants that like cold water did very well. This was the time, difficult to imagine now, when species like the little purslane, *Koenigia islandica*, or the delicate deep-water *Najas flexilis*, so rare and unfamiliar today that only recently were they given common names, were the 'weeds'. Frost-hardy plants were not then confined to mountain ledges but grew throughout Britain as long as the climate remained too cold for the more competitive grasses and trees to flourish. The rare and beautiful Woolly Willow (*Salix lanata*) probably colonized wet, newly exposed moraines at the edge of the glaciers, as it does in Scandinavia. It was probably the heyday for Drooping Saxifrage in Britain. Today this arctic species produces fertile pollen and seed regularly only at points north of 66°, far to the north of the British Isles. In the present-day Scottish Highlands the poor Saxifrage has virtually given up on sex and relies instead on vegetative bulbils. Finding a decent flower to photograph can be a frustrating exercise. Uncompetitive by nature, it is also unable to spread far from its very restricted sites on mountain summits.

The Late-glacial period of between 14 and 10 000 years ago was an eventful time for plants. According to Martin Ingrouille,[88] 'The bleak tundra landscape filled with flowers'. As grassland succeeded tundra, the alpine flora retreated northwards and upwards to the hills, where heavy grazing eventually forced them back to cliffs and corries. The reason our mountain flora lives on mountains is not only climatic. It is because these plants are not competitive in a lowland environment where grass and trees are the big winners. Most British alpines will grow happily in a lowland rockery, and even set seed there. The alpine 'meadows' full of tall herbs and ferns which succeeded them are easier to imagine because we still have some fine ones today, in the Pennine Dales and some of the more luscious Welsh valleys and Scottish glens. In places they may have survived little changed through succeeding millennia, maintained by grazing and mowing. There is a probable 'indicator species' for these ancient 'meadows': the Jacob's Ladder (*Polemonium caeruleum*). As anyone that has seen a true wild Jacob's Ladder will know, the garden plant is but a feeble imitation of the native plant, which has big, clear blue flowers with bright orange stamens, and graceful, narrow, 'ladderey' leaves. The pollen record shows that Jacob's Ladder was widespread in England and Wales between the melting of the ice and the coming of the forests. It likes damp soil, and the plants are often found rooted in thick moss on overgrown scree, usually on north-facing hillsides. It is a characteristic member of a type of herb-rich vegetation dominated by Meadow-sweet (*Filipendula ulmaria*) and False Oat-grass (*Arrhenatherum elatius*) which is believed to be a rare example of natural grassland.[146] Dramatic proof of its ancient nature was provided recently from Wharfedale, where 12 000-year-old pollen of Jacob's Ladder was excavated very close to where the plant grows today. The discovery of natural meadow grassland surprised ecologists, since we were brought up to believe that Britain was at one time a huge forest ('the Wild Wood'), and that all grasslands are artifacts of farmers. If it were so, plants like Jacob's Ladder could not have survived. What British botanists under-rate, because, after all, they are botanists, and live in a land which has lost its major wild herbivores, is the ecological impact of wild animals. The original Jacob's Ladder meadows were grazed by wild cattle and wild horses, and kept open by browsing deer. They were probably not so much open lawns of grass as lightly wooded valleys of the sort one sees in the Derbyshire Dales today. The dense 'Wild Wood' probably established itself in the interval between the killing off of the bigger wild herbivores and the introduction of domesticated cattle, sheep and horses – a period uniquely free of big herbivores.

Britain remained joined to present-day France, Belgium, Holland and Denmark until about 8500 years ago, when melting arctic ice flooded over the low plains where the Dogger Bank is today, and breached the valley between Kent and Calais, finally separating the Thames from the Rhine. The rising of the sea turned Britain's flora into a self-selecting one of hardy, pioneering plants. Of about 5000 species of flowers, ferns and trees in the nearest parts of the continent, fewer than 2000 had time to colonize Britain, and fewer than half that had time to reach Ireland. By then much of Britain was forest and the great advance of the plants had slowed down. Most woodland flowers are temperate species and so late arrivals to Britain. But at least some woodland flowers were there before the wood. For example, pollen of Wood Anemone (*Anemone nemorosa*) and Dog's Mercury (*Mercurialis perennis*) is known from Scotland well before the woods arrived.[70] On the glacial sands of the Highland fringe, the Wood Anemone still survives in what was probably its ancestral habitat, open heathland. In the context of rare plants, what is more important than woodland are the places the woods missed. Many woodland flowers bloom in open spaces within a wood, not in the wood itself.

The beautiful wild Jacob's Ladder (*Polemonium caeruleum*) may be a signpost of natural meadows, or at any rate places with exceptional long-term stability. [Peter Roworth]

What saved many flowers from the suffocating shade of the forest trees was limestone. A striking number of rare flowers are confined today to headlands and crags of hard, white limestone, or pavements inland, none of which were ever densely wooded. These became refuges for flowers which were intolerant of shading or were poor competitors. Most of them probably found their refuges at the time when the Wild Wood was closing in, and there they have stayed ever since, like pensioners seeking the warm dry places. Hence species like the Hair-leaved Goldilocks (*Aster linosyris*) at the Great Orme, or Cheddar Pink (*Dianthus gratianopolitanus*) in a Mendip gorge, or Limestone Buckler-fern (*Dryopteris submontana*) in the pavement grikes at Ingleborough have probably been rare and restricted for thousands of years – they owe their rarity to trees, not to humans. These isolated colonies may be very stable. Derek Ratcliffe[156] found that populations of Hutchinsia, for example, had hardly changed in a century, despite regular seed production. As with so many limestone plants, you tend to find a lot of it, or none of it. Hutchinsia just sits tight, year in, year out. Just occasionally, there is direct evidence of the extraordinary continuity of these isolated limestone plant communities. At Craig y Cilau in Brecon, there is, by good luck, a deep peat deposit next to a limestone cliff. There we find the pollen of Small and Large-leaved Lime at every level in the peat right down to the bottom of the deposit, about 6000 years old. And the rare Large-leaved Lime survives there still on cliff ledges where it is accompanied by a collection of rare whitebeams, including the endemic *Sorbus minima*, which probably evolved there. One finds a similar association of rare species on cliffs, headlands and gorges throughout the country, each one effectively an island in a former sea of tall forest. This may help to explain why so many rare species grow near the sea without being particularly adapted to a coastal environment. Flowers of the softer chalk, on the other hand, tend to be more widespread, which might suggest that the more shallow chalk soils were never very densely wooded and that the plants were not driven out to refugia to the same extent. However, even on the chalk there are areas unusually rich in rare flowers, like the Mole Gap in Surrey and the Medway Gap in Kent. Some chalk flowers, like Pasqueflower, must also be restricted in range for subtle climatic reasons. They too may have reached roughly their present-day distributions long ago, although in all too many cases their density has become much thinner through habitat destruction.

In the case of the Breckland of East Anglia, a unique flora survives because of extreme dryness. The rare flowers include species that are commonest in central and eastern Europe. They are relics of a time when Britain was still joined to France, Holland and Denmark and so this area was once a long way from the sea. Though the sea is nearer now, the Breck is still the driest part of Britain, and in the recent past it retained extensive areas of drifting sand – desert sand, not coastal sand. Most of the rarities are annuals, or at least short-lived plants, and poor competitors. Arable cultivation provided new opportunities for some of them. But today, the last refuges of flowers like Field Wormwood and Fingered Speedwell are on housing estates and waste land, so very different from the aboriginal open steppes of 10 000 years ago.

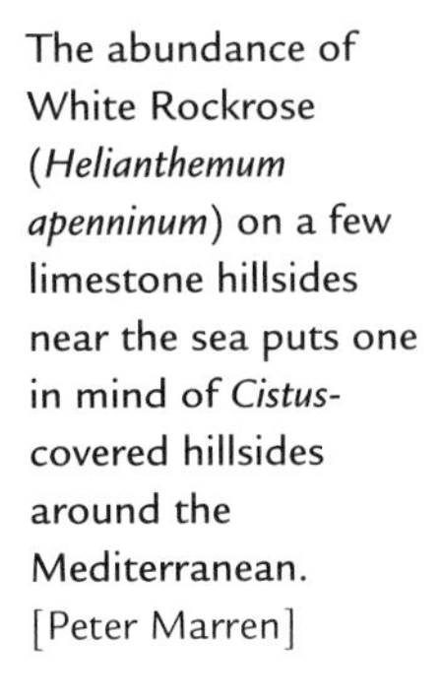

The abundance of White Rockrose (*Helianthemum apenninum*) on a few limestone hillsides near the sea puts one in mind of *Cistus*-covered hillsides around the Mediterranean. [Peter Marren]

Another refuge from the Wild Wood was in wetness, whether floodland, fen or bog. Drainage has long dried out most of the valley bottoms and coastal levels, where rare wetland plants thrived in the past. Bog plants probably reached their peak in the rainy Atlantic period, when peat mosses spread on waterlogged ground over much of upland Britain. The Rannoch Rush was common enough at this time to become an important peat-forming plant (see picture on p. 80). At Thorne Moors in Yorkshire you can pick up handfuls of it, preserved in the peat and easily identifiable by its rope-like rhizomes covered with old leaf-bases. Today its last stronghold in Britain is Rannoch Moor in Perthshire, to where you can squelch out to see it a few hundred yards from the loneliest railway halt in Britain. The fate of Rannoch Rush was shared by another major peat-forming species, *Sphagnum imbricatum*, now a relict species of scattered bogs in Scotland and northern England – an indication of how much environmental conditions have changed since the Atlantic heyday.

Finally, many rare plants escaped the Wild Wood because the trees could not reach them. Above the natural tree-line in Scotland, North Wales and a few parts of Northern England our glacial flora survives on cliff ledges, snow-patches and arid, windswept mountain ridges. But most of the rarities occur only where these conditions are combined with lime-rich soil. Some are in such small numbers that they may be near the survival line. Though safe from development in most cases, only slight environmental change may be needed to tip them over.

Many of our rare native plants, then, have been rare for a long time: they are naturally rare because they are survivors from an earlier stage in history when the climate and landscape were quite different. What is surprising is that, as far as we know, nearly all our original wild flowers and ferns have survived somewhere. This persistence is one of the most impressive aspects of the British flora. Conservationists worried about the survival of this or that species are looking at a time-scale of, at most, decades. But they have all seen a lot of history, and where there are dense knots of rare plants, like on the limestone headlands, they may be the places where change has been slowest, at least from the perspective of the plants. The species which have declined the most in recent times are not the naturally rare plants but the more 'successful', widespread ones of wet meadows, waysides, arable fields and commons. Paradoxically, farming would have given these a new lease of life – an escape from the Wild Wood. It must have been during the introduction of widespread arable farming, perhaps 4000 years ago, that many short-lived plants left their river banks and animal scrapes and entered the human domain as crop weeds – their Faustian pact with humankind. Grassland plants were preserved and increased by grazing, and the partition of land into pastures and meadows. Even woodland flowers probably increased in density through coppicing, hedging and timber management. Until about AD 1800, human activity was an opportunity for plant life, not a limitation.

TABLE 1 Presumed native species confined to a single locality or island in Great Britain (this excludes the Channel Isles and Ireland, and micro-species of brambles, hawkweeds and dandelions)

Species	Locality
Small Adderstongue *(Ophioglossum lusitanicum)*	A heath on St Agnes, Scilly
Lundy Cabbage *(Coincya wrightii)*	Cliffs on Lundy
Alpine Rock-cress *(Arabis alpina)*	Cliffs on Skye
Bristol Rock-cress *(A. scabra)*	Rocks in Avon Gorge, Bristol
Shetland Mouse-ear *(Cerastium nigrescens)*	Serpentine rocks on Unst, Shetland
Teesdale Sandwort *(Minuartia stricta)*	Streamsides on Widdybank Fell, Teesdale
Wild Cotoneaster *(Cotoneaster cambricus)*	Rocks on Great Orme, N.Wales
Arran Whitebeam *(Sorbus arranensis)*	Glens on Arran
Arran Service-tree *(Sorbus pseudofennica)*	Glens on Arran
Wilmott's Whitebeam *(Sorbus wilmottiana)*	Avon Gorge, Bristol
Bristol Whitebeam *(Sorbus bristoliensis)*	Avon Gorge, Bristol
Limonium paradoxum	Rocks at St David's Head, Pembs
L. transwallianum	Cliffs at Giltar Point, Pembs
L. loganicum	Cliffs near Land's End, Cornwall
Diapensia *(Diapensia lapponica)*	A mountain ridge near Glenfinnan
Euphrasia campbelliae	Coastal heaths on Lewis
Wall Germander *(Teucrium chamaedrys)*	Short grassland near Beachy Head
Wood Calamint *(Clinopodium menthifolium)*	Scrubby bank on Isle of Wight
Leafless Hawk's-beard *(Crepis praemorsa)*	A meadow in Cumbria
Radnor Lily *(Gagea bohemica)*	Rocks in Radnor
Sand Crocus *(Romulea columnae)*	Dawlish Warren, Devon

Perhaps this is a surprisingly small list, and some of these plants are endemics (see below) which probably evolved locally and never spread far. Perhaps equally surprisingly, few of them are threatened with extinction (though a substantial rise in sea-level would go hard with Small Adderstongue).

A few more, formerly more widespread native species have been forced back to a single locality. These include:

Species	Locality
Strapwort *(Corrigiola litoralis)*	Margin of Slapton Ley, Devon
Creeping Marshwort *(Apium repens)*	Port Meadow, Oxford
Fen Ragwort *(Senecio paludosus)*	A fen ditch near Ely
Lady's Slipper *(Cypripedium calceolus)*	'Somewhere in Yorkshire'
Triangular Club-rush *(Scirpus triqueter)*	Mud by the Tamar

The English Channel and the North Sea are wide enough to prevent the colonization of new species, with rare exceptions which can hitch a ride on migrating birds or float on ocean currents. We have gained a few species by evolution (our endemic species). We have lost a handful through recent extinction (see Lost Flowers). We have probably lost species which left no record, and, almost certainly, there are a few plants that have not been discovered yet. The most dramatic event in recent times, however, is the vast number of new plants that have arrived in Britain through trade and transport. As Martin Ingrouille points out,[88] the most dynamic aspect of our flora is its increasingly cosmopolitan nature. For example, while the second edition of Clapham, Tutin and Warburg's Flora (1962) included just four species

Above left: In some English bogs you can find fibrous lumps of peat made of this plant, Rannoch-rush (*Scheuchzeria palustris*), which is now confined to mid-Perthshire.
[Peter Roworth]

Above right: If Britain lay a couple of degrees further south, Sand Crocus (*Romulea columnae*) would be as common on the south coast as it is on the Channel Islands.
[Peter Marren]

of naturalized *Cotoneaster*, the first edition of Stace's Flora (1991) increased that number to over 50, and the second edition (1997) to 68. In 1908, Druce listed 1999 alien species; by 1994, Clement and Foster[32] had increased that number to 3586, and even that takes no account of some 580 non-native grasses. Today, the flowers you see on banks and verges from a speeding car or train are just as likely to be exotic as native. As Ingrouille enthuses, 'Not since the post-glacial colonization of the British Isles has the vegetation been exposed to such potential for change. We live in exciting times.'

Of course enthusiasts for rare native flowers tend to see things differently. Our concern is not so much with the advancing vanguard of exotic plants as with the native ones in the slow track, or left behind, the flowers which are going nowhere but down. Conservation is only the latest event in the long history of rare plants in Britain – a sign that society cares about the past as well as the future. We know that mere numbers deceive. Though we have a larger flora than ever before, we also have fewer flowers. Formerly common ones have become local, local ones rare, and rare ones endangered (actually some once fairly common ones are endangered too!). And what may soon be at threat is wildness itself.

Our endemic flowers and trees

Britain has been separated from the continental mainland for less than 10 000 years – the merest blink in evolutionary terms. Evolution normally takes longer than that to produce a stable new species. Hence, compared to islands that have been isolated for millions of years, like Crete or Madagascar, we have few endemic species, that is species which are found nowhere else in the world and which probably evolved here. As we have seen, the British flora is of a different kind – of hardy, pioneer plants which streamed north in the narrow time-gap between the last Ice Age and the formation of the English Channel. Nevertheless we do have some endemic plants – about 50 species of wild flowers and trees (see Table 2), more if you add local forms, subspecies and micro-species of hawkweeds, dandelions and brambles. Most of them are only *slightly* different from a close relative, and nearly all of them are rare. Indeed some endemic species are among the rarest native plants in our flora, with only a few dozen individuals in the world.

Almost by definition, British endemics belong to groups of plants which are actively evolving. Some are found on offshore islands or headlands or crags, like Lundy's endemic 'Cabbage' or Shetland's own Mouse-ear. Isolation speeds up evolution and produces new races or species adapted to local conditions. Many have breeding systems which strengthen genetic isolation. For example, a new variant derived from an outbreeding parent can be preserved by self-pollination or inbreeding. This is evidently what happened in the case of our most glamorous endemic flower, the Scottish Primrose (*Primula scotica*), which evolved from two parents, Bird's-eye Primrose (*Primula farinosa*) – happily still with us, though now separated from *P. scotica* by hundreds of miles – and the Norwegian Primrose (*Primula scandinavica*), which is not. The separate character of *Primula scotica* was probably preserved by self-fertilization.[111] It has a surprisingly low level of genetic variability which may possibly reflect long inbreeding. This, and its distribution in the far north of Scotland, must mean it is ill-equipped to survive climate change. Given the patriotic sensibilities aroused by *Primula scotica*, one can only hint, in whispers, that it did not necessarily evolve in Scotland, indeed that it almost certainly did not. It was probably widespread during the Ice Age, for its seed has been found preserved in peat in Cambridgeshire. Its present-day range merely shows the extent to which it has been pushed out by competitors.

The Scottish Primrose (*Primula scotica*), perhaps the best known British endemic plant, is confined to northern shores in Scotland and Orkney. [Andrew Gagg's Photoflora]

The Lundy Cabbage (*Coincya wrightii*) and its relative Isle of Man Cabbage (formerly a species, recently, alas, demoted to a subspecies, *Coincya monensis* ssp. *monensis*) are our two island 'Cabbages'. They look superficially similar with their clusters of big yellow flowers, but the Lundy one has hairy leaves and is perennial while the Isle of Man one has more divided hairless leaves and flowers only once. They both belong to a genus which has produced a group of closely related but geographically separated species, all confined to small areas on mountain ranges or the coast. The closest relatives to our two plants live in Spain, offering the amusing speculation that they did not actually evolve on Lundy or Man, but merely got stranded there. Perhaps competition was significantly less on the islands – though the Isle of Man Cabbage is quite widespread on the west coast (though, interestingly, it does not occur in Ireland).

The Early Gentian (*Gentianella anglica*) is one of the few fairly widespread endemic species. It almost certainly evolved from Autumn Gentian (*G. amarella*), a variable species which has produced at least two distinctive local forms, subspecies *septentrionalis*, found mainly in northern Scotland, and subspecies *hibernica*, endemic to Ireland. Early Gentian probably evolved from an early-flowering form, favoured by our relatively mild winters. Early-flowering would effectively isolate the form since it would preclude cross-pollination with Autumn Gentian. Yet it still seems surprising that although there are plenty of Autumn Gentians on very similar chalk downs on the other side of the Channel, not a single specimen of *Gentianella anglica* has ever been found outside Britain.

The most common cause of endemic plants in Britain is the abandonment of sex. Some plants, especially in the Rosaceae and Compositae, manage to produce fertile seed without fertilization. This is called apomixis. It means that the genetic character of the parent plant is passed on in identikit form to its offspring. In the case of the hawkweeds (*Hieracium*) and lady's-mantles (*Alchemilla*) the flowers produce little or no pollen, and so, effectively, all the plants are female. A group of sea-lavenders (*Limonium*), on the other hand, produce two types of pollen and have two types of stigmas, but among the various permutations, some are normally self-pollinated and retain their local character that way, while others are apomictic. This means you can have a lot of fun trying to name sea-lavenders. Apomixis is an evolutionary blind alley. No further development is possible since there are no exchanges of genes for natural selection to act on. The species carries on in its dead-end way until its environment changes, and at that point it presumably becomes globally extinct. Why should any species take this path? The obvious advantage of apomixis is that it enables a plant to reproduce quickly and efficiently. Once it starts reproducing in this feministic way, a plant is well on the road towards a separate and stable genetic identity, for local peculiarities will be perpetuated. Whether this makes them full species is debatable. They are much more similar to one another than sexual species, and are often hard to identify. Botanists often call them micro-species. They have fascinated specialists for over a century, but most of us leave them well alone.

Apomixis may have evolved in different ways. Probably the main one is through hybridization, and the resultant doubling of the number of chromosomes. The apomictic whitebeams (*Sorbus*) and lady's-mantles are all polyploids, that is, plants with a multiple set of chromosomes. Geographic isolation modifies characters like leaf-shape or the size of petals and sepals, producing a bewildering series of local species, some of which may be confined to an isolated headland, cliff or gorge, generally on limestone. As we have seen, the Avon Gorge near Bristol has two unique whitebeam trees. Two more, *S. arranensis* and *S. pseudofennica*, are confined to granite stream-banks on Arran, though are quite widespread there. *Sorbus leyana* is confined to crags on opposite sides of a valley near Merthyr Tydfil, *S. minima* to another couple of crags near Crickhowell. Some of these *Sorbi* occur in botanically rich localities which are probably refugia, suggesting that the whitebeams evolved a long time ago. (Though not all rare *Sorbi* are confined to remote crags. *Sorbus vagensis* was described by A.J. Wilmott from a tree growing 'inside Mrs Harris's tea garden' at Symonds Yat. Another yet-to-be-named form in North Devon is nicknamed the 'No Parking Tree' since it grows next to a 'no parking' sign!)

Another big group of endemics is found among the eyebrights (*Euphrasia*), those strange little half-parasitic plants with purple and yellow-veined flowers and small, saw-edged leaves. With them, sexual profligacy is the key and only habitat differences keep them from interbreeding all the time. (Euphrasias are named, incidentally, after the Greek Muse, Euterpe, who represented joy and pleasure.) There are about 20 'micro-species' (depending on whose classification you use) and they hybridize with one another in practically every combination available. Often there are more hybrids than parent plants. Nine species are endemic to the British Isles, and most of them probably originated as a hybrid which stabilized. For example, *Euphrasia rivularis*, confined to North Wales and the Lake District, probably stabilized out of a pairing of the much more widespread *E. rostkoviana* and *E. micrantha*. The endemic *Euphrasia anglica* replaces *E. rostkoviana* in southern Britain, and is itself one of the probable precursors of *E. vigursii*, confined to Cornwall and Devon. Some rare eyebrights are confined to islands on the northern fringes of Britain, like *E. campbelliae* and *E. marshallii* in the Hebrides and the northern isles.

These 'critical' plants are still in a state of taxonomic flux. Quite possibly they are destined to remain so, since in many cases it is an arbitrary decision where you draw the line between them. Present research suggests there are too many eyebrights and not enough whitebeams. In certain cases, a species may well be in the throes of evolution and has not yet become a stabilized species. A possible example is Newman's Lady-fern (*Athyrium flexile*), which is endemic to a few high cliffs and snow-pockets in the Grampian Highlands. Those (perhaps few) who have seen it say it is quite a distinct fern, with narrow, 'congested', rather rigid fronds and a long stalk bent

Above left: This alpine form of Hoary Rock-rose (*Helianthemum oelandicum* ssp. *levigatum*) with short stems and small, less hairy leaves is endemic to the summit of Cronkley Fell in Upper Teesdale. [Peter Marren]

Above right: There is debate about whether the Great Orme Berry or Wild Cotoneaster is a Welsh endemic (*Cotoneaster cambricus*) or an ancient introduction which has acquired local characteristics (*C. integerrimus*). [Peter Marren]

sharply in an elbow near the base to hold the blade in a characteristic horizontal plane. This unusual character is maintained in cultivation. The fern is evidently unable to compete with its close relative, the much commoner Alpine Lady-fern (*Athyrium distentifolium*), and so grows in even more extreme environments, among tumbled boulders on north-facing cliffs near the summits of some of Scotland's highest mountains. So Newman's Lady-fern seems to have a physical, geographical and ecological identity. Does that make it a species? Heather McHaffie, who recently investigated the ecology of this remotest of plants, found that the 20 or so known wild populations are not uniform. Some have broader fronds than others, and seem to be intermediate between Newman's and Alpine Lady-ferns. In other places only a single specimen has been found, suggesting the possibility of them arising from occasional mutations. One authority, Christopher Page,[131] believes that *Athyrium flexile* is an *incipient* species, a fern in the making, which has not yet stabilized. Another, Clive Stace,[180] considers it to represent either 'a chance combination of characters' or a mere variety of Alpine Lady-fern, that is, not a species at all. At the moment you can take your choice. Either way, it is an interesting fern which could provide insights into the evolutionary process.

Rare, endemic plants are by definition vulnerable, and most recent extinctions in the wild have been in this category. Two, Broad-leaved Centaury (*Centaurium latifolium*) and Interrupted Brome (*Bromus interruptus*) were British endemics (see Chapter 9, Lost Flowers). Fortunately most of our endemics seem reasonably safe on their islands and crags. Some are also in nature reserves, though this may mean little unless the manager knows about them and can recognize them; fine specimens of at least two endangered endemics, *Sorbus wilmottiana* and *Cotoneaster cambricus* have been accidentally cut down by conservation workers! One endemic species which may not last much longer, at least in recognizable form, is Plot's Elm (*Ulmus plotii*). As a tree, Plot's Elm can be recognized by its characteristically narrow outline and hanging branches, which contributed to the scenery in parts of the Midlands. But most of these trees were attacked by Dutch Elm Disease. This kills the mature trunks, but not the roots. However young Plot's Elm suckering from the roots is hard, if not impossible, to distinguish from other kinds of elm! Here, then, we have the strange situation of a plant which is not about to die out but which is no longer recognizable! I do not know a name for this phenomenon, but I offer 'optinction': it is there but you cannot see it.

TABLE 2 Endemic species or former species in Great Britain or the British Isles (NB I omit endemic subspecies except in a few cases where they have been treated as full species, and also endemic species of *Rubus* (brambles) and *Hieracium* (hawkweeds).)

Isle of Man Cabbage (*Coincya monensis ssp. monensis*)
Lundy Cabbage (*C. wrightii*)
Purple Ramping-fumitory (*Fumaria purpurea*). Also in Ireland.
Western Ramping-fumitory (*F. occidentalis*)
English Sandwort (*Arenaria norvegica* ssp. *anglica*)
Grey Mouse-ear (*Cerastium nigrescens*)
Mountain Scurvy-grass (*Cochlearia micacea*)
Boyd's Pearlwort (*Sagina boydii*)
Alchemilla minima (Lady's Mantle)
Wild Cotoneaster (*Cotoneaster cambricus*)
Perennial Flax (*Linum perenne ssp. anglicum*)
Early Gentian (*Gentianella anglica*)
Broad-leaved Centaury (*Centaurium latifolium*)
Scottish Primrose (*Primula scotica*)
Welsh Groundsel (*Senecio cambrensis*)
South Stack Fleawort (*Tephroseris integrifolia ssp. maritima*)
Shetland Mouse-ear Hawkweed (*Pilosella flagellaris ssp. bicapitata*)
Plot's Elm (*Ulmus plotii*)
Interrupted Brome (*Bromus interruptus*)
Scottish Small-reed (*Calamagrostis scotica*)
Young's Helleborine (*Epipactis youngiana*)
Newman's Lady-fern (*Athyrium flexile*)

Taraxacum (Dandelions)
37 micro-species, including the following Red-listed ones:
T. cherwelliense
T. clovense
T. geirhildae
T. serpenticola
T. tanylepis

Sorbus (Whitebeams)
Sorbus leptophylla
S. wilmottiana
S. eminens (possibly conspecific with Irish *Sorbus hibernica*)
S. porrigentiformis
S. lancastriensis
S. vexans
S. subcuneata
S. devoniensis. Also in Ireland.
S. bristoliensis
S. pseudofennica
S. arranensis
S. leyana
S. minima
S. anglica. Also in Ireland.

Euphrasia (Eyebrights)
Euphrasia rivularis
E. anglica. Also in Ireland.
E. vigursii
E. pseudokerneri. Also in Ireland.
E. cambrica
E. marshallii
E. rotundifolia
E. campbelliae
E. heslop-harrisonii

Limonium (Sea-Lavenders)
Limonium paradoxum
L. procerum (with a subspecies in Ireland)
L. britannicum
L. parvum
L. loganicum
L. transwallianum
L. dodartiforme
L. recurvum. Also in Ireland.

The hidden world of genes

You cannot see genes but they are the basis of life. Each species contains within it a usually large amount of genetic variation enabling it to adapt to changing circumstances. Where a species is very rare, however, genetic resources may be low, and individual plants may share exactly the same genes. This means that they are vulnerable to environmental change since they lack adaptability. Until recently the only way to study genetic variation was to grow plants in uniform laboratory conditions and measure characters like leaf shape or the presence or absence of a particular pigment. However it is now possible to study genes through their fingerprint chemicals (isoenzymes) or even directly, through DNA. The former are analysed using a process known as electrophoresis, which produces distinctive bands which differ from one genotype to the next. In the 1980s, more direct methods of assessing genetic variability were developed by comparing the actual genetic code contained in the plant's DNA. At first the process required more plant material than would be desirable in a very rare species, but the technique has since been refined to use only small quantities – a fragment of leaf or bark tissue. The method is known as 'amplified fragment length polymorphism' (AFLP). It produces a much larger number of 'markers' than isoenzyme studies, and therefore provides a more detailed and accurate idea of the amount of genetic variation present in a population.

This technique was used recently by the Jodrell Laboratory at Kew to elucidate the genetic variability in the famous colony of Monkey Orchids at Hartslock in Oxfordshire.[153] These delightful flowers with their passing resemblance to monkey acrobats are one of the successes of rare plant conservation. The Hartslock population was all but wiped out in 1950 when a farmer ploughed up most of the site. When the local wildlife trust purchased what was left of it for a nature reserve 25 years later, they could find only 8 plants. Since then this tiny population has been carefully nurtured and built up by hand-pollinating the flowers and managing their habitat until, by 1997, there were some 200 mature Monkey Orchids. Of these, a leaf was taken from 26 randomly chosen individuals, and their DNA extracted. The results showed that, as was expected, the genetic variability in these Hartslock orchids is extremely low. Many plants are more or less identical and the one which did differ slightly from the others grew at some distance from the main colony in an adjacent field. These results agreed with a morphometric (measurement-related) test made a few years earlier. Old herbarium specimens from Hartslock show there was once a far more healthy level of variation in this colony: it has gone through what geneticists call a bottleneck, having at one time been reduced to a handful of plants. The researchers drew no firm conclusions from the study, except to observe that there was nothing further to be gained from hand pollination since the genes are much the same: since the plants are self-incompatible, no fertilization is taking place. In the short term, the wildlife trust's policy of building up the population is the correct one. In the longer term, cross-pollination with Kentish or even French Monkey Orchids might help to restore some genetic variability – if that was considered ethically acceptable for what after all is a *wild* plant.

Genetic studies of the native flora using isoenzymes and DNA are scattered in scientific journals and reports, though much current work has not yet been published. DNA work has become scientifically fashionable, with its focus on molecular structure and expensive technology. In some cases, like that of the Monkey Orchid, the results bear out those of traditional, non-technological measurement-based work. They tend to prove what we fear, that small populations of plants have low genetic variability. It is not invariably so: the Shore Dock (*Rumex rupestris*), for example, shows excellent genetic health despite most of its populations being small and isolated. However its seeds are probably dispersed on tidal currents, allowing genetic exchange between one colony and another. On the other hand, a recent study of Twinflower (*Linnaea borealis*) has shown that many populations are genetically identical (clones), and have grown up purely vegetatively from a single individual. As a result these isolated populations are sterile: *Linnaea* is an out-breeding plant, and its pollen is ineffective on individuals of the same genetic stock. Given the localized nature of its colonies – here a patch in a pinewood, there another high up on a crag a kilometre or more away – the chances of any pollen reaching another population are low. That does not necessarily mean the Twinflower is in any immediate danger (there is a small patch of it in northern England that does not even *flower*, but it seems quite stable). But it certainly seems to be declining for reasons that cannot be attributed entirely to habitat destruction. Where an outbreeding plant is down to just a few small populations in the whole country, like the Blue Sow-thistle (*Cicerbita alpina*), genetic alarm bells may be ringing – though having seen most of its populations, I would guess that there are a few hundred years left even for this species.

Genetic studies have revealed that there can be geographically separated genetic races within an apparently uniform species. Professor John Parker of the University Botanic Garden, Cambridge has identified a form of Autumn Squill with 42 chromosomes instead of the usual 28. This form is restricted to western Cornwall and the Channel Islands. It is probably the original race of Autumn Squill in Britain, having colonized the country from Brittany while there was still a land-bridge between Britain and France. Unfortunately it has declined at some sites, apparently through competition with the commoner 28-chromosome plant. You might say 'so what?' – since they both look the same. But they are almost certainly not the same physiologically, and, as the older form, the 42-chromosome race might in the long term be the better adapted to local conditions. In the interests of the species, the preservation of both genotypes is desirable. How you achieve that is another matter.

Isoenzyme studies are also helping to elucidate the relationships between closely related species, like that troubling genus, the whitebeams. Since the rare species are all apomictic, and so have foregone any possibility of further genetic variation, one would expect them to have very low genetic variability. Using electrophoresis to isolate the isoenzyme peroxidase, Michael Proctor and colleagues have shown that this is indeed true of some species.[151] But other, generally more widespread ones, show two or more distinct genetic patterns that indicate separate genetic types (genotypes). In the context of their restrictive breeding arrangements, this implies that plants like *Sorbus eminens* and *S. porrigentiformis* are not uniform species at all, but aggregates of closely related forms.

It looks as though the whitebeams may be due for another overhaul, which is likely to produce a string of new species similar to the one we recently enjoyed among the Rock Sea-lavenders. One can only hope that any gene-based classifications will bear in mind the overriding need to identify species in the field, based on characters that one can see. Already, DNA studies are indicating that the way we classify wild orchids is partly artificial. The genus *Orchis*, for example, consists of not one but three groupings, showing that the species had more than one common ancestor. This means that *Orchis* as currently defined in our floras is an artificial genus. Some *Orchis* species are more closely related to Pyramidal or Dense-flowered Orchids than to other *Orchis*es. The Man Orchid, on the other hand, stands revealed as a perfectly good *Orchis*, though it is placed in another genus, *Aceras*, on the grounds of lacking a spur. Unfortunately it simply does not look like one. One dreads where such studies may be leading – perhaps to two sets of British floras, one with plants arranged traditionally and the other based on genes, with a wildly different structure. Most of us, I dare say, will go on using the first kind, even if we have to find them on an antique book stall.

DNA-testing has shown that the Monkey Orchid (*Orchis simia*), known for more than a century from a site in Oxfordshire, has low genetic diversity and may therefore be vulnerable to natural change.
[Peter Marren]

Chapter 5

A question of numbers

Starved Wood-sedge (*Carex depauperata*), named after its few widely separated flower-spikes, is one of the rarest British plants with fewer than 25 specimens known.
[Peter Wakely/English Nature]

The poet, priest and botanophile Andrew Young once asked George Claridge Druce if he had seen all the British plants. He replied that he had, though the last ten had given him a hard struggle. 'Are you the only botanist who has seen them all?' pursued Canon Young, but Druce merely smiled. When Young inquired about an orchid so very rare that it had been seen only a few times, Druce pointed out modestly that two of those times he had seen it himself.[213] Working in the Druce herbarium at Oxford for this book, I sought out the specimen that, I had read somewhere, had completed Druce's tour of the British flora. It was a dim and very rare little flower confined to Guernsey and Herm, called Guernsey Centaury or *Exaculum pusillum*. Druce had attached a typewritten note to say he had collected it on 31 July 1913, with the local botanist, E.D. Marquand. The plant had been rediscovered two years earlier after a lapse of many years by Lady Davy. 'It was fairly plentiful, over a very limited area,' noted Druce, as he picked a number of them to distribute among fellow botanists. And there were the proud words: 'My last British species'. I thought of this when I found *Exaculum* by the singing sands of Herm in May 1992. Alas, it was not *my* last British species by a long chalk.

Guernsey Centaury is among the most elusive wild flowers by virtue of its small size, capricious appearance and (for most of us) remote location. But there are flowers which, for most of the twentieth century, were even harder to track down. The famous last colony of the Lady's Slipper orchid was a well-kept secret, even to the great Ted Lousley who had seen nearly every other plant. Perhaps few botanists even today have set their eyes on

Hart's Saxifrage in its 'gully above the sea, called Polldoo on the island of Aranmore' off the coast of Donegal. When D.A. Webb was writing about the species in 1950,[199] only three collections were known, the most recent one having been gathered by the island's lighthouse keeper. David McClintock made a special journey to see it in 1956, but found the island unexpectably large, and the directions correspondingly vague. 'I walked around its coast getting half way each time,' he wrote 'and eventually, almost at the end of my circumambulation, found it in a small gully. So far as I know I am still the only living botanist who has seen it in the wild.'[115] Unfortunately Hart's Saxifrage has since then been demoted to a subspecies of Irish Saxifrage (*Saxifraga rosacea* ssp. *hartii*). Then there is Dwarf Millet (*Milium vernale*), aptly described by McClintock as 'small, prostrate and unobvious'. Someone found it on Guernsey in 1900, but forgot where, and assumed, wrongly, that it was from the cliff at Petit Bot. 'Hither went streams of hunters, bottoms up, but none could see it.' It was not refound until 1949, and not on the cliff, after all, but on dunes nearby.

Our rarest flowers

Which *is* our rarest wild flower? The question is often asked, but even if we agree on what 'rare' means, the answer will change from one year to the next. No one, as far as I know, saw a single wild Starfruit plant in 1995, though it sprang up again, phoenix-like, the following year, probably from buried seed. The orchid, *Ophrys bertolonii* must take the biscuit for the rarest ever flower, for there was only one, it lasted only a few weeks and it was never seen again. But that plant was only a stray from warmer shores, not an established British species. In terms of native wild flowers, the Lady's Slipper is a good stock answer; the clump of flowering stems on its Yorkshire hillside is a single genetic entity, and therefore you could say there is only one. But there are rumours of others, and even the Lady's Slipper could be considered less rare than the Ghost Orchid, which has not been seen at all since 1986. It is probably not extinct, since it has a habit of flowering only now and then, and taking us all by surprise when it does. The most *endangered* flower is another question. The Corn Cleavers (*Galium tricornutum*) has been just a gasp away from extinction for some years, but it has turned up in gardens before now and is probably under-recorded. But the Triangular Club-rush (*Schoenoplectus triqueter*) has been searched for most assiduously in its few known sites on muddy riverbanks, and, with only a single pure-bred population left seems to be on the brink of national extinction.

For certain rare wild flowers it is possible to estimate the actual numbers of individuals, although the population may vary wildly from year to year and I doubt whether any surveyor would claim to have seen *every* plant. But, treated with proper caution and scepticism, numbers are always fascinating. Some plants can evidently survive as extremely small, isolated populations. The world population of endemic Sorbi, like Ley's Whitebeam or Wilmott's Whitebeam is of the order of only a few dozen trees or bushes. The last count of the Wild Cotoneaster revealed only 14 bushes, although the latest thinking is that this attractive small shrub is an ancient introduction, and, if so, not an endemic species but an isolated colony of a continental one. Britain and Ireland have a lot of rare trees, often overlooked by field botanists who instinctively look downwards, not upwards. Perhaps the rarest of all is the true Wild Pear (*Pyrus pyraster*). Oliver Rackham[154] knew only four mature trees, one of which fell down and was made into a harpsichord, but its status is confused by wildling pear trees escaping from gardens and orchards. Plymouth Pear (*Pyrus cordata*), Service-tree (*Sorbus domestica*) and Large-leaved Lime are other trees which get by with very small populations.

The rarest plants in the British flora have national populations of less than 100 individuals. One of them is Starved Wood-sedge (*Carex depauperata*), which totalled only 20 plants in 1995 – though even that is a cause for celebration compared with the all-time low of just one plant in the 1970s! Of the same order is the Fen Ragwort, which produced 28 flowering stems in its roadside ditch in 1997, again an improvement on just five stems when it was first discovered in 1972, though some way below its best effort of 82 stems in 1984. Another desperately rare plant is the typical form of Perennial Knawel (*Scleranthus perennis* ssp. *perennis*) whose status at its only site, Stanner Rocks, is so precarious that, in the words of Ray

TABLE 3 Britain's rarest wild flowers?

Ghost Orchid (*Epipogium aphyllum*)	None seen since 1986
Lady's Slipper (*Cypripedium calceolus*)	Single large plant plus introductions
Triangular Club-rush (*Schoenoplectus triqueter*)	Single population, but situation confused by hybridization
Corn Cleavers (*Galium tricornutum*)	Only 4 plants seen in 1995 (and none at all in 1992 or 1994)
Fringed Gentian (*Gentianella ciliata*)	15 plants present 1994, but none seen 1996
Wild Cotoneaster (*Cotoneaster cambricus*)	About 14 bushes, plus introductions
Starved Wood-sedge (*Carex depauperata*)	About 20 plants in 2 localities (1995) plus introductions, but slowly moving up
Fen Ragwort (*Senecio paludosus*)	Less than 100 flowering stems - latest count 28 (1997)
Oxtongue Broomrape (*Orobanche artemisiae-campestris*)	About 40 plants in 4 sites
Prickly Sedge (*Carex muricata* ssp. muricata)	51 plants in 3 sites (1995)
Proliferous Pink (*Petrorhagia prolifera*)	About 75 plants seen in 1994
Alpine Rock-cress (*Arabis alpina*)	Last count was 83 plants in 3 places (1993)
Leafless Hawk's-beard (*Crepis praemorsa*)	Less than 200 plants all in one place
Wood Calamint (*Clinopodium menthifolium*)	A few hundred plants all in one place

Perennial Knawel (*Scleranthus perennis*) has two forms in Britain, one erect, the other prostrate. This is the prostrate one, endemic to a few sandy fields in Norfolk and Suffolk. [Peter Wakely/English Nature]

Woods, 'a misplaced boot could exterminate it'.[211] The prostrate form of this plant (*S. perennius* ssp. *prostratus*) in the Suffolk Breck is only slightly less precarious, relying mainly on old plough-lines. Given the length and detail of botanical recording in the Breck, we can be fairly sure that this plant has been rare ever since John Ray discovered it in the seventeenth century. Yet it may be fairly stable, and still occurs where he first found it.

There are many more species in the population range of 100 to 1000 individuals. These include some of our rare orchids – Monkey, Military, Late Spider and Red Helleborine, and also the East Anglian form of Fen Orchid, with only 197 plants, distributed in three sites, in 1997. The Creeping Marshwort (*Apium repens*) belongs here, with an estimated 500 or so plants living in the hoof-prints of cattle on Port Meadow, Oxford. So does the Field Wormwood with 465 plants in 1990 (not counting 'translocations'), most of them on a Breckland industrial estate. Interestingly, a large proportion of the species surveyed in recent years fall into the population range of 1000 to 10 000 individuals. At the bottom end are species like Ground Pine, Lundy Cabbage, Thistle Broomrape and Whorled Solomon's-seal

One of the most spectacular of our rare wild flowers, Meadow Clary (*Salvia pratensis*) occurs in shreds and patches across southern Britain but most commonly, for some reason, in Oxfordshire. [Peter Wakely/English Nature]

The Deptford Pink (*Dianthus armeria*) is a much decreased wild flower, possibly through lack of grazing on its favourite dry, grassy banks. It is doubtful whether it ever occurred at Deptford. [Peter Roworth]

(*Polygonatum verticillatum*), at least in their good years. Rather more comfortably placed is the Yorkshire Sandwort (*Arenaria norvegica* ssp. *anglica*), reported to be stable at 2000 to 4000 plants all in the Ingleborough area, but vulnerable to erosion since most of them grow by well-used footpaths. The Cotswold Pennycress (*Thlaspi perfoliatum*) and Meadow Clary (*Salvia pratensis*) are at the upper end of the range, but the majority of the plants are in one or two places. A few almost equally rare species fall outside this range of rarity only because their largest populations contains thousands of plants, like Cut-leaved Germander (*Teucrium botrys*). But numbers can deceive. Annual plants often respond spectacularly to disturbance, like the tens of thousands of Red-tipped Cudweeds (*Filago lutescens*) that appeared recently in two fields in Surrey, or the abundant Green Hound's-tongues (*Cynoglossum germanicum*) that sprang up on building sites and gale-damaged woods near Dorking. Such population explosions seldom last long, although unnaturally large numbers may be sustained for a while by clever management.

A striking aspect of so many of the species just mentioned is that a single site may hold more individual plants than the rest of Britain put together. In the case of Whorled Solomon's-seal, whose range is confined to the catchment of the River Tay, each isolated colony is considered to be part of a single genetic 'metapopulation'; in other words, the plant has only one 'site', but it happens to be a large and scattered one.[212] The intensely 'clumped' distribution of some species is hard to explain in terms of habitat: on the face of it, there is nothing unique about the Meadow Clary's favourite bank, or the Cudweed's chosen field. Their survival in these particular places may have as much to do with chance and history as with biology. This tendency continues with some of the species which number from 10 000 to 100 000 individuals. Alpine Catchfly is confined to only two widely separated British sites, but at one of them, Meikle Kilrannoch in Angus, an estimated 68 000 plants were present in 1993.[111] Even more spectacular was the discovery of a vast population of Grass-

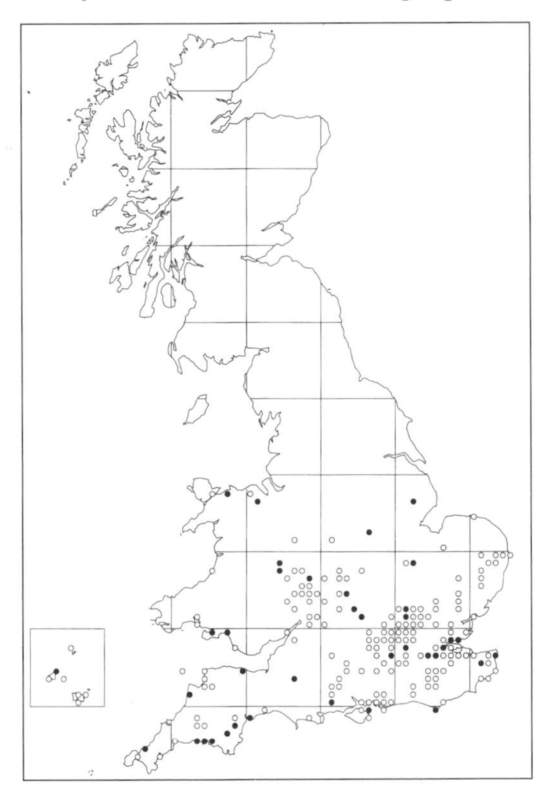

Distribution map of Deptford Pink. Solid dots are post-1950, open circles pre-1950. [Biological Records Centre]

poly around the swan lake at Slimbridge (see Chapter 12), which holds the bulk of the estimated national total of around 700 000 individual plants.

Some Red Data Book species would not be there at all if we assessed them by numbers instead of geographical cover. I would not like to have to count the Dorset Heaths (*Erica ciliaris*) in Dorset or Cornish Heaths on the Lizard, or White Rock-roses at Berry Head and Brean Down – for there must be many millions. If you live in these areas you might regard such species as common plants. If there was an arbitrary dividing line of, say, 100 000 plants, a great many species would be crossing from the Red Data Book to the Scarce Plant Atlas, and vice-versa – assuming we had the energy and funds to go out and count them all.

Of course mere numbers can deceive. The most threatened plants are not necessarily the rarest ones, but those which suddenly find themselves out of kilter with their environment. I would, for example, put more money on the prospects of the Lady's Slipper surviving the next century than mature Wych Elm. Alpine Catchfly or Dwarf Millet will probably get by on their specialized sites, but we are witnessing a catastrophic decline of once relatively common flowers like Corn Chamomile (*Anthemis arvensis*), Tower Mustard, Deptford Pink and Small-flowered Catchfly. Species tend to attract funds only if they are very rare and, preferably, pretty: hence the biology of some of our rarest plants is better known than most common ones. Field botanists are much more likely to 'adopt' a rare species than a common one, and who can blame them? The plants in the greatest danger of extinction within the next century will be among those which are currently ignored. The wise naturalist should regard Red Data lists as a stock list of the present, not a guide to the future.

How plants survive

Wild flowers live slowly. Normally we glimpse only a fragment of their lives. If you are patient you might see a flower being visited by an insect, or watch pollen or seed drifting in the wind. To find out what makes them tick, however, you need data. You can gather some data quite easily – the size of a colony, perhaps (once you agree on what constitutes an individual plant), the average number of flowers per plant, and perhaps seeds per fruit. But to see how each individual fares over the seasons and years you need to mark each one and record the same data at regular intervals. Two sets of data gives you a line on a graph – a sense of *direction*. After about twenty sets we usually begin to get an idea of what sort of plant it is and how it maintains itself. It sounds easy doesn't it? In the 1970s I had a brief flirtation with monitoring the rare flowers of Upper Teesdale. To plot the position of each plant accurately we used a special portable frame with a movable cross-bar. Each leg of the frame fitted into a nylon tube sunk into the ground. The position of each plant was marked on weatherproof drafting film, on which you also recorded any flowers, seeds or seedlings. Then, a couple of months later you went back and did it again.

Never mind the tedium (scientists are trained to be immune to boredom), consider the practicalities. First you have to find your permanent plots. It is no use putting up a marker. As we soon learned, Teesdale sheep eat anything, even, it seems, metal wire (either that or the wind blows them away), and in any case, a visible marker would risk drawing unwanted attention to the place. So we found our buried nylon tubes by taking a compass bearing. But the fell is a big bare place, and the plots were only a square metre or two. Until one was familiar with every tussock and rock, finding them took as long as the actual recording. Luckily there were few visitors to witness the moments when one's despairing wail rose, Lear-like, from the heath to join the plaintive chorus of the Plovers and the Dunlin. Then there was the Teesdale weather. This is one of the coldest, windiest, wettest parts of England, which suits all those Ice Age flowers very well, but it does challenge the constitution of *Homo sapiens*. As Margaret Bradshaw once remarked, 'a high degree of stamina is needed to maintain integrity under the climatic conditions . . . Indeed it may be an advantage if the human element (matches) the plants: a physiological arctic-alpine by heredity or naturalisation may be expected to perform most successfully.'[16] I performed abysmally on all counts, and it was at about that time that I realized I wasn't cut out to be a scientist.

The purpose behind those days on the fell was to try to answer often-asked questions like: Does sheep-grazing harm the Teesdale Violet? Does it matter if a Bird's-eye Primrose seldom sets seed? Or how long does a Spring Gentian live? To find out such life and death matters you need to follow the progress of a sample population over as long a period as possible. A population might decrease or increase or stay about the same. Or it may fluctuate around a mean, as most of the Teesdale flowers did. The Spring Gentian had its good and bad years, but although the graph bounces up and down, you perceive after a few years that these short-term changes mask a longer-term stability. The Bitter Milkwort (*Polygala amarella*) seemed at first to be an exception, showing a serious decline on its best site. We worried about it; we put wire cages over the best colonies. But eventually the Milkwort perked up again on its own accord. It evidently exhibits a very long-term bounce. The upswing in its numbers coincided with a decrease in rabbits.[29] It seems that rabbits like the Bitter Milkwort, despite its horrid taste, and that the plant's numbers go up as those of its main predator go down. During the bad times it probably puts most of its chips in the seed-bank.

Populations are stable when the rate of recruitment ('babies') is about the same as that of mortality ('tombstones'). During a baby boom, numbers go up; when mortality exceeds recruitment down they go again. It seems a miracle that most plant populations seem to remain balanced indefinitely on that knife-edge of stability – mathematically almost impossible, one might think. Given the unpredictable events in the life of a plant – the weather, the amount of rain, grazing animals, fire, flood, pathogens, plagues of aphids or voles – it is hardly surprising that the graph has peaks and troughs. But in normal circumstances, native wild flowers seem to overcome all such difficulties. They have had thousands of years to attune themselves to survival. The mistake of previous generations was to assume that rare plants rely mainly on seed to replenish their numbers. Some in fact produce little or no seed in Britain. The Small Cord-grass (*Spartina maritima*), for example, rarely if ever manages to ripen fruit. Instead, here, at the northernmost limit of its natural range, it spreads by means of tough, wiry rhizomes. These enable the grass to hang on in its patches of salt-marsh but do not allow it to colonize new sites (unless bits of the plant are transported by the tides). Hence this species is liable to become extinct locally when its habitat is reclaimed or eroded away by the sea. A more famous example of infertility is the Small-leaved Lime (*Tilia cordata*). Some particularly brilliant research by Jackie Huntley and Donald Pigott showed that in northern England the summers are no longer warm enough to provide enough time for the pollen tube to reach the ovary and achieve fertilization. What has saved the Lime from extinction in the north is its longevity. Some trees may live a thousand years or more.

It seems as though Spring Gentian, too, does not count on regular replenishment by seed. One of the problems we encountered was identifying individuals. The Gentian is a rosette plant which can produce satellite rosettes from underground rhizomes. We plotted them as separate individuals, but in most cases they were probably vegetative shoots or ramets. We noticed that genuine Gentian seedlings were rare. The flower stalk elongates after the plant has finished flowering, projecting the seed pods well above the turf – and there they are smartly nipped off by the sheep. We concluded that most of these small, dense patches of Spring Gentian were clones. Within a clone, individual 'ramets' may live up to 20 years or so. But the clone itself is potentially immortal. Hence, a single Gentian seed may provide a plant which survives for centuries by purely vegetative means. For the Spring Gentian, seed is almost irrelevant as a means of population recruitment. This means that the appetite of sheep for its flowers and seed pods cannot, in the short term, harm the plant. This largely vegetative way of life is not confined to rare flowers – some common woodland flowers, like Dog's Mercury, rarely produce seedlings, though they may produce plenty of seed.

Plants which live for a long time do not normally change their numbers dramatically from one year to the next. Some of them seem able to survive with small populations and low levels of recruitment. An example is the Spotted Cat's-ear (*Hypochaeris maculata*), a big robust yellow 'composite' with an enormous tap-root. At Humphrey Head in Cumbria a colony of only 30 or so plants has survived for 200 years, despite frequent predictions of its imminent extinction (as Lynne Farrell has remarked, the better Spotted Cat's-ear spotters spot more Spotted Cat's-ears). In one Suffolk site where it has been known since 1804, the plant has recently appeared again after a ten-year absence. In one Northamptonshire site, a colony of just four plants has persisted year in, year out without either increasing or decreasing. It is natural to assume that such a colony must be endangered – but perhaps it is just part of the life-style of a plant which is naturally rare.

Seed requires space to germinate, and this is not usually available in the close-packed turf of the dales and downs. Seeds are much more important for short-lived plants which grow in temporary open spaces or in naturally open habitats like sand-dunes or corn fields. Their problem is that most open spaces do not stay open for long – sooner or later they will grass over. Hence the maintenance of a sizeable bank of seed in the soil becomes vital. If you plotted numbers against time of a species like Ground Pine or Starfruit, the graph would look like the teeth of a saw.[82] Such flowers make their breaks when they can, producing a vast quantity of seed in a short time, before being forced underground again by the competition. It is of course a high risk strategy unless the ground is disturbed regularly and often. In the past annuals and other short-lived species could often rely on cattle or ponies or carts to provide it. But today, livestock are generally penned up on sown, artificially fertilized grassland with no room for common wild flowers, let alone rare ones. This is why so many rare flowers rely, paradoxically, on disturbance. They spring from the soil when someone lays a pipe, or digs a pit or a tree falls down revealing among its roots soil from a long-forgotten field.

In the right conditions, seed dormancy can last many decades. One of the most spectacular examples I can think of was the unexpected appearance of Ground Pine on land which had been unsuitable for it for over fifty years.

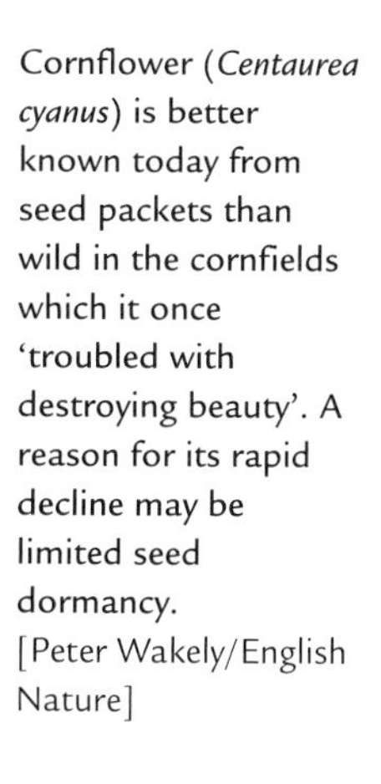

Cornflower (*Centaurea cyanus*) is better known today from seed packets than wild in the cornfields which it once 'troubled with destroying beauty'. A reason for its rapid decline may be limited seed dormancy. [Peter Wakely/English Nature]

That site, on the North Downs, had seen a form of shifting cultivation until 1940, when it was left to grass over and, after myxomatosis, to turn into scrub. By the 1980s it had become dense woodland. In the 1990s the National Trust began to restore downland in the area by felling and uprooting trees (having been given a head start by recent gales). It was during these activities that Ground Pine turned up – almost certainly from seed ripened back in the 1930s on open ground. Of course for seed of this age, buried quite deeply in the soil, considerable disturbance will usually be needed to raise it to the surface. You can envisage Ground Pine as not so much a pretty flower as a grain-silo of seeds, with just the occasional burst of flowering, when some of the seed is allowed to germinate. If we could detect seed, as well as flowers, we might well take a very different view of their conservation status. Since seed banks clearly play such a central role in the survival of rare short-lived plants, one does wonder how species without this faculty get by. Very badly seems to be the answer in the case of Corn Cleavers, which may be nearing extinction in Britain. Least Lettuce (*Lactuca saligna*), too, may prove dangerously vulnerable in its apparent reliance on cattle puddling on sea walls. This is a flower that has already lost its natural habitat along the upper edge of coastal salt marshes. One can only hope, for its own sake, that it may have a survival trick or two up its axils that we do not know about.

Unfortunately information on the seed dormancy of rare plants is very thin. Some Composites (the daisy family) and willows have short-lived seed, geared up to germinating straight away or not at all. The native Black Poplar (*Populus nigra* ssp. *betulifolia*) is a good example of a tree with no seed dormancy – and, perhaps in consequence, Black Poplar seed hardly ever produces a tree. DNA evidence has shown that most Black Poplars are clones, sharing the same genetic material. The usual practice was to plant them as cuttings on muddy streambanks, and, in the Upper Severn valley at any rate, one or two trees supplied the entire population. Otherwise, most available information is for annual arable weeds. One reason why the Corncockle is virtually extinct in Britain as a wild plant is that its large fleshy seeds cannot survive long – either they are eaten or they rot. Other species with limited dormancy, like Cornflower (*Centaurea cyanus*), Shepherd's Needle (*Scandix pecten-veneris*), Corn Buttercup (*Ranunculus arvensis*), Corn Gromwell (*Lithospermum arvensis*) and Greater Yellow-rattle (*Rhinanthus serotinus*) are faring very poorly outside protected sites managed for their needs.[207] And it comes as no surprise to learn that Interrupted Brome (*Bromus interruptus*), another near-extinct 'weed', has little or no seed dormancy. Indeed it seems that few native grasses do.

In many cases the potential investment of seed deposited in the 'bank' is enormous. An average wild plant of the Adderstongue Spearwort (*Ranunculus ophioglossifolius*), for example, produces about 2500 seeds (which is, incidentally, only a third of the capacity of greenhouse-grown plants). In the colony at Badgeworth, which covers just a few hundred square metres, an estimated 16 million seeds were produced between 1962 and 1980.[63] The Ribbon-leaved Water-plantain (*Alisma gramineum*) is even more fecund, producing up to 10 000 seeds per plant. For this species, however, flowering and ripening seed is a relatively rare event which happens only when the water level and temperature are just right. Even then, the flowers last for only a few hours, like those tropical arum-lilies that bloom only once in a hundred years, and then die. Most of the seeds go straight into the mud, and there they stay, awaiting the call. Perhaps one reason why we do not see many Ribbon-leaved Water-plantains in Britain is that when drainage engineers are at work they like to scoop out all this precious mud. By the Rhine, where the Water-plantain is more frequent, they apparently use gentler methods which remove encroaching vegetation while leaving the bottom mud intact.

Of course not all the shed seed will survive, and not all the seedlings will mature into plants. For most rare species, this is a little known area, for it is normally well outside the scope of shoestring conservation projects. The Milk-parsley (*Peucedanum palustre*) is one of the exceptions, mainly because it happens to be the sole food-plant of the glamorous Swallowtail butterfly. The butterfly once flew over Wicken Fen, but died out many years ago, and repeated attempts to reintroduce it have always failed. After a while it dawned on the scientists that there was not enough Milk-parsley to sustain the butterfly, and so, quite properly, they set out to discover how Milk-parsley lives, and what it likes. What was known already was that this is an umbelliferous flower (a distant relative of the carrot) of wet peaty fens, mainly in East Anglia. Each leaf rosette of Milk-parsley flowers only once, but the plant persists by developing new buds from the stem base. A typical shoot produces about 4500 seeds. The seeds can float and so should disperse when the water level is high, but probably not very far. The question was: how do we stack things in favour of Milk-parsley in order to benefit the butterfly? The ecology of all stages of the plant's life-cycle at Wicken were studied in detail by T.C. Meredith at Cambridge.[124] What he found out in some ways deepened the riddle. The Milk-parsley grows in two kinds of habitat: open fen, which is cut for thatch now and again, and in unmanaged, bushy fen. In the former, the plants grow more quickly and seed germinates more often. However, most of the seedlings are promptly eaten by slugs and snails, and a proportion of the mature plants are also, of course, lost every time the fen is cut. In the unmanaged parts, on the other hand, the plants live longer, but germination is poor, and most of the seed is eaten by voles and mice. In either case the Milk-parsley is in tune with its environment, with recruitment and mortality more or less in balance. But this does not, of course, help the Swallowtail. The suggested scenario for increasing Milk-parsley was to cut the fen just occasionally and late in the season, perhaps once in four years, to produce bursts of regeneration, and also to flood the ground in winter to get rid of the slugs and snails and voles. Probably the wetter the better in any case.

The intense blue stars of Alpine Gentian (*Centiana nivalis*) open only in sunshine. Its rarity in Britain may be due to our wet climate, though we also have too many sheep.
[Derek Ratcliffe]

Population biology produced a rather similar conundrum at the opposite end of Britain and for a much rarer flower, the Alpine Gentian (*Gentiana nivalis*). This attractive species with piercing blue flowers has a most unusual life-strategy for a mountain plant. It is an annual (though, as is so often the case with annuals, some plants survive beyond a year), and it therefore depends on seed. That means it must attract pollinators and ripen its pods, all in the fickle, uncertain weather of a Highland summer. That may indeed be the main reason why it is rare in Britain but common in Scandinavia and the Alps – we are that much cloudier and wetter (plus our hills are not high enough). Most of Britain's Alpine Gentians live on Ben Lawers in Perthshire, among perhaps the richest mountain flora in Britain. To survive in a cold, wet climate, the Gentian has a number of insurance policies. The intense blue flowers open and shut like clockwork as the weather changes, and in dull weather the petals furl tightly together exactly like an umbrella to keep the pollen inside dry. In an emergency, the plant can self-pollinate, and each plant produces up to 10 000 tiny seeds.[111] Moreover the seeds can remain viable in the soil for many years, and they can also pass unharmed through the gut of an animal. In Iceland and Sweden the Alpine Gentian is spread in cow-dung, and sometimes stars the grass right up to the door of the home farm.

With only two sites in Britain, the survival of the Alpine Gentian may be close to the line. It was with the very best intentions that the managers of the Ben Lawers nature reserve decided to fence off the best colonies, on steep, south-facing slopes. Their fear was that sheep were grazing the gentian's fruit pods before seed could ripen and disperse. But the enclosures, expensively erected with the help of helicopters, were a failure. No one at that time knew anything about the plant's population dynamics. At first the Gentians did indeed fruit more successfully. But within a few years, the former open, species-rich vegetation was outcompeted by grass, and there were fewer pods than ever. It looked like a no-win situation: too many sheep and the pods get eaten; too few and grasses take over. Perhaps light or rotational grazing would be most beneficial, but that would be hard to arrange on a remote mountainside with free-ranging sheep. But, as we now realize, the Alpine Gentian does not count on regular seed production. There is plenty of seed there already, locked up in the seed bank in the soil. *The Hitchhiker's Guide to the Galaxy*'s large friendly letters, 'don't panic', looks like good advice.

This kind of population study has potentially a great deal to offer to conservation managers. A range of options might be open to conserve a rare plant. In the case of Milk-parsley we could increase the area of cut sedge-fen to boost the carrying capacity of the site. Alternatively we could try to reduce the density of competing plants, as with the Alpine Gentian. For some annual plants we could try to maintain open ground by rotovation or peat cutting. We can slow down natural succession by cutting bushes or dredging ponds. Or we could try to increase the efficiency of seed dispersal: in the case of Milk-parsley one imagines that some sort of wind-pump to circulate the water might do the trick. Or we could wade right in, collect the seed ourselves and sow it somewhere promising. But whether population science really does aid nature conservation on a practical level depends on the degree of co-operation between researchers and managers. The scientists do not help matters when they write turgid papers in professional journals in which any possible practical application is ignored while the authors try to demonstrate their numeracy. Some conservationists, on the other hand, seem less and less inclined to read the literature at all, intelligible or not. The best chance of success is where the researcher works closely with the nature reserve people, as at Teesdale or Wicken Fen. But it has to be said that there is a far greater distance between practical conservation and theoretical science today than there was thirty or forty years ago.

The East Anglian typical form of Spiked Speedwell (*Veronica spicata* ssp. *spicata*) relied on free-range grazing to maintain short, open, bracken-free grassland. [Peter Wakely/English Nature]

The life and death of a colony

The British tradition for recording minute particulars is well seen in the loving attention lavished on some colonies of rare flowers. A dossier of observations spanning many decades can provide a peculiarly intimate glimpse into the life of a colony of plants, recording not only how the plant may fluctuate in numbers from one year to the next but also suggesting some of the reasons why – a kind of plant biography. This is rarely the result of a formal monitoring scheme: it just happened that way because naturalists have long made a habit of detailed field notes, and, thanks to societies and local museums, these notes have often survived their maker. In the case of some of the rare Breckland flowers, 'site histories' have been compiled by Gigi Crompton and others,[35] which trace the fortunes of a colony over a century or more. Particular groups of flowers have been known for up to 200 years, like the Spiked Speedwell which Sir John

Cullum had 'gathered in great beauty on the dryest and barrenest Part of Culford Heath' in the late eighteenth century. Here I want to tell the story of two plant colonies which eventually died out almost under the eyes of their good angels, the local botanists. If they have a moral, it is that things are not always what they seem. Wild flowers, even rare ones, can survive all kinds of upsets which in this case included military occupation, mowing, collecting and plagues of rabbits. What killed them off was insidious, long-term change which, by its nature, hardly anyone noticed, except the plant.

The Spiked Speedwell (*Veronica spicata*) occurs in two forms in Britain. The subspecies *hybrida* has a relatively tall, dense-flowered spike and grows on limestone gorges and cliffs in the west. Though nationally scarce, it can be locally common and is not endangered. The typical form, subspecies *spicata* is confined to the Breckland of East Anglia. It has a shorter looser flower spike and grows in open, sandy grassland, and its state is much more parlous. Unlike many Breckland rarities, Spiked Speedwell is a long-lived plant, which spreads vegetatively using creeping stems which can take root. The seed-set is evidently very low, and the plant typically grows in dense patches. In July 1910, at a place called Garboldisham (pronounced 'Garblesham') Heath, Suffolk, W.H. Burrell and W.G. Clarke found a strong colony 'as flourishing as any in the country'. Growing amongst bracken on undisturbed heathland, they reported, were several hundred plants, 'presenting a most beautiful appearance'. Some had tall, virile flower spikes with up to 100 separate florets, quite unlike the feeble depauperate forms you see today. The following year, there seemed to be even more, or perhaps they just searched harder: 'In an area of about 100 by 50 yards there were at least one thousand specimens of *V. spicata*. In just over a square yard, where the growth was most dense, we counted 180 plants'. Yet, by 1913, and with no signs of disturbance, they could find only a dozen flowering spikes. Evidently, they concluded, 'one season is much more favourable than another'. That year, G.C. Druce visited the colony, noting that 'as the plant is so very local, members [of his Botanical Exchange Club] must be content with a meagre gathering, which has been carefully made so as not to injure the living plants . . . The specimens were gathered on 30 August, in company with Dr C.E. Moss'.

Then came the First World War. The heath was used as a military rifle range, and the botanists feared disaster for *Veronica spicata*. 'Part of the area has been much cut up . . . I expect that this year the outlying plants are the only ones that stand much chance.' Their worst fears seemed realized when Clarke and Burrell failed to find a single flower in 1922. Then – a miracle! – a solitary plant appeared in 1923, and the following year 'there were several hundred plants in the colony . . . in grassy patches among the bracken'. Most had probably been there all the time, but had been cut before they could flower. The years between 1924 and 1940 were gloriously uneventful. The Speedwell was visited by such luminaries as T.J. Foggitt, J.E. Lousley, Noel Sandwith and A.S. Watt, who reported nothing alarming. A sketch-map made at this time shows four clumps of Spiked Speedwell, all close to an ancient earthwork of chalk rubble crossing the heath known as the Devil's Ditch. Access to the area was difficult during the Second World War, when again a red military flag flew over Garboldisham Heath. The estate was sold in 1944 – changes in ownership are a dicey time for rare plants – but the new owner left the heath much as it was. After the war was over, Francis Rose found 'a huge population over a wide area of level heathland', with at least 2000 flowering spikes in one place.

It seems to have been in 1953 that Garboldisham Heath was ploughed up. Subsequent visitors found a few plants at the edge of what was now an arable field, and more on Devil's Ditch which had been left intact and became the main stronghold of the Spiked Speedwell. By now it was harder to find, and specific directions were needed: '11 paces from 7th telegraph pole coming from East Harling'. Another plant turned up on the road verge 'where (the) excursion coach stopped and Audrey Disney trod on it'. By 1961, the colony seemed to be in danger. 'We had a prolonged hunt [for it] on the Devil's Ditch,' wrote Martin George of the Nature Conservancy, 'but were quite unsuccessful. It seems that all three plants that I saw last year have succumbed during the winter. It is difficult to find a definite reason for this . . .' In 1964 there is another note: 'Only a few plants survive along the top of the dyke and these are in danger of being overgrown by bracken and coarse grasses. Arrangements are in hand to autoscythe the dyke top to encourage the growth of *V. spicata*.'

The rest of the story is best given verbatim.

22.5.1964 'We . . . obtained permission to auto-scythe the top of the dyke where *V. spicata* has grown in the past. [The owner had been asked not to use herbicides near the dyke.] We then autoscythed an area about 10 yards wide and 20–25 yards long, clearing away bracken fronds, coarse grass, nettles etc.'

19.6.1964 I visited the area briefly . . . but could not find any plants of *V. spicata*. J.M. Schofield.

24.8.1965 3 specimens at East Harling. ('There were 30 here in 1960').

22.7.1971 'One flowering plant *c.* 20 yards from road . . . badly overgrown with coarser grasses and bracken. A small area around the plant was cleared.' P.W. Lambley.

1972 & 1973 'Failed to find the plant.'

11.8.1974 '14 vegetative shoots found in single square metre in a small clearing in the bracken . . .' G. Crompton.

20.8.1975 '5 flowering spikes. Bracken cleared for 20 yards around site, and in a former area nearby.' R.P. Libbey.

2.11.1976 '6 vegetative shoots. Bracken more open but grasses very dense after the rains. No sign of an old flower spike. Elder tree nearby appears to be dead.' G. Crompton & W.H. Butcher.

27.7.1977 '8 spikes in bud in group of a dozen plants . . . Bracken has invaded area again and was removed by hand.' G. Crompton & R. Payne.

After that the record falls silent. There seems little chance of the plant returning, for the area is now a pig farm.

What probably sustained the Spiked Speedwell was light open-range grazing, the form of pastoralism which sustained much of our wildlife for centuries, but which is now confined to a few commons and parks and special places like the New Forest. Grazing keeps the more aggressive plants like bracken and tall grasses at bay, and maintains an open sward in which rare flowers like this can survive – even if some of them get eaten. As far as one can tell, the Spiked Speedwell could cope with intervals of military training during two World Wars – since grazing no doubt continued, as it does on present-day army ranges. Even the ploughing of Garboldisham Heath was not immediately fatal, though it reduced the population to a dangerously low level and denied the plants the regular grazing they needed, except by rabbits – and by coincidence, the outbreak of myxomatosis temporarily wiped out the rabbits too. Inevitably bracken and other tall vegetation took over, possibly hastened by fertilizer drift from the new crop fields. Bracken can grow taller than a tall man in parts of the Breck. Mowing the ground now and again could never compensate for age-old rough grazing, and was clearly ineffective. That is the problem with conservation remedies: they are essentially short-term, often too localized and continue only so long as volunteers, money and enthusiasm are available. Spiked Speedwell has declined even more than most Breckland rarities because it can grow neither on arable land nor in tall, dense vegetation. Without old-fashioned pastoralism to sustain it, it depends almost entirely on the charity of conservationists.

Our other story concerns the Hairy or Downy Greenweed (*Genista pilosa*) in Ashdown Forest, Sussex. Like Spiked Speedwell, there are two forms of Downy Greenweed in Britain – a trailing plant with tiny downy leaves which grows near the sea in Cornwall and Pembroke, and a more robust form with more erect shoots that grew among heather in Ashdown Forest. This latter form resembled the Downy Greenweed of Spain and northern Germany more than the plants in western Britain. For many years it was one of the most celebrated of Sussex flowers. Fears were expressed about its survival as long ago as the First World War, when the best known site, at Gills Lap ('Galleon's Leap' in the Winnie-the-Pooh stories), was used as an army camp. Downy Greenweed has been known in this 'very wild and heathy' district since 1873. In some years it was reasonably plentiful, though easily passed over as Petty Whin (*Genista anglica*). Then came the Second

Hairy Greenweed (*Genista pilosa*) overlooking Mullion Cove, Cornwall. [Bob Gibbons]

World War. It was the same story: the decline of pastoralism followed by myxomatosis. Scrub began to invade the open heath. The heather became older, drier and more prone to fires. A big fire during flowering time in 1960 virtually wiped out the colony; next year only a single survivor was found, close to a path. After a long absence, a suspected plant appeared in 1980, but before it could flower the place had been mown flat.[164] Fires have been blamed for the extinction of this colony, but Downy Greenweed can survive light burning in Cornwall. The underlying cause was again the decline in grazing, changing the formerly open grass and heath of Ashdown Forest into scrub and bracken. Photographs of the same view, decades apart, show this dramatically. But the change has been gradual and evolutionary; and the casual visitor does not notice it.

A second colony, in heathland opposite a layby, seems to have suffered a series of misadventures which gradually weakened it. In 1971 the site was burned, and later part of it was ploughed up as a firebreak. A few plants still survived. Mary Briggs saw one in 1974; 'on a perfect June evening, (we) walked onto the heath in the sunset – *G. anglica* was in full flower and standing out brightly in the evening sun, but a little further on (we saw) a low clump of deeper yellow . . . *G. pilosa*!' This plant, a fine upright one, was photographed by Peter Wakely of the NCC, little suspecting that he was taking a historic picture. The sun was indeed going down on *Genista pilosa*. On Mary's next visit, the plants were almost hidden under a pile of logs constructed for a horse jump. The year after that they were roasted by another fire – one of the many heathland blazes of summer 1976 – which penetrated to within a few inches of the colony and then stopped. Perhaps the firebreak had worked. The last person to see the plant, in June 1977, wrote that 'we found three plants this time, two very prostrate and hardly growing well with no flowers, the third on the edge of the ride just a few inches from last year's burnt ground – this flowering sparsely'. The *Genista pilosa* seems to be in a precarious situation, he added. It was. David Coombe searched for it in vain in 1979, noting 'much erosion and pony jumps'. Several search parties in the 1980s failed to find it either, and since the detailed recording for the *Flora of Ashdown Forest*[164] in the 1990s also drew a blank, it looks as though the Forest's most famous flower is now extinct, and that we can say so with unusual certainty. Conservation came too late to save it, and focusing narrowly on restoring this one rare species would in any case be a waste of time. The message of the Downy Greenweed is that the historic landscape of Ashdown Forest must somehow find a way of restoring its ancient grazing customs. If a Downy Greenweed could speak, that was probably what it was trying to say.

TWENTY YEARS IN THE LIFE OF A RARE FLOWER

'Survival for one of Britain's rarest orchids has been secured after a round-the-clock security operation by lovers of the country's flora' (*Daily Telegraph*, 1998). The management of the only known wild plant of the Lady's Slipper has long been the responsibility of a special committee, and more recently under English Nature's Species Recovery Programme. Hence this is probably the most closely monitored wild flower in Britain. These are just a few highlights from bulletins published in BSBI News or English Nature's newsletter.

1979 'Visitor pressure on the Lady's Slipper is increasing (and) has caused erosion in the immediate area . . . PLEASE DO NOT VISIT THE PLANT IN 1980.'

1982 'Numbers of visitors had decreased considerably and the immediate surroundings . . . were recovering well. 5 flowers bloomed.'

1984 Good news: a seedling has appeared.

1987 'The single wild plant continued to produce good growth and 6 flowers bloomed. Seed from this plant has been used in germination experiments at Leeds University and Kew. The latter has successfully raised protocorms . . .'

1989 'The wild native plant is healthy and produced 7 flowers. Some natural seedlings have been observed in the vicinity. The wild site is still very fragile and members are again urged NOT to try and see the plant.'

1991 Alarm. The orchid has suffered 'a vicious attack' by voles which bit through several shoots while the flowers were still in bud. 3 flowers were produced, and were hand-pollinated.

1994 The plant flowered again and seems to have recovered from the voles. 'One bloom was produced on the first seedling to flower. The laboratory grown seedlings introduced some time ago continue to survive and are regularly monitored . . . It is pleasing to note that visitor numbers continue to fall.'

1995 The 11-year-old seedling has again produced a flower! There are now 60 shoots present from seedling planting and the scattering of hand-pollinated seed.

1996 Press release: Lady's Slipper seedlings have been reintroduced to former sites, open woods on the side of the Pennines under English Nature's Species Recovery Programme. At Ingleton Glens, near Settle, some seedlings have been planted in a protective enclosure next to the footpath, so the public can see them. If all goes well, they should flower by the year 2000.

1998 Latest report: some 14 flowers bloomed on the original site, where some 60 seedlings are now growing in protective wire cages. Of 11 seed pods produced, 9 were removed and taken to Kew for micro-propagation, and the remaining 2 left to disperse naturally. Young plants are also being raised by members of the Hardy Orchid Society and Alpine Gardener's Society. Some 16 wild sites now contain young Lady's Slipper plants. Two more are to be made open to the public gaze: at Castle Eden Dene and what may possibly be another wild plant in Silverdale, Lancs.

Chapter 6

Living on a tightrope:
hybrids and predators

Hybrids are a mixed blessing for plants. They are the raw material of evolution, where the genetic cards are reshuffled and another round in the endless game of natural selection begins. On the other hand, hybridization may come to threaten the survival of a species. We have a number of plant species which probably originated through hybridization, such as the eyebright, *Euphrasia campbelliae*, a flower found nowhere else in the world except the Hebridean island of Lewis, where it probably evolved. The rare Alpine Woodsia fern (*Woodsia alpina*) is believed to have evolved by the mixing of genes from Oblong Woodsia (*Woodsia ilvensis*) and an arctic species not found in Britain, *Woodsia glabella*. This particular event happened long ago, and probably not in Britain. At least one species, however, has evolved very recently, not on a mountain top, like *Woodsia* but near a market town in Wales. This is Welsh Groundsel (*Senecio cambrensis*), the product of the genes of native Groundsel (*S. vulgaris*) and the introduced Oxford Ragwort (*S. squalidus*). Normally plant hybrids are sterile, or at least have low fertility. Though hybrids may be common and a population of them may last hundreds or even thousands of years, they stay hybrids. In the case of Welsh Groundsel, however, the hybrid was fully fertile and stabilized into a new species. This event must have happened very recently in evolutionary terms, since Oxford Ragwort was unknown in Britain before 1794.[87] Perhaps it evolved on the very spot that it was found, on a road verge near Ffrith in Flintshire. Since then it has somehow turned up on building sites in Edinburgh, though building sites being what they are its new life north of the border may prove a short one.

There are some 700 known hybrids in the British flora. Some of them are very occasional, the product of parents who do not normally communicate, like the strange orchids that result from two very different-looking parents.

One of these is the hybrid of the Bee Orchid and the Fly Orchid, known scientifically as *Ophrys x pietzschii*, a strange-looking thing with the pink sepals and petals of the Bee but the long, dark lip of the Fly. Like most of these occasional oddities it is sterile. The single known plant first appeared in 1968 in a quarry at Leigh Woods near Bristol and died of old age about 20 years later. In that time a hazel bush had grown up and shaded the plant, which might have shortened its life. I missed it, unfortunately, but I have been shown the exact spot. There were Fly Orchids all over the place, but no Bee Orchids. They grow in another quarry some distance away and so the two orchids are separated by distance as well as genetic incompatibility. No other wild specimens of *Ophrys x pietzschii* have ever been found.

Welsh Groundsel (*Senecio cambrensis*) must have evolved recently, for one of its parents, Oxford Ragwort (*Senecio squalidus*), was unknown in Britain before 1794. [Andrew Gagg's Photoflora]

Other hybrids are very common, though in most cases only a botanist would spot them. A well-known one is the 'Pink Campion', *Silene x hampeana*, which forms readily whenever its parents, the Red and the White Campions come into contact, and sometimes replaces the red one altogether. Since the Pink Campion is often fully fertile, it persists happily enough on its own. Possibly the Pink Campion is a product of farming and hedges, which have brought Red Campion, a woodland plant, together with White Campion, which prefers open places in full sun. In other words, it is human activity which in this case has increased the amount of hybridization. In other cases, hybrids have probably occurred and persisted among their parent plants for as long as the landscape itself. On Widdybank Fell in Upper Teesdale the rare Teesdale Violet (*Viola rupestris*) and the upland form of the Common Dog-violet (*V. riviniana*) occur together and occasionally produce a distinctive hybrid, *Viola x burnatii*. The hybrid is sterile, but it can spread vegetatively and form local patches which are practically immortal. Yet their ability to hybridize does not seem to harm the prospects of either parent. Their breeding mechanisms ensure that hybrids are the exception, not the rule, and the majority of Teesdale Violets on Widdybank Fell contain not so much as a whiff of the genes of their common relative.

In certain cases, long-lived hybrids *can* become more common locally than the parents, and even persist in their absence. A good example is the Least Water-lily, *Nuphar pumila*, which, as a pure-bred species is confined mainly to the sheltered bays of lochs and broad rivers in Scotland. In northern England, however, another small yellow water-lily has been much confused with it. This one is slightly larger with a differently-shaped stigma. It is called *Nuphar x spenneriana*, and is the hybrid between Least Water-lily and the common Yellow Water-lily, *Nuphar lutea*. The interesting thing is that the parents hardly ever see one another nowadays, and so have no opportunity to hybridize anymore. Chris Preston, an authority on water plants, believes that most populations of *Nuphar x spenneriana* were formed in the distant past, possibly as long ago as the Late Glacial period.[148] What may have happened is that *Nuphar pumila* got here first, since it tolerates cooler conditions than *Nuphar lutea*, and was then gradually displaced in the south by the larger species as the climate grew warmer. For a time the two water-lilies were in close contact, and it was then that they produced hybrids. Today it seems that the hybrid has a competitive advantage over *Nuphar lutea* in the cool lakes of northern England and the central lowlands of Scotland. At any rate it often occurs on its own, behaving as though it were a species in its own right. Rosemary FitzGerald tells me that a hybrid water crowfoot she is studying, *Ranunculus tripartitus x aquatilis* may be similarly ancient. It has long been thought that the rarer parent, Three-lobed Crowfoot, *Ranunculus tripartitus*, was ousted from much of southern England through hybridization with the more aggressive Common Water-crowfoot, *Ranunculus aquatilis*. But old herbarium specimens of supposedly pure-bred *R. tripartitus* from the New Forest and elsewhere turn out on closer inspection to be this same hybrid. An alternative hypothesis is therefore possible: that *Ranunculus tripartitus* was never a fully-evolved species except in western Britain (where it is more isolated), and that the hybrid, far from being a modern threat, is an ancient plant. If so, this of course, has fundamental repercussions for the conservation of the Red-listed Three-lobed Crowfoot, a priority species in the Biodiversity Action Plan.

An even more overtaken plant is the Alpine Enchanter's Nightshade (*Circaea alpina*), which is much less common than its hybrid with common Enchanter's Nightshade, *Circaea x intermedia*. This last is quite common in upland Britain, while genuine *Circaea alpina* is very local indeed, confined mainly to central Cumbria and the Isle of Arran, with scattered sites in the Scottish Highlands and in Wales. Compared with the more robust hybrid, Alpine Enchanter's Nightshade is a delicate plant, typical of seepages on wet rocks or wooded slopes, and with markedly less vigorous powers of vegetative propagation. The beautiful coloured maps reproduced in Geoffrey Halliday's *Flora of Cumbria* (1997) show how the hybrid is widespread in wooded valleys throughout the county, while thoroughbred Alpine Enchanter's Nightshade is restricted to the heart of the Lake District, mainly in undisturbed woods at the heads of Borrowdale, Haweswater and at one end of Ullswater.[78] Interestingly it is relatively common on Arran where it presumably benefits from a greater degree of isolation. Alpine Enchanter's Nightshade is probably another Glacial relict species, which lost out when woodland containing the more aggressive *Circaea lutetiana* spread over Britain as the climate grew warmer. Its range probably continues to contract to this day. However, old records of *Circaea alpina* are unreliable since for a long time the hybrid was mistaken for the real thing.

The species tug o' war

While recent research is showing that some hybrid populations are ancient, and so probably stable, there are signs that other hybridization events are recent, and in some cases are threatening rare flowers with genetic dilution. One such case is the Fen Violet (*Viola persicifolia*) at Otmoor, Oxfordshire. Though recorded as quite abundant there in the past, more recent searchers have found only a hybrid violet, *Viola x ritschliana*, which has the pale blue colours of the other parent, Heath Dog-violet (*Viola canina*). This hybrid is probably the result of the drainage of this formerly wet fen, which endangered the Fen Violet and brought it into closer proximity with the Heath Dog-violet. The hybrid is sterile, but can spread vegetatively to form dense masses of flowers (which are – irrelevant, I know – very pretty). In this case, there is a twist in the tale. A few years ago, disturbance from scrub clearance brought up long dormant seed from the days when the Fen Violet ruled Otmoor, and, hey presto, they promptly germinated into pure-bred pale lilac Fen Violets.[99] Hence the seed bank may yet renew the genes of a species apparently lost through hybridization.

The same thing may be happening to some populations of the Dorset Heath, *Erica ciliaris*, the only flower that shares its name with its habitat. When I first began to explore the wonderful heaths of Purbeck, I had difficulty finding a good-looking typical plant among swarms of *Erica x watsonii*, its hybrid with Cross-leaved Heath (*E. tetralix*). It has even been suggested that past hybridization explains the limited range of Dorset Heath, mainly on the Purbeck Heaths and in West Cornwall.[26] In my experience, the hybrid is commonest on the drier periphery of bogs, and, if so, perhaps drainage and natural succession have brought the two species together more often. Fortunately Dorset Heath seems more than capable of holding its own on home ground.

The future of the Tuberous Thistle (*Cirsium tuberosum*) seems far less assured, and many former sites now show only the hybrid with Dwarf Thistle, *Cirsium x medium*, proving that in this case mass hybridization is recent and ongoing. It is doubly regrettable, for undiluted Tuberous Thistle is a gallant cossack of a thistle, whose military air is enhanced by the openness of its downland setting, not to mention the tanks and other military vehicles scudding along on some of its sites. The hybrid is a watered down version, a feebler thistle, with narrower flowers and more prickly leaves. If in doubt, the way to separate them is to look at the hairs on the bulbous bit of the flower through a hand lens: Tuberous Thistle has cottony ('arachnoid') hairs, Dwarf Thistle crooked ('jointed') hairs, and the hybrid a combination of the two. Once seen, real Tuberous Thistle is never forgotten, but the hybrid can be confused with a tall-grass form of Dwarf Thistle (var. *caulescens*) with a long flower stem.

In 1957, Donald Grose listed all the then sites of Tuberous Thistle he knew in its main stronghold in Wiltshire.[76] There were about 30, but already the hybrid was present and increasing in many, while others were being ploughed up. Thirty years later, Ron Porley and Sue Everett revisited many of Grose's sites, and discovered that the hybrid had increased much further at the expense of the parent plant. In only six places were pure plants found, and even there hybrids, too, were present. On another nine sites there were hybrids only, including the well-known one on the earthworks of Avebury Ring, the source of many herbarium specimens of Tuberous Thistle. The best surviving colonies were on army land on Salisbury Plain, where there are still several thousand plants scattered over the northern half of the Plain. (Getting access to them is another matter. Ron Porley once returned to his car after a hard day's thistle-counting on one of the army ranges to find a crumpled wreck with the doors hanging off it. A tank driver seems to have been overcome by short-sightedness.)

What has caused this disastrous situation? The most likely explanation is that the two thistles were previously separated ecologically, the Dwarf Thistle favouring short grass, heavily grazed by sheep or rabbits, and the Tuberous Thistle deeper soils with tall grass, which crowds out turf-hugging plants. Everett[52] has suggested that it was sudden changes in grassland management that brought the two together, for example, where grazing has increased on hitherto neglected or lightly grazed grassland. Ironically, precisely that kind of change has often happened on nature reserves, to promote the short swards rich in downland species – with the result that several Wiltshire nature reserves now have large stands of hybrid thistles, and not a single pure-bred Tuberous Thistle left. It is also possible – and, as a walker on the Wiltshire Downs it is certainly my impression – that the stemmed variety of Dwarf Thistle, which survives better in tall grass, has increased, and it seems to be the form which hybridizes most readily. Only on the army ranges, where things are more ecologically stable despite the crack of firearms and the blast of shells, and on one or two privately owned downs, does the Tuberous Thistle look reasonably secure for the moment.

Another plant whose willingness to hybridize poses a problem to conservationists is the Scottish Dock, *Rumex aquaticus*. The Scottish Dock can form hybrids with at least four other docks, providing a lot of fun for specialists. Above all, it hybridizes with the one that resembles it most, the Broad-leaved Dock, *Rumex obtusifolius*, hated weed of gardeners and graziers. You distinguish them and their hybrid by looking at the fruits: Scottish Dock's are triangular and tapering, Broad-leaved Dock's are heart-shaped,

Above left: Pure Dorset Heath (*Erica ciliaris*) like this is sometimes outnumbered by the less distinctive hybrid with Cross-leaved Heath.
[Bob Gibbons]

Above right: The last stand of Tuberous Thistle (*Cirsium tuberosum*) is in Wiltshire, especially on Salisbury Plain. You can identify it with your bottom. You can sit on a true-bred Tuberous Thistle in comfort. The hybrids, on the other hand, will make you yelp.
[Bob Gibbons]

and the hybrid's something in between with little teeth at the base.[163] Pure-bred Scottish Dock can be a big strapping dock as tall as a man, with dense yellow and red panicles and distinctly triangular basal leaves. It is confined to the shores and bonnie banks of Loch Lomond, especially around the incoming River Endrick. It was much more widespread in the Late-glacial period, for its remains have been found in widely separated parts of Britain. On the other hand it might have been even more rare in the more recent past. The floodplain of the River Endrick was drained in the eighteenth century, and only when the drains fell into disuse did the area become suitably wet again. That might account for its late discovery, for the species was unknown in Britain before 1935, when the Glasgow botanist Bob Mackechnie found a puzzling dock at Balmaha, on the eastern side of Loch Lomond, and sent specimens to J.E. Lousley for identification.

The good news is that most Scottish Docks live on a National Nature Reserve. The bad news is that the hybrid is flourishing, especially where the two docks find one another along old field drains or in wet meadows. I see, to my annoyance, that my own photograph of it, taken years ago, seems to be this hybrid, evidently a rather nondescript and shapeless plant compared to the Gothic spires and pinnacles of true Scottish Dock. All the nature reserve people can do is try to keep them apart by maintaining the water

level and pulling up Broad-leaved Dock wherever they find it. Another danger, it seems to me, may come from conservation volunteers or contractors uninstructed in the finer points of dockology, and so inclined to view the whole tribe of them as undesirable.

In the past few years a lot of interest has focused on a little-known and undeniably unspectacular plant called the Triangular Club-rush, *Schoenoplectus triqueter*. The reason is that it seems to be on the brink of extinction in Britain (and not far back from it in mainland Europe either), and one of the reasons for that seems to be hybridization. Triangular Club-rush lives on the mud banks of tidal rivers and is more easily noticed from a boat than from the bank. You can spot it from some distance by the bluish-green foliage, from closer up by its sharply triangular (triquetrous) stems and, very close up, by its smooth round glumes (a close relative has bumpy ones). In times gone by, it grew by the Thames in London, Westminster and Hammersmith, and quite likely would have sprouted on the mud we saw so much of in the BBC's recent television adaptation of *Our Mutual Friend*. Its extinction on the Thames can be dated with

Above: Pure Scottish Dock (*Rumex aquaticus*) can grow to six feet tall with stately spires of speckled red-and-yellow fruits. Unfortunately it seems all too ready to hybridize with its numerous vulgar relations.
[Andrew Gagg's Photoflora]

Right: These are the enfeebled non-flowering remains of a once healthy stand of Triangular Club-rush (*Scirpus triqueter*) by the River Tamar. Some former populations now consist only of hybrids.
[Peter Wakely/English Nature]

unusual precision: the last site disappeared in 1946, when the river wall by Kew Bridge was rebuilt, destroying what was said to be the last patch of tidal mud in London.[20] Apart from the Arun, where it has not been seen for many years, the only other known sites for Triangular Club-rush are on the Medway in Kent and the Tamar on the border of Cornwall and Devon. And it is in a bad way in both of them.

Triangular Club-rush is not a very easy plant to survey. You need a boat, and should you disembark to examine its finer points you are liable to disappear up to the waist. Moreover, there is only a small window near low tide when the plants are easy to see, and steering a boat to the right place with one eye on the incoming tide needs time and non-botanical skill. In 1987, Rosemary FitzGerald and Eric Philp were able to survey the Medway banks with binoculars from the comparative comfort of a cabin cruiser, keeping a rubber dinghy in tow for closer investigation. They travelled for miles past towering reeds and inaccessible marshlands, without any sign of their quarry, though there was plenty of Grey Club-rush (*Schoenoplectus tabernaemontani*) locally, which shares the Triangular one's bluish-green hue and hence can cause palpitations in the unwary. Then, as they reached the bends between Snodland and Burham Marshes, 'two clumps of a noticeably different nature were found, shorter and more slender, darker green, barely coming into flower, and of a rather floppy habit quite different from the stiff clumps of *S. tabernaemontani*'. Their excitement grew when Eric Philp launched the dinghy and returned brandishing specimens with a three-angled stem.

Was it Triangular Club-rush? The jury is still out, but it seems more likely that it was a hybrid, or at least a population of not quite pure *Schoenoplectus triqueter* introgressed with the genes of its close relative. The stems were triangular, but perhaps not sufficiently sharply so. There were suspicions about some slight bumpiness on the glumes. But this is the closest that anyone has come to refinding the plant on the Medway in the past forty years. Since much of the tidal area is inaccessible, the pure-bred club-rush may yet survive somewhere, but most of the once open mudbanks beloved by this species are now under dense vegetation.

On the Tamar, the species was once locally common. When first found in 1857 it was 'growing most copiously on a mud-bank ... on the Calstock side', and was easily visible from a passing paddle steamer. Yet here too it has suffered a prolonged decline and the small remaining stand may be threatened by hybridization. A survey by inflatable boat by Rosemary FitzGerald, David Holyoak and Nick Stewart in 1997 found only two clumps of apparently pure-bred plants where there had been four only eight years before.[60] A third clump, found in 1995, had gone. In other places they found only the hybrid between Grey and Triangular Club-rush, and not much of that either. What has happened? Old photographs of the area collected by Rosemary show that the area has greatly changed since the turn of the century. Ironically it looks more natural now, with shady woods and reedbeds on formerly open banks, and no sign of the steamers, barges and wharfs that cluttered up the place when mining was at its height, a hundred years ago. What we cannot see is the difference in the river mud, but the huge increase in reeds suggest that it has become much more eutrophic through decades of run-off from farm fields. There is, in fact, comparatively little open mud left. It seems that Triangular Club-rush actually flourished in areas of heavy boat-traffic and general disturbance – which should not surprise us all that much when we recall that it was first found in the City of London. As on the Medway, changing conditions have brought the taller, more aggressive Grey Club-rush into closer contact with the Triangular one at a time when the latter's numbers were shrinking through the replacement of open mud banks with reedbeds. A further clue may be the fact that herbarium specimens of Triangular Club-rush examined by Tim Rich all seemed to be sterile. The species could not have been producing much seed. These interlinked circumstances – hybridization, habitat loss, sterility – may have combined to propel the species to within a whisker of extinction in Britain.

Port Meadow's flower

Meanwhile, on the pleasant pastures and meads of Oxford, the fortunes of another very rare flower have moved in the opposite direction – what was thought to be a hybrid population turns out to be pure-bred after all. This is Creeping Marshwort (*Apium repens*), a tiny umbellifer which grows in wet muddy places in old meadows, especially where cows or ponies have disturbed the ground. Ecologically speaking, it is confined to wet, neutral grassland which floods in winter, of the sort that is dominated by Creeping Bent (*Agrostis stolonifera*) and Marsh Foxtail (*Alopecurus geniculatus*), and known to vegetation surveyors as MG13. Geographically, Creeping Marshwort is even more restricted. It used to occur in several unimproved meadows on the floodplain of the River Thames, near Oxford, but today is found only on Port Meadow. Even at Port Meadow it is hard to find unless you know where to look and what to look for. I have walked there often without noticing any Creeping Marshwort, until Camilla Lambrick showed me a plant the size of a teaspoon inside the hoof-print of a cow, and looking nowhere near as glamorous as it does in the drawings by Stella Ross-Craig or Ann Davies. As I lay on my tummy in the mud with my nose in the ground, Camilla pointed out its salient points: the characteristically divided leaf segments, the creeping stem with its procession of leaves and flowers and the pleasant scent (and taste) of fresh parsley. There was for a long while doubt whether a pure population of *Apium repens* still existed in Britain.[73] Some had taken convincing pieces home to cultivate, only to see them growing up into hybrids, or worse, into Fool's Water-cress, *Apium nodiflorum* (a moment, one imagines, when the Latin name was preferable). Fortunately, growing trials at Kew and genetic fingerprinting at the Royal Botanic Gardens in Edinburgh has demonstrated that the species does survive, and that the hybrids are confined to drier ground. It is nonetheless one of our rarest plants, with only a few hundred individuals in all, the best patch of which is monitored each year by Camilla and members of the Cotswolds Rare Plants Group. On its soggy home ground, Creeping Marshwort seems to breed true – insofar as it breeds at all, for no viable seeds have been found.

This tiny, parsley-smelling umbellifer in the hoof-print of a pony is Creeping Marshwort (*Apium repens*). Grown in a plant-pot it will look more like its portrait in the floras. [Peter Wakely/English Nature]

The secret both to the survival and the extreme rarity of Creeping Marshwort lies in its adaptation to heavy grazing. It has a long flowering period, from July to October, and determinably keeps pushing up flowers after the first are grazed off. The flowering shoots (peduncles) are flexible – and though normally very short at Port Meadow, can shoot up to 10 cm or more, given the chance. However, its creeping habit, so necessary in an environment where seeding is hazardous, does confine it to open habitats. Only regular winter flooding will keep the competition at bay on these naturally rich alluvial soils. Clearly, though, Creeping Marshwort is not flourishing. Why should a species which is widespread in Europe, and which grows in an apparently 'ordinary' habitat, be confined to a single meadow in Britain? The probable answer is that Port Meadow is unique as a big ancient flood-meadow which has always been grazed all year round. Although other big flood-meadows have survived, nearly all of them are historically 'shut-up' in spring to grow hay. This period of tall growth prevents the survival of tiny flowers of wet mud like Creeping Marshwort. There is only one heavily grazed field with a history of absolutely stable management over the past thousand years, and that is Port Meadow. Maybe it is a bit over-grazed at the moment, maybe it is drier than it was, but at least conditions there allow the plant to survive. Though Creeping Marshwort has no fossil history, it was probably much more widespread on river floodplains in the distant past. Its presence today is testament to the unique character of Port Meadow.

Missing parent mysteries

The little Creeping Spearwort (*Ranunculus reptans*) is one of Britain's mystery plants. Around the sandy margins of certain lakes in northern England and Scotland there is a tiny yellow 'buttercup', anchored to the ground by a tuft of root at each node, with the merest green slivers standing in as leaves. There has long been speculation about what these plants are. They seem to be at an intermediate stage between the true *Ranunculus reptans* of Scandinavia and Iceland, and the narrow-leaved, northern form of our Lesser Spearwort (*Ranunculus flammula*). It was Druce who first suggested that the plants are hybrids, and this view was upheld by R.J. Gornall,[71] who named them *Ranunculus* x *levenensis*, after Loch Leven, where the plant was discovered as long ago as 1764. But if our plants are hybrids, what happened to one of the parents, *R. reptans*? Gornall's view is that this parent is sometimes brought to Britain by duck and geese on migration from their breeding grounds in Scandinavia. Experiments have shown that the seeds can remain in the stomach of a mallard or a goose for up to three days – more than long enough for a bird to eat *Ranunculus reptans* by an Icelandic lake, fly across the North Sea, and defecate in a Scottish loch. Gornall suggested the following scenario at Loch Leven. Each year thousands of Pink-footed Geese arrive in massed aerial squadrons and there they rest for a few days, loafing about on mud banks, and excreting the remains of their last meal in Greenland or Iceland. Sometimes they wade into deeper water to splash about and preen themselves. During these few days, any undigested achenes of Creeping Spearwort will wash into shallows and end up in the mud. As the water draws down the following summer, some of these seeds may ger-

Below left: Creeping Spearwort (*Ranunculus reptans*) is confined to sandy lake-margins in northern Britain. It may depend on regular reinforcement from the guts of migrating geese. [Peter Roworth]

Below Right: Moore's Horsetail (*Equisetum x moorei*) is a distinctive hybrid plant that can survive in the absence of either parent. Though it can colonize new areas vegetatively, it cannot evolve further since its spores are sterile. [Christopher Page]

minate. By then, of course, the geese will have long since returned to their midge-infested nesting grounds, their unconscious role in the life-cycle of Creeping Spearwort successfully accomplished.

There are problems with this attractive idea. Most plants introduced in this way would be pure *Ranunculus reptans*, for the hybrid is relatively scarce outside Britain. Moreover the hybrid occurs in some quantity around Ullswater and other Cumbrian lakes, which do *not* receive large numbers of migrant wildfowl. And would pure *R. reptans* be so overcome by our native Lesser Spearwort that the entire population hybridized more or less spontaneously? Its biology would suggest otherwise, for it is a perennial, capable of spreading rapidly by the use of stolons, and is partly self-fertile. Complete hybridization would surely need a much longer time-scale, and there may after all be something in the older theory, that Creeping Spearwort is a native plant that colonized Britain back in the late Ice Age, and gradually succumbed to hybridization with the more successful partner. It may be that both theories are true, and also that pure *carte blanche* Creeping Spearwort does sometimes occur with us, but has been overlooked by collectors (could they have collected the more robust plants, consciously or not?). Most herbarium material is from the long-established hybrid populations at Loch Leven and Ullswater, but plants found on the shore of the Loch of Strathbeg, on the north-eastern tip of Scotland, resemble closely the real thing (the difference is often slight – the hybrid is a touch larger, the beak of its achene a touch shorter – but you may need to grow it on in pots to be sure). DNA testing may help to resolve the situation. Creeping Spearwort is a sweet little 'buttercup', and it would be nice to have a pure strain of it somewhere, even if it was squirted out of a duck's bottom.

Two or possibly three more parents are missing from among the known hybrids of horsetails and clubmosses. Horsetails are apt to hybridize with one another in almost every conceivable combination. The progeny is often fertile and behaves as if it was itself a true species, with its own ecology and behaviour. Clubmosses are less prone to genetic experiments, but we do have one very odd one called, prosaically enough, the Hybrid Alpine Clubmoss (*Diphasiastrum* x *issleri*). There is no doubt about one of its parents. It is the familiar Alpine Clubmoss (*D. alpinum*), from which the hybrid differs in its larger, looser habit, and more flattened trailing shoots which usually have a prickly look. The hybrid is rare but widely distributed, and usually found at a lower elevation than Alpine Clubmoss in dry heathland, from the Malverns up to Caithness. But what is the other parent? None of the other British clubmosses fit the bill, and the traditional candidate is a continental species called *Diphasiastrum complanatum*, which occurs in open pine forests throughout northern Europe, with the interesting exception of Britain and Ireland. Christopher Page[131] offers a second contender, *Diphasiastrum tristachyum*, which occurs at middling altitudes in Europe, roughly halfway between *D. alpinum* on the high moors and *D. complanatum* down on the wooded valleys. The English material of Hybrid Alpine Clubmoss tends to be more robust than the Scottish, and occurs on more base-rich soils than the shallow, granitic soils preferred in Scotland. They may well have different parentage. Page offers this scenario: that both *D. complanatum* and *D. tristachyum* were once British, probably growing in native pine forests. They declined as the climate grew warmer and wetter during the Atlantic period, when moorland replaced many woods, and in the process they were outcompeted by the eventual winner, *D. alpinum*. As the more temperate species, *D. complanatum* might have died out long ago with the last English pine forests, leaving only a genetic echo in this mysterious hybrid of the Malverns and a few other places. The Scottish form is more likely to be a hybrid of *D. alpinum* and *D. tristachyum*. DNA testing may be able to prove or disprove this idea, but in the meantime, if I still lived in north-east Scotland, I would be looking very closely at any Alpine Clubmoss I found in a native pinewood.

All but one of the hybrid horsetails have parents that happily survive somewhere in Great Britain. The exception is Moore's Horsetail (*Equisetum x moorei*), which is confined to the coast of South-east Ireland, in Counties Wexford and Wicklow. One of its parents is the so-called Dutch Rush (*Equisetum hyemale*), a widely distributed native species. The other is the Branched or Ramose Horsetail (*E. ramossissimum*), which does not occur in Ireland, and only as a very rare introduction in Britain. Yet Moore's Horsetail is perfectly at home over some 50 km of warm, sunny shoreline in the lee of the Wicklow Mountains, often within the spray zone of the Irish Sea. In its characteristic habitat of clay banks and dune slacks, it is a plant of some presence – a mass of half-metre length yellow-green quills, with bands of black, tapering teeth, and often a little cone on top. Like other horsetails, you can pull it to pieces, like popper beads, plant each bit in a pot, and watch them take root and grow, like willow cuttings. But Moore's Horsetail is unusually tolerant of salt water, and, naturally enough, its hollow stems float. This useful combination of talents suggests how Moore's Horsetail may have established and spread in this very mild corner of Ireland. Page[130] conjures up the impression of a tough plant, washed into the sea on eroding banks, floating for several days in the current and then being deposited upstream, like a botanical Robinson Crusoe, and putting down roots. Could fragments of Horsetail have survived an ocean journey from Brittany or Bordeaux? And, if so, did it arrive as a ready-made hybrid, or was the Branched or Ramose Horsetail once an Irish plant? What seems clear is that the hybrid is more cold hardy than its southern parent, inheriting these qualities from Dutch Rush, whose northern range lies well within the Arctic Circle. Perhaps it also inherited some of its vegetative vigour from the absent parent.

Animal life on rare flowers

One argument in favour of conserving rare flowers – used mainly by zoologists – is that they might be eaten by rare animals. Any animal that really subsisted entirely on rare flowers would indeed be rare – for it could scarcely be more numerous than its food supply. On the face of it, this would look like a recipe for evolutionary suicide, like the poor Bread-and-butter-fly in *Alice Through the Looking-Glass*:

> 'Crawling at your feet,' said the Gnat (Alice drew her feet back in some alarm), 'you may observe a Bread-and-butter-fly. Its wings are thin slices of bread-and-butter, its body is a crust, and its head is a lump of sugar.'
>
> 'And what does it live on?'
>
> 'Weak tea with cream in it.'
>
> A new difficulty came into Alice's head. 'Supposing it couldn't find any?' she suggested.
>
> 'Then it would die, of course.'
>
> 'But that must happen very often,' Alice remarked thoughtfully.
>
> 'It always happens,' said the Gnat.

Downy Woundwort (*Stachys germanica*) is mysteriously confined to Oxfordshire, where it occurs in hedgerows, scrubby banks and wood borders, mainly within the historic boundary of Wychwood Forest. Woodmice and voles have a taste for the young plants. [Jo Dunn]

This sad story came true in 1977 for another improbable insect, the Viper's Bugloss moth (*Hadena irregularis*), which lives not on Viper's Bugloss, as was once thought, but on the rare Spanish Catchfly (*Silene otites*). Spanish Catchfly is confined to sandy places in Norfolk and Suffolk, and so, therefore, was the moth. Never very common, it finally seemed to run out of Catchflies, and so died out. From my own unusually intimate acquaintance with Spanish Catchfly (it grows all round one of our insect trapping sites in the Breck), I suspect that most present-day Spanish Catchflies are too small and spindly to support the solid-looking larva of this moth. They would simply bend over, like straws. Perhaps the plant was more robust in the past when its habitat was less beleaguered, as it is on the continent.

Rare flowers are of course grazed, nibbled or mined every bit as much as any other wild plant. A few seem particularly susceptible, like the

Epipogon or Ghost Orchid, which provides an unusual and apparently irresistible treat for slugs. The Lundy Cabbage is a favourite snack for practically all the island's large herbivores – sheep, goats, rabbits, ponies and Sika deer – despite tasting 'horribly cabbagey ... like essence of Brussels sprouts' (pers. comm. Roger Key, who bravely sampled it, in the interests of science. Tim Rich tells me it is better when cooked.). In her close study of Downy Woundwort, Jo Dunn[49] found a wide range of wild life feeding on various parts of this tall hedgerow flower, ignoring its protective coat of long silky hair. Rabbits, hares and deer graze the whole plant, especially the tender shoots. The overwintering leaf rosettes attract slugs and snails. The ripe seed pods are eagerly filleted and their contents swallowed by woodmice and bank voles. Greenfinches have attacked ripe Downy Woundworts grown in gardens. At a more insidious level, the larvae of a leaf-rolling moth, *Cnephasia interjectana* have infested the stems and leaves, and severely damage young plants. And the shield-bug *Eysarcoris fabricii*, which normally feeds on the seeds of Hedge Woundwort (*Stachys sylvatica*) and dead-nettles, has been seen perched on Downy Woundwort leaves, eyeing it speculatively. On top of all that, we could add the specimen of *Homo sapiens* who one year stole all the seed-heads of the best colony. I hope whoever it was got a garden full of leaf-rolling moths and shield-bugs for their pains.

None of these predators confine themselves to Downy Woundwort; to them the rare plant was probably just another dead-nettle. It would be a very foolish insect that specialized in this species, for its appearances are very uncertain, and in some years it may not appear at all. Yet some insects, and also certain parasitical microfungi, do rely more or less exclusively on rare flowers. One reason why Touch-me-not Balsam is considered a native species is that it is the sole host plant for the caterpillars of two moths, the Netted Carpet (Geometridae), which eats the leaves and flowers, and a micro-moth called *Pristerognatha penthinana* (Tortricidae), which mines the stem and roots. (It is not absolute proof, for another little moth, the Balsam Carpet, feeds mainly on *introduced* balsams.) Their specialist strategy may, in this case, be hazardous, for *Pristerognatha* was last recorded in 1914, and the Netted Carpet is in the Red Data Book, classed as Vulnerable.

Most of these specialist 'bread-and-butter-flies' are associated with perennial plants that are restricted in range rather than numerically rare. Only a small proportion of our rare flora has a specific associated insect. One of them is Nottingham Catchfly, the sole host of a pretty little Noctuid moth called the White-spot. However, the White-spot is much rarer than the Catchfly, being confined to the south-eastern coastline, mainly on shingle beaches, where its host-plant occurs in profusion. The White-spot is one of a group of moths that specialize on feeding on the flowers and fruit capsules of campions (one of these moths is actually called The Campion). In Britain, the White-spot would probably be outcompeted by its relatives on common

Red or Sea Campions, but on Nottingham Catchfly it seems to hold its own, even though the other moths will eat the plant too. The White-spot is among a number of insects which are at the edge of their range in Britain and have adopted a much smaller range of host plants than they do near the middle of their range. The best known example is the Swallowtail which is restricted to the scarce but locally abundant Milk-parsley in Britain (see Chapter 5), though in Europe it will use a whole range of plants, including the foliage of garden carrots.

A related flower, the Hog's Fennel, has an equally exclusive clientele. Rarer than the Milk-parsley, it is restricted to the salt-marshes and sea-walls around the Thames estuary and northwards on the Essex coast as far as Harwich. Despite its rarity, it has acquired a rust, *Puccinia rugulosa*, and two moths: *Agonopterix putridella* (Oecophoridae), which 'spins together the ends of leaves into distorted clumpy bunches',[155] and *Gortyna borelii* or Fisher's Estuarine Moth (Noctuidae) whose substantial larva mines the roots. The latter can infest up to 10 per cent of a population, causing stems to drop and flowers to fall off – a useful sign to those hunting for this desirable moth. But Fisher's Estuarine Moth is unlikely to threaten the survival of its host-plant. The two have probably co-existed for thousands of years, and no doubt factors like parasitism keep the moth in check. However, there are dangers. In Sweden, the Sticky Catchfly suffers from anther smut (*Ustilago violacea*), which does not kill the plant but replaces its pollen with the purple spores of the smut, and thus prevents the flower from pollinating and setting seed. Severe infestations can, it seems, wipe out a population by preventing recruitment.[205] This is a potentially serious matter for the Sticky Catchfly in Britain, for it occurs in small and widely scattered colonies and once gone, cannot return. Fortunately, although the smut is common in Britain, the Sticky Catchfly seems to be smut-free.

Partnerships of rare plants and even rarer insects probably go back a long way. Once they must have made more sense. For example, Dwarf Birch (*Betula nana*), which supports a range of insects, including two exclusive ones, was widespread in the Late-glacial period, though it is now confined mainly to northern Scotland. Its little ecosystem probably evolved when the plant was common. Similarly, Mountain Avens (*Dryas octopetala*) has three exclusive leaf-mining moths, as well as a special rust, and it too was extremely common in the cooler distant past, so much so that its remains are used in dating pollen samples, like distinctive pot-shards or coins. Had it not been so, these insects would not have staked their futures on it.

Other associations might have evolved when the climate was warmer and drier than at present. A squash bug, *Gonocerus acuteangulatus*, is confined to Box Hill where, appropriately enough, it lives on Box (*Buxus sempervirens*). In France it prefers Alder Buckthorn (*Frangula alnus*), while in Sicily it is a minor pest of stored fruit and nuts. Remarkably, the very rare

Field Wormwood (*Artemisia campestris*) has two apparently exclusive micro-moths. Like Box, this plant was common during the warm Boreal period, and it may even have been grown as a crop by Neolithic farmers. But now it is confined to a handful of suburban sites in the Breckland, and its family of little moths is probably doomed unless they find another foodplant. Another specialist is the plume-moth, *Stenoptilia graphodactyla*, which can make a mess of Marsh Gentians (*Gentiana pneumonanthe*). It has two generations, the first of which mines the developing shoots, while the second feeds on the seeds, and occasionally the flowers. Marsh Gentian fans may be pleased to learn that some caterpillars living inside the blue trumpets may be 'drowned by a shower of rain filling the flower with water'.[10]

TABLE 4 Rare flowers and their insect diners

This list is based on a report for the NCC by Mark Parsons,[136] with a few additions from my own browsing.

Plant	Insects
Wild Cabbage	*Selenia leplastriana* (micro-moth)
Lundy Cabbage	*Psylloides luridipennis* (a flea beetle)
Nottingham Catchfly	White-spot moth (*Hadena albimaculata*), *Coleophora otitae* (case-moth)
Spanish Catchfly	Viper's Bugloss moth (*Hadena irregularis*)
Dwarf Mouse-ear	*Caryocolum blanduella* (micro-moth)
Marsh Mallow	Marsh Mallow Moth (*Hydraecia osseola*), *Pexicopia malvella* (micro-moth)
Touch-me-not Balsam	Netted Carpet (*Eustroma reticulatum*), *Pristerognatha penthinana* (micro-moth)
Downy Greenweed	*Phyllynorycter staintonella, Syncopacma suecicella* (micro-moths)
Marsh Pea	*Rhyncopacha tetrapunctella*, *Ancylis paludana* (micro-moths), *Adelphocoris ticinensis* (Capsid bug)
Wild Liquorice	*Cydia pallifrontana* (micro-moth)
Sea Pea	*Epischnia boisduvaliella* (micro-moth)
Hog's Fennell	Fisher's Estuarine Moth (*Gortyna borelii*), *Agonopterix putridella* (micro-moth)
Milk-parsley	Swallowtail butterfly (*Papilio machaon*)
Spring Cinquefoil	*Ortholomus punctipennis* (Ground-bug)
Mountain Avens	*Stigmella dryadella* (Nepticulidae), *Parornix alpicola*, *P. leucostola* (micro-moths)
Wild Service Tree	*Stigmella mespilicola* (+ other *Sorbi*), *S. torminalis*, *Parornix anglicella* (+ hawthorn) (micro-moths)
Box	*Gonocerus acuteangulatus* (squash bug), *Anthocaris butleri* (flower bug)
Field Wormwood	*Sophronia humerella*, *Bucculatrix artemesiella* (micro-moths)
Bastard Toadflax	*Epermenia insecurella* (micro-moth), *Sehirus dubius* (shield bug)
Marsh Gentian	*Stenoptilia graphodactyla* (plume-moth)
Hoary Mullein	*Nothris verbascella* (micro-moth)
White Mullein	*Cionus longicollis* (weevil)
White Horehound	*Pterophorus spilodactylus* (plume-moth), *Meligethes nanus* (beetle), *Haplothrips marrubicola* (thrip)
Downy Willow	*Pontania crassipes* (sawfly)
Net-leaved Willow	*Empria alpina* (sawfly)
Water Soldier	*Bagous binodulus* (weevil)

Perhaps the most intriguing of all these associations is the Lundy Cabbage (*Coincya wrightii*) with its three client beetles: the Lundy Cabbage Flea Beetle, the Lundy Cabbage Weevil, and the Lundy race of the Cabbage Stem Flea Beetle. The Cabbage is confined to Lundy, and since the beetles live on nothing else, so are they. None of them would win a beauty prize even in a beetle competition. One, the weevil, is small and dark, with a glint of bronze. It mines the stems, and to find it you need to rend apart your cabbage, which is technically illegal. (Apparently a pair of these weevils can completely strip a plant in six weeks.) Lundy Cabbage Flea Beetle (*Psylloides luridipennis*) is equally small and dark with a hint of blue, and eats the leaves. It, too, is rather elusive, for at the first sign of trouble the beetles leap into the air like their namesakes and disappear into the shrubbery. The traditional way to catch them is to use a long-handled sweep net and wave it vigorously to and fro through the 'cabbage patch'. Because of these unobtrusive but interesting beetles, the Lundy Cabbage has always been an entomologists' plant. Appropriately enough, it has become the study-object of Roger Key, English Nature's chief entomologist, who has suggested that 'this little community of insects presents a unique opportunity in Britain to study evolution in action – our own equivalent of the Galapagos Islands'.[98]

Lundy Cabbage (*Coincya wrightii*) is confined to the eastern shore of Lundy, where it was discovered by a Devon doctor in 1936. At least two kinds of beetles seem to eat nothing else. [Peter Marren]

Chapter 7

The lust for rarities:

did collecting damage our wild flora?

About one hundred and sixty years ago, a botanical pilgrim stood beneath the waterfall by Torc Mountain in Killarney, shielding his eyes from the dazzle and seeking a way across the torrent at his feet. It had been a long journey for Edward Newman, who lived in London, and, having crossed the seas and the wild paths of Ireland by a combination of rail, steamer and horse-and-trap, he now faced an awkward scramble. Leaving the seat where ordinary tourists could admire the fall in safety, Newman worked his way upstream, 'leaping from stone to stone along the bed of the torrent, which, in times of flood, as happened to be the case when I paid this visit, is rather an exciting and ticklish operation'.[120] At length, he reached the rocks at the foot of the fall. Almost deafened by its roar, and drenched in spray, Newman clambered up the side using roots and branches as holds until he reached his goal, a particular projecting rock with 'a little platform at the top, where (you) can stand very comfortably'. At last, level with his eyes, in its elemental setting of water, rock and twisting roots, was the object he had come so far to see. Newman described the half-shaded rock wall as 'robed' with that most beautiful of filmy ferns, *Trichomanes*

The Killarney Fern (*Trichomanes speciosum*), prize of the Victorian fern-hunter. [Peter Wakely/ English Nature]

speciosum, 'the dark green fronds hanging down, dripping with wet and begemmed with sparkling drops'. It was 'well worth the wet stockings', he decided, and, while on the subject of worthiness, he added that the misty lakeland beyond the fall, with its pastel blues, greens and browns, was 'well worthy of the rare fern which it cherishes in its bosom'.

Trichomanes was first discovered in Ireland at this very place in 1804. It was already known from near Bingley in Yorkshire 'at the head of a remarkable spring', where it was collected to eventual extinction, and was later to turn up in Scotland and Wales. Nowhere, however, was it as abundant as in south-west Ireland, and so it has been known ever since as Killarney Fern. By the 1830s, when Edward Newman arrived to pay his respects, Torc Mountain had become *the* place to see it. The many visitors included naturalists, gardeners, walkers and admirers of romantic scenery. At that time there probably seemed no harm in removing a piece of the fern for one's rockery or front room; indeed the developing science of 'pteridology' was based on nurture and cultivation, with a keen eye for distinctive 'varieties' to grow on. Fern lovers were more concerned with conservatories than conservation. Ward, the inventor of the glass case for growing delicate ferns like *Trichomanes*, had passed this way. So had a Mr Robson, who no doubt thought he was performing a great service when he removed great strips of the fern with the intention of transplanting it on Valencia Island and other scenic places in south-west Ireland. A Mr Andrews, a friend of Edward Newman's, had discovered a particularly large and beautiful variety with graceful lanceolate fronds, 'in a wild and romantic cave covered with a drapery of the overlapping fronds, hundreds of which, hanging gracefully down, formed a pendulous mass of the loveliest green'. Newman named it '*Andrewsii*' in his honour, and it became the most desirable form of all. Many visitors to the Irish lakes helped themselves to *Trichomanes*. There was no difficulty in stripping it from its underhangs and caves; indeed the problem was to avoid pulling up too much. The bristly black rhizomes clasp the wet rock like ivy, sometimes in lines like strawberry runners, sometimes in a fishing net of overlapping roots. One good tug and you might strip away several yards of it, as Newman himself did, the fronds all hanging down like flags on a string.

Half a century later, Reginald W. Scully, author of the Flora of Kerry, made his way to the same waterfall at Torc Mountain, and sat there for a while watching the light sparkle in the spray. He knew the area well, and had witnessed how the once profuse 'robing' of *Trichomanes* had been reduced to tatters, year by year, until, on his last visit in 1892, only a single shred remained on 'a huge mossy boulder'.[176] And now, he found, even that last shred had been looted. The fern survived in remote places higher on the mountain where, as Scully put it, they were 'safe from anything less destructive than a charge of dynamite'. But here, in its best-known locality, it seemed to have been collected out of existence. In desperation, Scully took to examining the boxes of ferns that were being touted from hotel to hotel by local tinkers. The last time he had passed this way he had been offered 'a large box filled with the roots and fronds of the *Trichomanes*'. At first, the 'peasant', as Scully described him, had asked £5 for the entire box. Failing to find a purchaser 'he ultimately disposed of the contents in lots for which he only too readily obtained sums varying from 2s. 6d. to 7s. 6d.'. That box, Scully supposed, 'represented the total destruction of another station, the growth probably of ages'. A few years later, there was no more *Trichomanes* for sale, only sad trays of spleenworts, maidenhairs and lesser filmy-ferns, still rooted in moss. It was from about this time that naturalists began to be more circumspect when talking and writing about the Killarney Fern. That justified caution continues to this day, and *Trichomanes* is one of only three species for which grid references are never given, not even on confidential computer records at the Biological Records Centre. Not without reason did one botanist say of this species, 'O *Trichomanes*, Breaker of Friendships, Destroyer of Confidences', such are the fallen reputations and broken promises strewed in its wake. Tell it not in Kerry; publish it not in the streets of Killarney!

There can be no doubt that the collector had an impact on the Killarney Fern. The object of their desire possessed a collection of attributes which made it unusually vulnerable. It was rare. It was pretty. It was associated with wild, romantic scenery. It became a fashionable plant for Victorian conservatories: the 'Wardian Case', which allowed non-specialists to grow delicate ferns, might have been designed for it. All this helped to put a price on its head. Unfortunately the Killarney Fern is a slow-growing plant which seldom strays far from permanently moist hollows and caves. Newman was one of the first to realize that cultivated Killarney Ferns thrive only in isolation: they seem unable to compete with other ferns, and only hang their heads and wilt in mixed company. And at that time even the healthiest fronds grown in ideal conditions showed little inclination to produce fertile spores. Those who have monitored wild colonies of the plant over long intervals have been struck by how little change there is from year to year. Derek Ratcliffe told me that 'it just sits there'. The reason for this apparently unenterprising behaviour probably lies in prehistory. While most of our flora consists of relatively recent arrivals, which by definition are species capable of colonizing new places fairly quickly, the Killarney Fern may well be a much older Tertiary plant, with memories of lush forests teeming with extinct animals, which managed to survive the Ice Age in moist, frost-free pockets near the warming sea. This theory is consistent with the present range of the sporophyte (frond-producing) plant. It behaves like a jungle plant, holding fast, well adapted to its half-lit, humid world, and neither increasing much, nor, until the collectors arrived, decreasing significantly. In his great book

Ferns,[130] Chris Page summed up the situation:

> Sat squarely upon a smooth-washed stone, one cannot but help reflect on the grandeur of such habitats, with the immense erosive forces of fast-moving water, reverberating between cliff-faces, hung with the delicate tracery of fern fronds waving slightly in the passing cool air. We admire a situation – a geology, a landscape, and a natural history – which has probably changed but little in many a million year.

There is, however, a surprising twist to this story, for there is more to a fern than a frond. But for the recent revelation of unsuspected Killarney Ferns-in-disguise I must refer you to Chapter 12.

Is the sad tale of the Killarney Fern an isolated case, or were other rare plants hunted almost to the point of extinction by collectors? Collecting, like hunting, is an emotive subject. You often find collecting being cited as a cause of decline in the conservation literature, though generally without much evidence. But the amassing of rarities for a plant press or a garden is more of a nineteenth-century fad than a modern danger, and has long gone out of fashion. Our flora has had a long time to recover from past excesses, and the fear of collectors blinded people to the real and increasing danger that all wild plants face: wholesale habitat destruction. The most vulnerable group of plants to collecting were the ferns. David Allen called the mid-Victorian fashion for ferns the The Victorian Fern Craze.[2] Another epithet is 'pteridomania'. Many of the ferns grown in Victorian conservatories, or in 'Wardian cases' on the parlour table, came from nurseries, but the literature of the period suggests widespread looting of wild places. This was not done in a sneaking, guilty spirit, but as a fashionable amusement – you could argue that dragging nature indoors, dead or alive, was what Victorian natural history was all about. Prints from contemporary books and journals show formally dressed gentlemen armed with hooked poles, pulling down tufts of spleenwort, or family parties in various ferny nooks and dells, busily filling their boxes and hampers. Kilvert's diary for 1870 mentions his helping some ladies to collect ferns on sea-cliffs in Cornwall: 'Not very successful ... H. had got her some much finer ones, but she did not despise mine, though they were very poor little ones in comparison.'[112] There was an artlessness about some collectors, often reflected in accounts of their exploits. Here, for example, is a Mr C. Barter, describing his encounter with the elusive Dickie's Fern:

> By a small rill that fell over the rocks, I managed to creep down and was gratified to find [the fern] in profusion ... I gathered fronds from six to eight inches long ... From the luxuriant specimens gathered

> here and the abundance of plants noticed, *I presume no ruthless hand had been plant-gathering here of late*, whilst the difficulty of reaching the spot will always afford it protection from invaders, excepting perhaps those affected with the Fern-mania. *I filled my box with plants and fronds*, leaving abundance for those who choose to follow by venturing the same road [my italics].[7]

Notice that it is always others that have the 'ruthless hands', never the author. The motto for fern collectors everywhere was provided by William Sutherland, who liked to mix his descriptions of plant hunting with sentimental musings about water nymphs and fairy dells. 'Let "Excelsior!" [Higher] be your cry,' he wrote, 'and the filling of your vasculum with the rarity your object.'[188] It is pleasant to imagine these middle-aged adolescents scrambling up the hillsides, gathering as they went, snatching up a tuft of some rare alpine by the roots with the exultant cry, 'Excelsior!'.

By combing the literature, it is not hard to find instances where fern gathering did lead to the permanent reduction or even extinction of a particular colony. The most heavily collected areas were in the west, the Cornish coast, the mountains of Snowdon and the Lake District and in well-visited beauty spots like Castle Eden Dene or High Rocks near Tunbridge Wells. Here are a few examples of what went on.

Cornwall

In his journal of a plant-gathering expedition to Cornwall, John Ray remarked that 'nothing is more common than *Osmunda* [*Osmunda regalis*, the Royal Fern] about springs and rivulets in this country'. Today, you could walk a long way without finding a single frond. Habitat destruction would account for much of the loss, but contemporaries blamed it on the plundering of this large, conspicuous fern during the Victorian era. The Royal Fern suffered doubly, both as an attractive fern and as a source of compost – 'Osmunda fibre' – ideal for the hothouse orchids which were coming into vogue at the turn of the century. The collecting became so notorious that Cornwall County Council eventually made a by-law prohibiting its removal – one of the first to protect a wild plant (whether the law was enforced is another matter). Edward Step (1908) gave one example from the parish of Tregear.

> A well-to-do woman, who was neither a dealer nor a collector, made several excursions in a carriage and pair to a bog where the Royal Fern grew, and would not rest until she and her maid had removed every plant to her own home, because she knew the fern to be rare in the

> district. That was the sole point that appealed to her: the Royal Fern was rare. Thereafter, to her wonted boast of relationship to an archbishop, she was able to add a claim to social distinction on the ground that she possessed all the Royal Ferns that grew in that section of the county! Our attempt to make her ashamed by a little sarcasm was thrown away. We might as well have sought to stir up a crocodile by pelting it with paper confetti.[183]

Virginia Woolf recalled collecting the roots of *Osmunda* from Halestown bog near St Ives at about that time, which she gathered each autumn to make into pen holders. An old botanist, C.B. Clarke remarked to her father, 'All you young botanists like *Osmunda*' (Virginia Woolf, *Moments of Being*, 1976). In those days, if you liked something you brought it home with you.

Somerset

In the nineteenth century, the Forked Spleenwort (*Asplenium septentrionale*) was plentiful on old walls at Culbone and Oare, and on loose shale near a path above Porlock Weir. But 'when these localities became known, a fern collector called Potter collected hundreds, if not thousands, of roots for sale, and by 1893 it had become very rare, and the exact place where it still survived was kept secret'.[74] It was last seen there in 1939.

Durham

Oblong Woodsia (*Woodsia ilvensis*) was the target of pteridophiles visiting the cliff known as Falcon Clints in Upper Teesdale. Specimens collected between 1821–95 exist in most of the regional herbaria in Britain. One sheet shows eight whole plants collected on a single visit. Particularly assiduous collectors were Samuel Simpson and S. King, who roundly boasted of their exploits in the pages of *The Phytologist* in 1844. By about 1900, every last Woodsia had been looted.[72] For good measure, these collectors also took away most of the only population of Holly Fern in Durham. It survives there, just, with the help of a reintroduction programme.

Elsewhere in the county, Royal Fern was again the victim of an acquisitive lady. Graham relates how most of the plants on the Derwent banks near Lintzford were 'dug up by a lady's gardener after he had revealed the site to the lady in question'.[72] The fern is now very rare in the county. Another lost fern was the Forked Spleenwort, which 'grew at one time in considerable quantity on dry bare rock, exposed to the sun at Beldon above Blanchland'. Part of the site was destroyed when a lead-mine was opened, and a large section of the rock on which the Spleenwort grew was removed to make way for a pumping engine. But, according to a contemporary, once its presence had become known to collectors, it was 'utterly exterminated'. The species is now extinct in Durham.

Snowdonia

The rare ferns and alpines of Snowdon and its neighbours were a magnet for collectors for more than two centuries. One can date the heyday of plant-hunting with unusual precision: from the 1680s, when Edward Lhuyd recruited local inhabitants in his quest for alpines, to 1896 when the Mountain Railway was built and tourists suddenly seemed to lose interest. In mid-Victorian times substantial sums were paid by wealthy collectors for rooted specimens of choice species like Oblong Woodsia and Tufted Saxifrage, whose localities were kept secret, partly for conservation reasons, but also to secure a monopoly for the supplier. For plant connoisseurs, mountain guides made themselves available in the lobbies of the main hotels to help their clients find what they were looking for. The ordinary tourists, meanwhile, were sold common spleenworts and parsley ferns as great rarities by wayside traders. The standard equipment for the collector of *Woodsia* and other rock plants was a long pole with a hook or 'radicator' at the far end. 'Eradicator' might have been a better name. Samuel Brewer used one of these to separate *Woodsia alpina* from Snowdon in 1726, and during the Victorian heyday we hear of an annual outing of local doctors to Cwm Idwal, each of them with a twenty-foot pole over their shoulder, like medieval pikemen.

The competitive element in this craze for alpines led certain unscrupulous collectors to strip away every last specimen, presumably to deny the site to 'rival' collectors. For example, an Oblong Woodsia colony on Glyder Fawr was reduced to bare rock in 1844, and C.C. Babington found a similar ruthlessness at work with its equally rare relative, Alpine Woodsia. 'Extirpation is the rule in Wales with tourists and collectors who call themselves botanists,' he wrote in his diary. And rarities were 'subject to constant solicitations from botanical tourists'.[93] As usual, the Killarney Fern was the most desirable species of all, despite its ready availability from nurseries. There was a colony in a rocky hollow on Snowdon, reputed to be the finest in Britain, growing 'in the form of a beautiful curtain, down which the water is constantly trickling; the whole having much the appearance of a crystal screen'.[93] Not for long though. A second colony was also stripped away to apparent extinction, but 24 years later it had recovered, to the finder's surprise and delight. Fortunately these ferns still occur in the area, albeit much reduced. But the Interrupted Clubmoss (*Lycopodium annotinum*) has not been seen since 1836 on its most southerly British site on the Glyders.

The Lake District

The Lake District suffered from the same cupidity for ferns as Snowdonia, but for a shorter period. The most significant victim was Mountain Bladder-fern (*Cystopteris montana*), eradicated from its only known English site, on Helvellyn, soon after its discovery there in 1880. Other ferns destroyed or permanently reduced by collectors were Royal Fern, Holly Fern, Oblong Woodsia, Hay-scented Buckler-fern, Killarney Fern and Interrupted Clubmoss (although in some cases sheep-grazing was a contributory cause).

Oblong Woodsia (*Woodsia ilvensis*). This is possibly the best patch of it in Britain. [Derek Ratcliffe]

Dumfries

The Moffat Hills had become a well-known district for rare ferns by the 1870s, and here the Innerleithen Alpine Club came to the aid of visiting collectors. Between them they wiped out more Interrupted Clubmoss, and nearly put paid to another colony of Oblong Woodsia. Just a few fronds of the latter remained when G.C. Druce paid it his respects. He 'wished it well', then thought again and took another bit home with him.[115]

Perthshire

A well-known site for Alpine Woodsia near Ben Lawers was heavily collected from 1828 onwards, and Martin Rickard has tracked down 29 herbarium collections from this one spot.[166] They far exceed the number of fronds left there today. In 1900, F.C. Crawford had removed 19 complete plants, and a few years earlier, a W.B. Wareterfall came away with enough fronds to supply four public herbaria. The site was supplying private herbaria as late as the 1950s.

Fern collecting could be a hazardous occupation. Several well-known collectors suffered a near miss, as they slipped on wet rocks, and at least one professional botanist's guide, William Williams, plunged to his death on the cliffs of Snowdon, reputedly while gathering fronds of Alpine Woodsia for a client. His body was found beneath the very place where Lhuyd had discovered this attractive little fern in 1690.

It seems that rarity, not beauty, was the attraction for fern collectors, for the once desired Forked Spleenwort must be about the least lovely fern in

the flora. The Woodsias and Killarney Fern suffered disproportionately *because* they were rare. However, there were other ferns equally rare and exploited equally ruthlessly that have fully recovered their former numbers. I once attempted to track down all the specimens of Dickie's Fern (*Cystopteris dickiena*) that I could find in the main public herbaria, knowing that most of them were taken from the same cave on the coast near Aberdeen.[119] There were scores of sheets and hundreds of specimens, and this cave is only the size of a large garage. Professor Dickie himself was sure that collecting had eradicated the plant. Yet, if you return there now, you find tufts of this delicate fern in a luxurious arc across the cave roof, and more among the fallen rocks beneath. The reason soon becomes apparent. The mature fronds often carry ripe sporangia and the damp rocks shelter many young, sporeling ferns. Conditions in this cave, with its permanently cool, moist interior, result in rapid growth and the population has long since recovered from the raids of 140 years ago. The Maidenhair Fern, another popular plant in Victorian greenhouses, is a similarly fecund and fast-growing species. Not only has it recovered well at many of its old, heavily collected sites, but it has become an effective colonist of walls, coastal defence works, wishing wells, railway ballast and even, in one case, a front doorstep – anywhere, in fact, so long as water can percolate through the stones and there are no severe frosts.

Alpine Woodsia (*Woodsia alpina*), a rare relict of the late glacial flora. [Derek Ratcliffe]

Yet the inability of a few rare ferns to recover from over-collecting requires some explanation. One likely reason why they are rare in the first place is that they are slow-growing and unable to colonize new sites. Ferns and fern-allies come, of course, from ancient stock. Unlike flowers they have alternate generations, and, like frogs, they need damp places for at least one of them. Most have a very precise ecological niche – even so common a plant as Great Horsetail (*Equisetum telmateia*) has a habitat so exact that you could describe it in a few words – a place where water trickles on the surface between limestone and clay. Oddly enough, the

Dickie's Fern (*Cystopteris dickieana*), a mysterious fern of shady sea-caves, once believed to have been 'extirpated' by collectors. [Derek Ratcliffe]

two Woodsia ferns often live in what look like 'ordinary' places, and are seldom closely associated with a rich alpine flora. Both grow in rock crevices or scree of the kind that is available by the acre in Snowdonia, the Lake District and the Highlands. And yet all but the very best colonies are small, and many of them seem to lack vigour – I remember one colony of Alpine Woodsia in particular that seems to 'cringe' into the rock crevice as though it was deeply unhappy with its lot. Some published photographs of it have the same look. In Scandinavia, neither Alpine nor Oblong Woodsia are like this – they are widespread and locally quite common, and often grow vigorously, like little green shuttlecocks. What has gone wrong in Britain? Lusby and Wright (1996) suggest that the Woodsias are relict species which have been declining naturally for a long time, as Britain's climate becomes less and less like that of Sweden.[111] They may now have reached a point, hastened by unrestrained collecting, where their small isolated colonies have become vulnerable to chance events, like rockfalls or drought. Recruitment seems to be very slow, and, like the Killarney Fern, the two Woodsias rely not so much on spores as on tough rhizomes, pushing through crannies in the rock. Observation suggests that they have only a limited ability to recolonize lost ground, and none at all for exploiting opportunities on a neighbouring cliff. Slow natural decline is the lot of these Ice Age plants, and their survival depends to an unusual degree on their being left undisturbed. Now they have climate change to look forward to, and I wish them luck.

It is hard to find evidence of similar lasting effects of collecting damage on rare wild flowers. The Lady's Slipper is the famous example of a flower that was all but collected out of existence (see later in this chapter). Orchids do tend to bring out the worst in us. Famous localities, like Castle Eden Dene and Kilnsey Crag seem to have been picked as bare of rare orchids as they were of rare ferns. Near the turn of the present century one naturalist wrote of the orchid-gathering industry on the downs of Kent, mentioning 'armfulls' of Fly Orchids, and a terrible old man who made it his business to uproot every Lizard Orchid he could find and sell them to gardeners at ten shillings each. The virtual disappearance of the Military and Monkey Orchids at this time has been blamed on collectors, though in this case I think they are being maligned – the rabbit is a much more likely culprit. Apart from the Lady's Slipper, the two species that were most seriously reduced by collecting were the Summer Lady's-tresses in the New Forest and the Fen Orchid in East Anglia. These orchids grow in Sphagnum bogs, and are easily uprooted. Here is a short extract from a local club outing to Burwell Fen, Cambridgeshire, in 1835:

> We had very good sport both in plants and insects. *Ophrys loeselii* [Fen Orchid] was found in great plenty. Between four hundred and five hundred specimens were brought home. It was growing in the grass

> and moss among the pits where they cut turf. There were two bulbs to each plant, and the bulbs were scarcely in the ground at all, so that we picked them out easily with our fingers.

Such accounts make it look as though collectors destroyed the Fen Orchid. But when they stopped digging turf or peat, the Fen Orchid died out anyway, through loss of habitat.

The other main target group were alpine plants, which were at one time heavily raided by nurserymen, although whether these did much long-term damage may be doubted. Ben Lawers was a magnet for nurserymen and collectors, but its alpine flora still seems about as luxuriant as it could be, or would be if the sheep left it alone. The Alpine Catchfly still spangles the lonely knoll of Meikle Kilrannoch with its pink, scented flowers, though you could paper a hallway with herbarium sheets labelled from this spot. In the Lake District it has fared less well, but the non-human culprits are all around, woolly, black-faced and permanently munching. Scone Castle, in Perthshire, was once a well-known locality for *Moneses uniflora*, the One-flowered Wintergreen, not only exquisite to look at but with a scent described by Reginald Farrer as 'of orange blossom, so poignant in its deliciousness as to be almost a pain to remember'.[111] The colony was plundered ruthlessly from its first notice in 1825. By 1858, John Sim thought it 'almost extirpated, and no wonder: the Edinburgh students devour every green thing'. These university parties could be large – Professor Balfour took 110 students here on 29 June 1861. The wintergreen also suffered from private collectors, including Colonel Henry Maurice Drummond Hay who mounted 25 specimens on a single sheet. The last Scone plants were dug up in 1883.[111] Other heavily collected localities were crags in or near Edinburgh at Castle Rock, Blackford Hill, and, most of all, Arthur's Seat in Holyrood Park. The attractive and vulnerable Sticky Catchfly was all but eradicated, and survives today as just four plants (currently being assisted by planting experiments). The Sow-bread or Wild Cyclamen (*Cyclamen hederifolium*) was another over-popular rare plant. According to Hanbury and Marshall (1899), there was a trade in the plant at Alderden Wood in Kent, which 'almost extirpated it, digging up the roots and selling them for transplantation into shrubberies etc'.[81]

Collectors fell into three main groups: the nurserymen and dealers, supplying plants to a wealthy or specialist clientele; botanists and gardeners who sought specimens for their private herbaria or garden, and the wider public who just picked flowers. Nurserymen supplied plants that were in fashion, and although these once included native alpines and ferns, the trade was well on the wane by the late nineteenth century as cultivated and exotic plants became more widely available. Botanical collectors often took more specimens than they needed, partly to exchange with other botanists, partly because they desired the finest possible collection, and unfortunately that often meant the biggest. You find many examples of excessive hoarding in the literature, and very entertaining they are too. For example, G.C. Druce once complained about the 'wholesale raids' of Allan Octavian Hume, whom he dubbed 'a most reckless collector': 'He told me he had made about 10 000 gatherings. He did not unfortunately confine himself to the flowers but liked roots as well. After that admission I never helped him to another locality, especially as he had just ravaged our Fritillary meadows and dug up bulbs galore'.[8] On the other hand, some considered that Druce himself had itchy green fingers. W.H. Mills once took him to see some marsh orchids in Cambridgeshire and 'was horrified when Druce pulled specimens up by the armful; he would have hit him, had not their companion, A.H. Evans, forcibly restrained him.'[4]

The botanical world at that time was not particularly conservation-minded. Granted that Druce, as the leading light of the Botanical Society and Exchange Club, paid lip service to the need to avoid reckless gathering, and was also active in the Society for the Promotion of Nature Reserves, his advice was often ignored, even by himself. No doubt, as David Allen suggests, some of the old guard agreed with the Gloucestershire botanist who reasoned that the God of nature, who caused flowers to grow where they did, could be relied upon to preserve them in perpetuity.[3] Disciples of St Thomas Aquinas could argue, with scriptural authority, that God gave us flowers for our enjoyment, and, that having done so, He would arrange things so that there were always plenty more. A most consoling doctrine.

For the wider public it was the most attractive rare flowers that were in demand, especially when they grew in a favourite beauty spot like Cheddar Gorge or Dovedale. Among those which were dug up and touted to tourists were Cheddar Pinks in the Mendips, Spring Gentians in Teesdale, and Marsh Gentians at Studland in Dorset, which Cyril Diver saw 'being sold by the trug-full'. Spring was the main season for picking. Popular favourites included the Pasqueflower – in 1929, one witness watched a party of 30 young women picking every specimen they could find on Royston Heath, in Hertfordshire. Picking Fritillaries was a local custom in some places, with special open days and an invitation to 'pick your own'. When Mickfield Meadow in Suffolk was made a nature reserve in 1938, and the (probably harmless) custom banned, notices were torn down and the pickers carried on regardless. For Wild Daffodil fans, the Great Western Railway even ran a special service to the 'Golden Triangle' in the woods around Newent in Gloucestershire – though there the Wordsworthian experience of wandering through the golden hosts might have mattered more than picking a bunch to take home.

There were rare cases where a farmer resented visitors so much that he spitefully destroyed a rare flower himself. This seems to have been the fate of the Blue Iris (*Iris spuria*) in Dorset. The flowers grew in wet meadows and were well known locally long before they reached the ears of the botanical world. Someone spotted them in a vase at a 'wild flower competition' in the local school during the Second World War, and was told that it grew wild in the fields, and that children had been in the habit of picking it for years. An old man remembered they had called it 'childweed'. Perhaps the farmer did not like children. He ploughed up the last iris field in the early 1970s. 'None of you buggers are going to see it again,' was his promise. He was 'fed up with botanists'. He was not the only one. A colony of Pasqueflowers on Fleam Dyke, Cambridgeshire, was dug up by a farmer to prevent its admirers from wandering across his land. More recently the same thing happened to one of the few sites left for Burnt Orchid in northern England 'to counter the attention shown to it by botanists'.[61]

Fears that indiscriminate collecting might be harming the wild flora were being voiced as early as the 1840s. However it was the growth of public transport, soon followed by private cars, and the ease in which people could now reach their favourite wild flowers, that prompted the first nationwide poster campaign to save wild flowers, in 1925. The Society for the Promotion of Nature Reserves (SPNR), the ancestor of today's Wildlife Trusts, was convinced that picking flowers was undesirable for a number of reasons. To begin with, 'if picked, the flowers last but a little while, and unless a sufficient number of them is left to seed, the flowers will disappear' (this is not, in fact, strictly true but population science was then unknown). More tellingly, the Society considered that picking flowers was an anti-social act: 'The beauty of the countryside would be sadly marred were no flowers to bloom on the banks or in the woods ... Plants and trees, as Nature placed them, are a delight to the eye: let all who pass by enjoy them.' Conservation bodies repeat such sentiments today, though seldom with the same elegance. However, the SPNR was not hopeful of success: 'It is uncertain whether the appeal will entirely achieve the object in view because a far stronger deterrent is required to restrain the unscrupulous collecting hog from uprooting all the rare plants within his reach.' In other words they saw the ultimate remedy in by-laws or legislation.

Rarities under the lens: imaging the Military Orchid. There was a queue to photograph this fine specimen, and note the compacted bare soil. [Peter Marren]

With the advantage of hindsight, we can see that the Society was barking up the wrong tree altogether. By perceiving picking and collecting as the main danger to wild flowers they were in danger of overlooking the real threat of habitat destruction. Fear of the collector is a hangover from Victorian times. In a famous experiment, the horticulturalist, John Gilmour, tested the effect of picking on the bluebells at Kew Gardens. It turned out that picking did not harm the plants at all; such damage as there was came not from picking but from trampling.[69] But in any case, even the most zealous pickers could manage to remove scarcely a tithe of the flowers eaten by sheep, deer or voles. Gilmour's conclusions apply to many of the most popoular flowers – orchids, lilies, alpines – most of which have perennating organs like corms, bulbs and woody rootstocks, and are no more damaged by picking than a fruit tree is by a harvest. (Ironically, the conservers of the Lady's Slipper used to pick every shoot before it flowered. This made the plant less conspicuous, and, they thought, would also help the plant save energy and therefore prolong its life.) Even rare annuals are unlikely to be permanently harmed by picking, since most of them have taken out life insurance in the form of seed banks preserved in the soil.

And yet picking wild flowers is regarded as anti-social behaviour today to an extent that the conservation pioneers could hardly have imagined. It is commonly thought to be against the law (though, except for named 'endangered' species, it isn't). On nature reserves and National Trust land one is invariably told to leave the flowers alone for others to enjoy, sometimes with the legal backing of a by-law. Those bold enough to pick cowslips or bluebells in a public place are setting themselves up for, at the least, unfavourable comment. Admittedly, it is hard not to feel furious at the sight of someone gathering bunches of wild daffodils or orchids, especially when you find them dumped by the car park, the pickers having discovered belatedly that wild flowers soon wilt. But have we not lost something of the past familiarity with wild flowers by this 'hands off' attitude? – and spoiled a harmless enjoyment that previous generations took for granted? It may seem surprising that some of Britain's leading botanists actually advocate picking flowers, precisely for that reason. David Pearman, for example, believes that 'all references to threats of collecting should be routinely excised from any publication' – since they are almost always fanciful – and he was President of the BSBI! We feel that ignorance is a far greater threat to our flora. The posy of flowers in a jam-jar, or the specimen drying on the herbarium sheet, makes you really look at the plant and appreciate its character and subtleties. We all know the buttercup – those yellow dots in a green field – but how many of us have admired the way the petals catch the light like slivers of mica, or watch the tiny insects dining on the corona of pollen inside the golden cup? Probably only the people that stoop to pick them!

Paradoxically, the old culture of picking and gathering wild flowers and ferns, and some of the folklore and anecdote attached to it, has moved on to wild fungi. It is edible toadstools now that are widely collected in such numbers that we hear again the inevitable warnings and cries for legislation. Picking wild mushrooms is providing many people with a half-forgotten pleasure. Like all personal feelings, it defies description and to those who dismiss feelings as unscientific may make no sense at all. But Andrew Young had a good go at capturing the pleasure of picking in his evergreen *Prospect of Flowers* (1945):

> He dropped me at a field-gate where a woman sat collecting money. Paying my pence I entered the field, maroon-coloured with the drooping heads of Fritillaries. People moved slowly about, stooping to pick those flowers that looked like reluctant serpents. All was so unexpected and strange that I had the feeling I was in heaven; I was even troubled to think that I was not engaged like the others. Picking flowers seemed the only occupation in heaven. But if in heaven, why not on earth? Perhaps something can be said for picking wild flowers.[213]

Undeniably, collecting does still pose a threat to a few wild flowers and ferns, generally those with a price on their head. Wild bluebell bulbs have a lucrative market internationally since Britain happens to have most of the world's bluebells. In 1998, the sale of wild bluebell bulbs was made illegal. Among rare plants, the main targets are wild orchids. In 1979 it was reported that someone was willing to pay £300 for a single specimen of the Ghost Orchid – which is hard to understand as the plant is not rare in Europe, and our plants are no different from theirs. There was a spate of orchid thefts at that time, and again in the mid-1980s, suggesting from its periodicity that it might be the work of a single trader. Recent examples cited in the press sound like the work of traders, not botanists – the looting of Water-soldier (*Stratiotes aloides*) from fen dykes in the Norfolk Broads, for example. But such instances seem to be quite rare. A more frequently heard complaint is the activities of photographers. No photographer deliberately sets out to damage a plant, but the impact of a succession of people, with their gear and heavy tripods, plus their need to 'garden' the best specimen, can nonetheless have that effect and, worse, they may compact the soil around it. A rare hybrid orchid that appeared on Noar Hill, Hampshire, a few years ago was soon surrounded by trodden grass and impacted soil which probably shortened its life and put paid to any chances of it seeding itself. The pressure is at its greatest with a newly discovered wild flower, like the Fen Ragwort in 1972, or the tongue-orchid *Serapias parviflora* in 1989. The Lady's Slipper was much visited after the secret was broken in the 1960s, and the site was showing serious signs of wear and tear from photographers by 1979. Its admirers have managed to restrain themselves since then, perhaps as the result of appeals to keep away, or possibly because most of those who wanted to see it had done so already. Another hazard is that all this attention may be observed with curiosity and, such is human nature, creep up afterwards and pick the plant. I remember once being watched by several boys while I photographed the Bristol Rock-cress. Like an ass, I told them it was very rare, and that hardened their resolve to pick it. The result was that I had to remain there, prepared to defend the Rock-cress with my fists if necessary, until they eventually got bored and after a few routine insults wandered off.

For every rare flower that has people queueing like penguins to see it, there are many thousands more that bloom more or less unseen. I have found very few recent examples of over-collecting or the theft of endangered species – orchids excepted. It is surely more of a fear than a fact. And the most impressive fact is that in 23 years of protective legislation for endangered flowers there has not been a single prosecution. Botanists no longer collect wild flowers, only images of them.

Protecting wild plants

No British wild plant was protected throughout the land until 1975. A few, like the Plymouth Thistle and Royal Fern, had benefitted from previous local by-laws preventing picking and uprooting, but these were applicable only on public property. In 1931, the Council for the Preservation of Rural England (CPRE) set up an advisory board on 'wild plant conservation'. One of its aims was a model by-law forbidding the uprooting of 'ferns, primroses or other plants' on public highways, commons and other places of public access. No fewer than 24 counties agreed to adopt the by-law, but in most places it was no more than a gesture. Seldom was the by-law displayed prominently, no one was keen to act as an informer, and policemen had more urgent matters on their plate. As one of them commented, 'it is extremely unlikely that any person is going to be so stupid as to begin rooting up wild plants with a police constable looking on, and the fact that a constable sees a fern in somebody's car would not justify his making a charge against the owner unless the latter was foolish enough to admit that he had recently dug up the fern unlawfully.'[178] Though the by-law remained in force, most people were unaware it existed.

Plymouth Thistle (*Carduus pycnocephalus*), protected under a local by-law since the 1930s. [Andrew Gagg's Photoflora]

The rights and wrongs of uprooting primroses and wayside ferns was more of a social matter than a scientific one. However some local authorities did request a list of any plants which might be threatened if picked in large numbers and so needed special protection. In response, H.W. Pugsley drew up county 'schedules' which were distributed to magistrates, police headquarters and schools in 1934 'for information and guidance'. At the same time, the board considered the wisdom of sponsoring a Bill which would entirely ban the uprooting of a small number of very rare species without the landowner's consent. But the attempt foundered, partly because it seemed as though the defendant would have to prove himself innocent. In other words, it seemed unworkable.

Things stayed as they were until 1967, when the BSBI, the SPNR and the Council for Nature jointly sponsored a Bill to protect all wild flowers from uprooting, and certain rare ones also from picking and sale. Their first attempt failed, but a BSBI report showing the disastrous decline of so many rare plants since 1930 presented a convincing case for protection at some future date. The team had another go, this time by grafting wild plants onto a similar Bill to protect wild animals sponsored in the Lords by the Earl of Cranbrook. By great good fortune, they found Peter Hardy MP to pilot it through the Commons as a Private Members' Bill. Fortunately (and rather surprisingly) it proved non-controversial. The Conservation of Wild Creatures and Wild Plants Bill went through without hindrance and became law in 1975. Twenty-one wild plants were given full protection, every last seed and rootling of them, on penalty of 'a fine not exceeding £100' – which was not much, even in 1975. The species chosen were considered to be endangered,

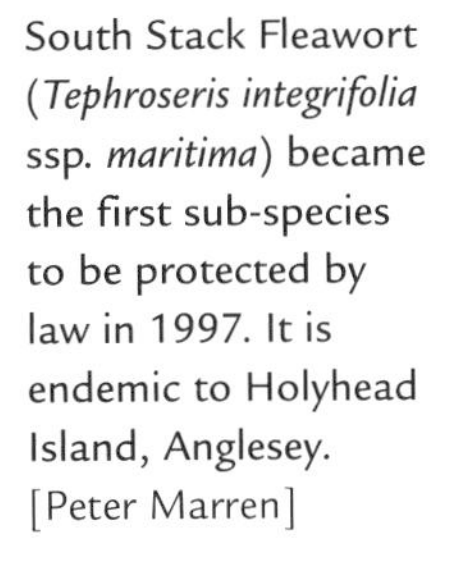

South Stack Fleawort (*Tephroseris integrifolia* ssp. *maritima*) became the first sub-species to be protected by law in 1997. It is endemic to Holyhead Island, Anglesey. [Peter Marren]

and, being mostly attractive, also vulnerable to collecting. Not surprisingly, a high proportion of them were orchids and alpines. This was the list:

Alpine Gentian
Alpine Sow-thistle
Alpine Woodsia
Blue Heath
Cheddar Pink
Diapensia
Drooping Saxifrage
Ghost Orchid
Killarney Fern
Lady's Slipper
Mezereon
Military Orchid
Monkey Orchid
Oblong Woodsia
Red Helleborine
Snowdon Lily
Spiked Speedwell
Spring Gentian
Teesdale Sandwort
Tufted Saxifrage
Wild Gladiolus

All well and good, but the question was, or should have been: what were these plants being protected *against*? There was little evidence of recent damage from deliberate uprooting or sale. And the law failed to protect them from the real threat – habitat destruction – since it specifically exempted 'any operation which was carried out in accordance with agricultural or forestry practice'. This meant that it was all right to plough up, or plant fir trees on top of, a Monkey Orchid, but if you took so much as a single one of its dust-like seeds you were in trouble. The sponsors understood that prosecutions were unlikely for the reasons outlined by that policeman back in 1934. However, as Peter Hardy had explained during the Commons debate, 'legislation in this field (has) a real effect, perhaps largely because of the educative influence that it had on society'. It was hoped that the new Act would change attitudes to wild plants in the way that the Protection of Birds Act had achieved for wild birds. Equally the botanists hoped to draw attention to the vulnerability of wild flowers and the need to conserve them for the enjoyment of future generations. Whether it is necessary to resort to legislation to improve public education is arguable. (Conservationists tend to be pragmatists – to them the end justifies the means.)

The real value of the Wild Creatures and Wild Plants Act was that it existed, and so could be built on. An important clause allowed the then Nature Conservancy Council (NCC) to review the protected list every five years and advise the minister of any other species that had become endangered in the meantime. The nature and degree of the threat to our rarest plants was refined two years later in the first Red Data Book, which listed more than 21 species 'in danger of becoming extinct'. The landmark Wildlife and Countryside Act of 1981 absorbed the relevant parts of the Wild Creatures and Wild Plants Act without amendment (except for a stiffer fine). Since it coincided, however, with the latter's first five-year review, the NCC successfully pressed the case for the inclusion of a much longer list of endangered plants. More species were added, and one was taken off. The one that got away was Mezereon (*Daphne mezereum*), which had proved to be less rare than was thought. Most other species that find their way on the list tend to stay on, since extreme rarity is their natural lot. Three more reviews on, there are now 111 fully protected wild flowers and ferns on the list (see Appendix 2), plus an assortment of mosses, lichens and fungi. The latest additions (1998) are Deptford Pink, Dwarf Spike-rush, Cut-grass and South Stack Fleawort. Although it is time-consuming to keep adding new plant species to the protected list, since a separate and detailed case must be made for each one, such advice is rarely challenged by politicians since no vital interests are affected by it. (It was quite otherwise with species like Basking Shark or Pearl Mussel whose protection affected individual livelihoods.) Since, however, the avowed aim of the legislation is educational, it might reasonably be asked whether any useful purpose is served by continually revising the list. After all the number of people who could even identify a Dwarf Spike-rush (to say nothing of a Bearded Stonewort or an Ear-lobed Dog-lichen) must be quite small.

The benefits of protecting a rare plant are mainly indirect. Their sites are more likely to be made nature reserves or SSSIs, and money is often made available for research on listed species. Mere protection is, however, a blunt instrument that does not achieve much on its own. Much more productive is the Biodiversity Action Plan (BAP) which has a planned recovery programme for each species with targets that define what recovery means. The BAP works because government and conservation charities have signed up to it and are committed to making it work; it taps into the popular sentiment in favour of saving rare species at home as well as abroad. Its more positive attributes, steered through schemes like English Nature's Species Recovery Programme and Scottish Natural Heritage's Rare Plant Project, make the protected list as presently constituted look increasingly redundant. Plantlife is campaigning for more meaningful legislation which would prevent damage to a plant's habitat as well as to the plant itself. This would, for example, make it illegal to use fertilizer or plant trees on a colony of rare plants, an entirely logical extension of the law, one might have thought, but one which has yet to be made.

A tale of lost Slippers

The Lady's Slipper orchid (*Cypripedium calceolus*) is the prodigy, if not the monstrosity, of our wild flora. Its huge exotic bloom – those clashing colours, a golden clog or slipper held in the clasp of four purple banners – is a triumph of natural bad taste. It looks like some archetypal jungle plant, or the sort of thing a romantic poet might invent after a session on the opium. The

Lady's Slipper has something in common with King Henry V: it became 'too famous to live long'. From its first mention in print, 'Calceolus Mariae', Our Lady's Slipper, was dug up and dug up again, until none but a single plant was left. It is not a very uplifting story, and the more you delve into it the worse it gets, but the near-extinction of our most dramatic wild flower does at least convey a message: that gorgeous blooms and densely populated countries do not mix. This flower is in decline all over Europe, and for once the reason is not habitat destruction. For all but the most self-controlled it is a matter of: see it, pick it. It seems that the trick of nature that produced an ingeniously designed but necessarily conspicuous flower has backfired badly.

I only saw the Lady's Slipper once, and something tells me I will never go there again. It took a couple of years to work its secret location out of the old gentleman with whom I went on orchid rambles in my youth. When he finally surrendered, I remember his solemn warning: ' Remember, *Mum's the word*. I'll be in the doghouse if they find out I told you.' He was half convinced there was a curse on the site. A year or two back he had visited the place with two companions, one of whom had the latest, most expensive, reflex camera on the market. I think it was called a Rolleiflex XL. While kneeling before the orchid, busy with his lenses and filters, he had

Below left: Lady's Slipper: the last British plant. [Bob Gibbons]

Below right: The map plots the historical distribution of Lady's Slipper. For most of the twentieth century only one site was known, but the orchid has recently been reintroduced to several of its old localities. A solid dot is a post-1930 record; open circles are pre-1930. [Biological Records Centre]

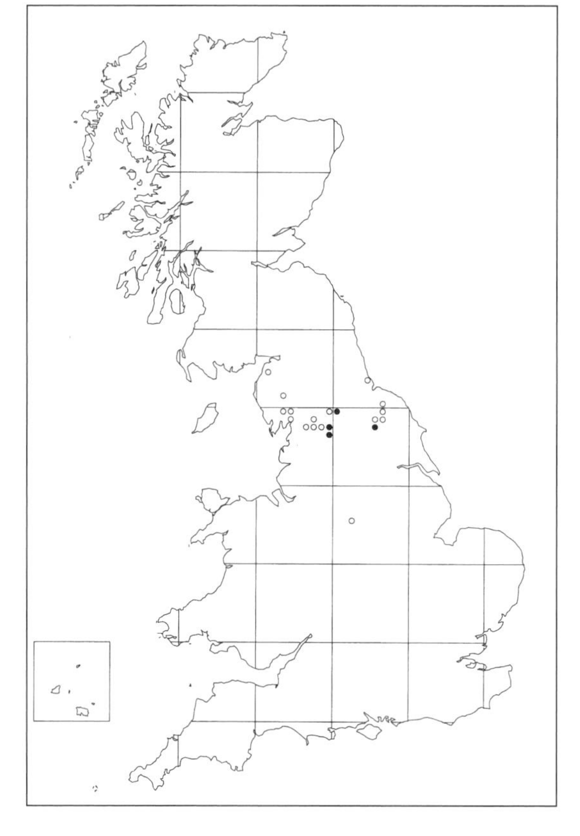

dropped the camera. Downhill rolled the Rolleiflex, bouncing off roots and bumping from rock to rock with sickening metallic thunking noises. It would be amusing to report that it finally rolled down a rabbit hole, but I believe that the camera was recovered, intact if somewhat battered, and has gone on to snap many another rare flower. As for my kind old friend, he seemed rather subdued about the whole business, quite unlike his tales of Encounters with The Lizard and his singlehanded Detection of The Monkey. Later, when I too went there, hanging on to my Nikon like grim death, I realized why. The Lady's Slipper is not disappointing – it is every bit as improbable and bulging as you expected – but, well, travelling is better than arriving. The fun lay in the detective work, the anticipation and the sense of discovery. Actually seeing it in the flesh was a bit of an anticlimax.

Another reason was the warden. I could probably have found the spot blindfolded. I had drawn a coloured Treasure Island-style map of the place, with every rock and tree marked, together with exact instructions, 100 paces from the gate, turning right at the third rowan by the wall and so on. I need not have bothered since, not far from the hallowed ground, there was a pitched tent with a guardian inside, and he insisted on taking us there himself for fear we might tread on it. I was given a form to fill in. *Who told you?* it asked. Evidently they were trying to trace a network of who told who, and, fearing that the net might close around my dear old mentor, I left it blank. One could imagine him being pilloried in the botanical press as the rat who grassed on the orchid. As I say, I doubt whether I would want to go through all that again, even for *Cypripedium*, and we are indeed regularly asked to keep away by the orchid's own private committee.

Before word got around, the Lady's Slipper had been virtually lost to the botanical world. The last record in print was 1937, and it was deliberately left open to speculation whether the plant survived at all. Victor Summerhayes, author of the standard work on British orchids, was careful not to commit himself. The Wild Flower Society magazine tells of a Miss Vachell who had two great ambitions – one was to carry her purse in a Highland sporran, and the other was 'to see a Cypripedium'.[204] Alas, it seems she never did. Even the great Ted Lousley, who had seen almost every other wildflower, failed to track down the Lady's Slipper until late in life. The *Atlas of the British Flora* included all its historical localities, not distinguishing past and present. It had been effectively airbrushed out of the picture, rather like Trotsky after his exile from Russia.

If this seems an extreme reaction, it is worth considering the historical context. This was the least safe plant in Britain, and there had been a price on its head since the reign of Charles the First, if not earlier. The first mention in print of the Lady's Slipper as a native wild flower – it was already in gardens – was by John Parkinson in his book *Paradisus in Sole*, an extravagant account of 'all sorts of pleasant flowers which our English ayre will permit to be noursed up', along with tips on their 'right ordering, planting and preserving of them'. Preservation, though, was not exactly what Parkinson had in mind for Our Lady's Slipper, for he spilled the beans with exactitude:

> It groweth ... neare upon the border of Yorkshire in a wood or place called the Helkes, which is three miles from Ingleton as I am enformed by a courteous Gentlewoman, a great lover of these delights, called Mistriss Thomasin Tunstall, who dwelleth at Bull-banke, neare Hornby Castle in those parts, and who hath often sent mee up the rootes to London, which have borne faire flowers in my garden.

And not only in Parkinson's garden either. A generation earlier both Gerard and Tradescant grew Lady's Slippers in their gardens, though where they got them from is uncertain. Mistriss Tunstall has come in for her share of censure for supplying roots to all and sundry, but she seems to have had more sense than to kill the goose that laid such golden eggs. It was not in her time but much later, in the 1790s, that a gardener from Ingleton robbed Helkes Wood of its last Lady's Slippers. As Reginald Farrer tells it, in his book *My Rock Garden* (1920):

> After Mistriss Thomasin had long been dead as the Cypripediums she sent up to Parkinson, there came a market gardener, a base soul animated only by the love of lucre (and thus be damned to a far lower Hell than the worthy if over-zealous gentlewoman), who grubbed up all the Cypripediums that she had left, and potted them up for sale. The Helkes Wood, now, is an oyster forever robbed of its pearl.

Tales of the Lady's Slipper often take on this flavour of good and evil. At a valley near Arncliffe, where the flowers blossomed beneath a whitebeam tree, the good angel was the local vicar, who 'kept careful watch over it, and went every year to pluck the flowers and so keep the plant safe, for without the flower you might, if uninstructed, take the plant for Lily-of-the-Valley'. Then one year he fell ill. Enter the demon, in the form of 'a professor from the north' who said he would pay so much for every rooted plant brought to him. And 'the valley was accordingly swept bare'. Farrer concluded with a curse worthy of Dante: 'Accursed for evermore, into the lowest of the Eight Hot Hells, be all reckless uprooters of rarities, from professors downwards.'

The essential point is that all this happened a long time ago. The sale of Lady's Slippers in the market place of Settle took place in the 1780s, well before the excesses of mid-Victorian plant hunting. The allure of the Lady's Slipper went beyond mere botany: it seems to have become a symbol of local pride, like Yorkshire beer or the White Rose. The plant was given titles like 'the great Flower Prize of Craven' or 'the Pride of our Northern Flora'. That being so, every country landowner or proud squire in his great house wanted a row of golden slippers in his garden. Specimens plundered from Helkes Wood were still flowering at Hipping Hall in 1835, long after the wood itself had been looted out. They also adorned the flowerbeds of Edge End near Ulverstone in 1857, its proprietor noting that 'the botanist who supplied it from Settle had destroyed the station'.[127] Settle seems to have been the centre of the trade. In 1781, William Curtis was told of a man who had 40 of them to sell, presumably all with their roots and tubers intact. 'Transplanted into dale gardens, as it had been many a time,' wrote the author of *The Vegetation of Craven*, 'it lives on for many a year, until some manuring, or acidic intervention upon its basic calc requirement compasses its dissolution.'

By late Victorian times, the Lady's Slipper was already more a garden flower than a wild one. The entry in *The Flora of West Yorkshire* (1888) is little more than a record of extinctions: 'Helks Wood. Not to be found when I was there,' 'Now hardly ever met with,' 'One plant found and carried off,' etc.[103] A Flora of Derbyshire of 1864 claimed it was 'native' on the Heights of Abraham overlooking Matlock, but a subsequent county flora of 1889 is careful not to mention the plant. Botanists were becoming deliberately vague on the subject of Lady's Slipper, and details of where and when it disappeared from its last north country sites are hazy (and this has made the identification of former sites for reintroductions a difficult matter). But in a curious twist to the tale, it seems that at least one landowner was nurturing it in his garden and greenhouses for returning to the wild. This seems to have happened at Castle Eden Dene, a popular resort for seekers of natural wonders, where the orchid was said to be 'more a propagated or nursed plant than a true wilding'.[72] Greenhouse Lady's Slippers are also said to have been introduced into the nearby Hawthorn Dene, where they survived well into the twentieth century. Thus the dividing line between wild and cultivated plants was becoming blurred even as the plant neared extinction. The sole remaining plant is as manicured as any film star, with its own little industry of wardens, administrators and scientists, and a back-up team busy producing 'test tube' seedlings on plates of agar jelly. It has been saved, but is it any longer wild?

What does one make of this record of a species plundered almost beyond recovery? Evidently it was not the passing rambler plucking a bloom for his hat that caused the rot, but the trade in the species. One nineteenth-century author, who was certainly not a plunderer, even explained how to pick the flower without harming the plant: 'Cut the flower stem with two or three leaves below it, and leave the rest severely alone; nought lost by so doing – a hundred to one it will be there again for your pleasure next season.' He knew one resident who 'over seven years cut twenty flower stems from one clump, six in one year, and Our Lady's Slipper is there yet'. It was not casual pickers nor even botanists that were to blame, but nurserymen and gardeners, and above all, its special association with Yorkshire that made it such an attraction. Perhaps if it was easier to propagate, the orchid might have had a better chance of survival.

All this focus on wicked professors and selfish gardeners might have overlooked more fundamental matters, like habitat change. Many of the places in which the Lady's Slipper used to grow are still wild and undeveloped – some indeed are nature reserves. But in most of them the plant would not occur today even if it had been left alone. They are now too shaded, or over-run by brambles, bracken and other aggressive plants. Others are grazed flat or planted with crop trees. What struck me most on visiting the locality of the last Lady's Slipper was how unlike the rest of the dale it is. The site has a pristine feel to it, with just the right amount of light and shade, and a rich associated flora not yet hoovered out of existence by the ubiquitous dales sheep. It may be that even the Ladys Slipper was not so much destroyed by collecting as reduced by unfavourable natural changes to the point where it became vulnerable to chance events, like a passing gardener.

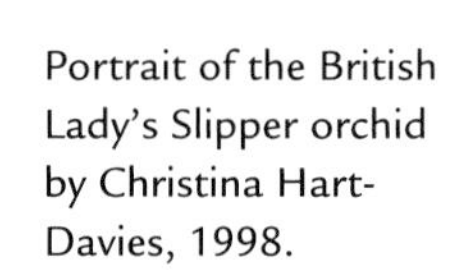

Portrait of the British Lady's Slipper orchid by Christina Hart-Davies, 1998.

Chapter 8

A matter of origins

When I was about seven, and starting to take an interest in natural history, someone gave me a copy of *The Observer's Book of Wild Flowers*. This was an old-fashioned book even in its day, with its paint-by-numbers descriptions and tiny pictures based on *Sowerby's English Botany*, but it made an ideal 'tickers' companion which you could slip into your pocket and deface with a big biroed tick whenever you found a new flower. But one flower I never ticked. This was the Corncockle (*Agrostemma githago*). And that was a puzzle, because according to *The Observer's Book*, 'Wandering through or round our cornfields any time during the summer, one is almost sure to find

Left: Corncockle (*Agrostemma githago*) was once a serious agricultural pest. Since its seeds could not easily be separated from wheat, they were often ground and baked in the bread, giving it a bitter taste and causing upset stomachs. But no living farmer has had to worry about Corncockle. [Bob Gibbons]

Right: History inverted: Corncockle being grown as a crop on a Berkshire farm for wildflower seed mixtures. [Peter Marren]

this beautiful flower'.[185] Never mind that where I came from you were liable to receive a thick ear if you tried wandering through a cornfield. Assuming that our corn must be uniquely cockle-free, I asked my mother whether she had found the missing flower when she wandered through cornfields, dodging irate farmers, in years gone by. No, she hadn't (though she remembered Corn Marigolds), and my grandmother didn't seem to know it either. And in nearly forty years of further wandering, I have yet to find a single cockle in a cornfield, though like everybody else I find it now and again in sown wild flower meadows and in gardens. Plainly, that confident statement in *The Observer's Book* was not true.

This is where a little literary research comes in. Although that edition of *The Observer's Book* was on sale in the 1950s and 1960s, the offending sentence had been borrowed from another book, written at the turn of the century by Edward Step (who died in 1931).[183] But even Step may not have been writing from firsthand experience. He was a Londoner, and according to *The Historical Flora of Middlesex* (1975), Corncockles were already few and far between around London by then.[97] Step, it transpires, had himself borrowed freely from an even earlier book, written in 1852 by grocer's daughter, Anne Pratt.[147] And Anne Pratt was not entirely original either; for example her often-quoted story about Pheasant's Eye (*Adonis annua*) being sold in Covent Garden as 'Red Morocco', was culled from William Curtis' *Flora Londinensis*,[36] the first instalment of which was published in 1775! Thus can a story travel through time in that well tried-and-tested vehicle, literary plagiarism!

The Observer's Book compiler could be forgiven for getting it wrong about Corncockles when the principal reference book of the day, *The Flora of the British Isles* (1952) by Clapham, Tutin and Warburg, was still describing it as 'common, but decreasing'. Decreasing perhaps, but common it wasn't, and had not been so for a long time. Professor Dudley Stamp, who was born in 1898, remarked that he 'was not old enough to remember the Corncockle in abundance',[181] though E.J. Salisbury, twelve years older, could just remember it as 'a not uncommon weed' in wheatfields in his home county of Hertfordshire.[174] A trawl through the local floras shows that the Cockle was already becoming scarce by the start of the twentieth century, though it could still occasionally make a nuisance of itself. Hence the Corncockle was not a victim of modern farming, as is so often stated. It was a victim of *nineteenth-century* agriculture. Its survival strategy is simple and unsophisticated: Cockles have little or no seed dormancy and depend on being resown with the crop. Their large, fleshy seeds are similar in size to a grain of wheat (their rounded shape and knobbly surface provide the flower with its molluscan name), and so were impossible to eradicate by sifting. Once more sophisticated methods of screening seed became available, however, the Cockle became an overnight casualty of progress.[55] Though, according to Donald Grose, it enjoyed a brief resurgent burst in Wiltshire during and after the Second World War, when it appeared in some sixty places after old pasture was ploughed, it did not last long.[76] Geoffrey Grigson, who knew the county well, said he last saw a Wiltshire Cockle in 1955. Today the only cornfield Cockles that survive have been deliberately sown. The site with the best pedigree is a field corner at Burwell, by the Devil's Dyke in Cambridgeshire, where a farmer has lovingly nurtured his Cockles for the past thirty years using locally collected seed. It is effectively extinct as a 'wild' flower, though, paradoxically, it is more familiar today than for many years past, thanks to its popularity as a garden plant.

Did the British Corncockle ever have an existence independent of the farmer? Probably not. It has been with us for a long time, for its characteristic seeds and pollen grains have been found near Roman settlements. But none have been detected before the days of Julius Caesar, and it therefore seems likely that Cockle seed first arrived here by ship in sacks of grain imported from the Mediterranean. It may have continued to rely on regular imports; E.J. Salisbury (1961) mentions that nineteenth-century supplies of grain from Russia were not infrequently rotten with Cockle seed.[174] It seems, therefore, that the Corncockle was one of nature's hitchhikers, an import from the Mediterranean and Ancient Rome, that rode around Europe in a sack.

Contrast the story of the Corncockle with that of another rare arable weed, Corn Parsley (*Petroselinum segetum*). This, too, grows at the edges of arable fields, often on light calcareous soil. It, too, has declined, and not only in Britain but throughout its European range – indeed, Britain probably has as much Corn Parsley as anywhere. But in other ways the two 'weeds' are poles apart. To begin with, Corn Parsley exhibits far more 'geography'. Whereas the Cockle is generally distributed, Corn Parsley is a distinctly western, oceanic plant, commonest along the Atlantic and west Mediterranean coast. England is towards the northern limit of its range, and, as one would expect, it occurs mainly in the warmer areas of south and east, and along the coast of South Wales. Interestingly, it never seems to have reached Ireland. Also unlike the Cockle, Corn Parsley has a 'wild' habitat, in dry, rather open grassland, typically on crumbly banks by rivers and dykes, on grass verges and near the sea. It is especially characteristic of floodland by the brackish parts of the Thames estuary, and on dry, rocky places on the coast of South Wales. In its better known habitat, arable fields, Corn Parsley shows a firm preference for well-drained soil, especially on the chalk and the Jurassic limestones. Its geography suggests a plant that has settled into the landscape. One might adduce from this that it is a native species, and this is confirmed by the fossil evidence. Its pollen has been found in post-glacial layers in many sites, and also from three interglacial periods. Corn Parsley has been around a long time, and 2000-year-old pollen found preserved in Cumbria suggests that it was more widespread in ancient times than it is today.[141]

Corn Parsley (*Petroselinum segetum*), perhaps a natural plant of coasts and river banks that took advantage of arable farming. [Peter Marren]

This firm grounding in the landscape works in favour of a plant's survival. Corn Parsley has a strategic advantage over Corncockle in its built-in life insurance policies. Its relatively small, light seeds retain their viability, tiding the plant over during the cold, wet seasons when seed production is low. It does not therefore rely on being resown with the crop. It is a monocarpic species, with a first-year rosette stage, followed by flowers in the second year, very like its close relative, garden parsley (*Petroselinum crispum*). Biennials develop strong deep roots and can be stubborn colonizers of waste and fallow land. Experiments suggest that it does best after a late autumn sowing, surviving over winter as a small rosette. What comes across is a reasonably flexible plant, and to some degree an opportunist, not a one-shot wonder like the Corncockle. Corn Parsley can certainly appear in unexpected places. A friend once showed me a clump growing beneath his garden sundial which had evidently arrived there by natural means. Possibly it was a recent arrival, but just as likely it may have been there, active or dormant, for longer than the sundial; perhaps longer even than the garden. Its reward for tenaciousness and opportunism is that Corn Parsley, though now rather rare, as most of the 'traditional' arable weeds have become, is not endangered. It is unlikely to die out completely. With modern chemical and technological farming, it has either 'retreated' to the wild sites, or hung on wherever the landowner has been kind enough to allow it to survive. Grassy field headlands, for example, allow the Corn Parsley to thrive in places on the Oxford limestone: here it owes its chance to the partridge. Probably the advent of widespread arable cultivation enabled the plant to colonize much of lowland Britain from its aboriginal base on the coast and riverbanks, thus blurring its 'natural' pattern of distribution. Today this advance has been reversed by progress. It is our loss. Corn Parsley lacks the scruffy pink glamour of the Corncockle, but, when well grown, it is a lovely plant with its tiny 'puffs' of white flowers on a green scaffolding of flower stalks. The books fail to do it justice. You could not mistake it for anything else, except possibly the Stone Parsley (*Sison amomum*), but that plant has an extraordinary pong of petrol, whereas the Corn Parsley smells, well, exactly like parsley.

Naturalization and rarity

Although it is extremely difficult to do so consistently, British nature conservationists try to maintain a clear distinction between native and naturalized plants. Native species deserve our protection, the rule runs, but naturalized ones must take their chances. The first Red Data Book (1977)[142] did in fact include many rare naturalized plants, but most of them have since been weeded out. Only native species qualify for direct legal protection (though one or two non-native ones have crept in all the same). Although there is logic in this approach, which is based on long botanical tradition, it implies that non-native species have no conservation value, and this can occasionally lead to ludicrous situations. Twelve years ago I tried to persuade the NCC to protect the Crocus Field at Inkpen, Berks, as a Site of Special Scientific Interest. The Field was on the market, but although the local wildlife trust was keen to buy it, they had no money and no grant would be forthcoming unless the place was made an SSSI. There was a very good case for doing so. The Field was the best site in Britain for a naturalized species, Spring Crocus (*Crocus vernus*), which was then in the Red Data Book. People came from far and near to admire the beautiful display of purple and white flowers in early spring. Moreover, the Field was itself a rarity, an undrained, unfertilized old meadow, full of declining flowers like Pignut, Ragged Robin and Sneezewort. Yet it was the Crocus that was the sticking point: it was not a native species, and therefore its presence was seen as polluting, as if it was no more significant than a casual weed or a garden escape. Floral fashion had swung a long way since the early years of the century, when the Crocus Field was on the very first 'want list' of national nature reserves precisely *because* of its rare Crocus. In logic, it seemed the only way to protect the Field now was to get rid of the Crocus and restore the pasture to its pristine pre-Crocus state. Fortunately better councils pre-

Spring Crocus (*Crocus vernus*), commonly grown in gardens, also grows wild in a few meadows and churchyards. [Jane Smart]

vailed, the Trust did receive its grant and the Crocus Field is now an SSSI and a nature reserve. The Crocus showed its gratitude by rapidly doubling its numbers. But it seemed touch and go at the time.

Surely the time has come to make a distinction between long-naturalized flowers and recent arrivals. Plants which established themselves in the wild before, say, the year 1600, and have survived by their own devices ever since are ecologically much closer to native species than garden escapes. Often, indeed, it is hard if not impossible to tell the difference. There is a good argument for basing conservation assessments on longevity rather than a notional 'nativeness', especially where there is a significant historical dimension involved. English Heritage, as well as English Nature, might be interested in the Spring Crocus at Inkpen, with its possible links with a near-by Knights Templar foundation, or in the Birthwort (*Aristolochia clematitis*) growing suggestively among the ruins of Godstow Nunnery near Oxford. Some naturalized plants, like the wild pinks on the walls of Rochester Castle and Fountains Abbey, or the Tower Cress (*Arabis turrita*) on crumbling college walls in Cambridge, could be regarded as 'heritage' in themselves. But at present, such plants enjoy little or no formal protection: they are in conservation's limbo land, neither garden plants nor wild ones.

In terms of plant hunting, rare naturalized plants are sought after just as heartily as native ones. They are given a starring role in local floras and get their picture into Keble Martin. Their persistence in a particular place can be remarkable. The Starry Clover (*Trifolium stellatum*), for example, has grown on the shingle at Shoreham Beach in West Sussex since at least 1804 – probably for much longer. Were it not for its Mediterranean distribution and known performance as a casual elsewhere on the south coast, no one would have doubted its nativeness. Prostrate Toadflax (*Linaria supina*), which grows on sandy ground near Par in Cornwall was until recently Red-listed as a possibly native species. Yet its claims to be one are even weaker than Starry Clover's. The local botanist Frederick Hamilton Davey observed it colonizing the harbour area along with the 'rare adventives of several continents' just a century ago. He practically watched it leaving the ship.[8] The point, surely, is that even as an introduced plant, the Prostrate Toadflax's naturalization in Cornwall is of scientific interest, and, since it is eagerly sought after and photographed, it must have cultural interest too. Who are we to say it is unimportant? One could visualize the botanical character of modern Britain not so much as a ghetto of isolated island plants but as an increasingly cosmopolitan flora, with wild plants coming and going with trade and transport, an essentially dynamic flora which changes from year to year. Most of the incomers last for only a few seasons, and just a few become serious pests. But from time to time a species may establish itself quietly in a corner of the country, usually somewhere warm and probably not far from their original garden or port, and pretend to be native. Preserving the ephemeral alien flora would be a hopeless aim, especially as its interest lies in its dynamism and unpredictability. But these long-naturalized plants are part of Britain's local character, a product of history and geography as well as biology. The unique presence of the Starry Clover says something particular about Shoreham Beach, even if it is only that, as it happens, it offers the best environment available in Britain for the Starry Clover.

Long-naturalized plants confined to one or two places are perhaps the most endangered species in our flora. Fans of Keble Martin may remember the pretty pink and yellow flower in the corner of Plate 4, rather overshaded by the Wild Peony. This is Barrenwort (*Epimedium alpinum*), which is in all the floras from Hooker onwards as a naturalized plant of wild places in northern England. When I knew nothing about rare flowers except what I read in books, I thought Barrenwort had a peculiar presence of its own and looked forward to exploring the 'subalpine woods' where it was supposed to grow (rendered all the more exotic because one didn't know what a subalpine wood was either, though you thought vaguely of willow-pattern). But you would probably have to travel back to the Victorian age to find Barrenwort in a convincingly wild place. No one grows that particular species anymore, and so the source in country gardens which supplied the seed that somehow escaped to the subalpine woods no longer exists. Stace[180] notes that 'most modern records [of *Epimediums*] probably refer to other taxa', like the hybrid *E. x versicolor*, which is the red-flowered one you see in cottage gardens today. So one small ambition of mine may never be fulfilled. Some other 'traditional' species of British floras are in an equally bad way. The Tuberous or Fyfield Pea (*Lathyrus tuberosus*), a famous Essex plant, was all but extinct by 1970, thanks to agricultural improvements in the 1950s and 1960s (though there is a big patch of it by The Fleet in Dorset – should it be renamed the Weymouth Pea?). This plant is well known to entomologists as the supposed foodplant of the Wood White butterfly (wrongly, fortunately for the butterfly). Perhaps we should have tried harder to preserve it, especially near Fyfield. If John Ray could come back to earth, he would probably express displeasure to learn that Hairy Yellow-vetch (*Vicia hybrida*) no longer grows on Glastonbury Tor, or that Hairy Vetchling (*Lathyrus hirsutus*) was destroyed on Hadleigh Castle (though I gather that English Heritage have reintroduced it).

The original urban wild flower was the now very rare London Rocket, the seventeenth equivalent of Rosebay Willowherb. Individually it is a somewhat nondescript looking 'cress' with raggle-taggle leaves and a tuft of narrow pods and yellow florets. Like flax, it looks better *en masse*. Londoners got their chance to see the plant at its best in 1667, after much of their City had been devastated by the Great Fire. The leading botanist at Oxford, Robert Morison, had a theory to account for the sudden abundance of *Sisymbrium irio*. They were 'hot, bitter' plants, which proved that they were 'produced spontaneously without seed by the ashes of the fires mixed with salt and lime'. Their natural habitat was, appropriately enough, on Mount Vesuvius. Morison's interesting account continued as follows:

> The spring after the conflagration at London all the ruins were overgrown with an herb or two, but especially one with a yellow flower: and on the south side of St Paul's Church [= Cathedral] it grew as thick as could be; nay on the very top of the tower. The herbalists call it *Ericoleoris neopolitana* – Small Bank Cresses of Naples; which plant Thos. Willisel told me he knew before but in one place about the town and that was at Battle Bridge [= King's Cross] by the 'Pindar of Wakefield' and that in no great quantity.

London Rocket's day was soon over, but it lingered as a 'casual' weed, occasionally turning up in some funny places, like the Snake Pit at London Zoo. Right on cue, it reappeared in some quantity in 1945, after the City was again devastated, this time by German bombs, but not in its former numbers.[20] For the past 50 years it has flowered regularly in gardens on Tower Hill, where it used to receive visitors from the Wild Flower Society each year before the Society's annual London tea party. It has appeared in other towns, hanging around a car-park in Taunton, for example, and popping up on the pavement outside a fancy-dress shop. It would be as sad to lose *Sisymbrium irio* at the Tower of London as to lose an orchid in a meadow. It may not be a link with the Ice Age, but it is a souvenir of '1666 and all that', which has gathered to it a fair amount of history and folk-lore.

Several other British cities have a special rare flower, but only two have a special *flora*. One, at Bristol, we have met already. It is a wild flora, surviving on the slopes and crags of Avon Gorge, though some species have escaped onto nearby road verges and flower beds. The unique flora of Plymouth is less easy to categorize. Two of its rare flowers, Plymouth Thistle (*Carduus pycnocephalus*) and 'Plymouth Campion' (*Silene vulgaris* ssp. *macrocarpa*) are naturalized Mediterranean plants which probably got there in ships. Plymouth Pear (*Pyrus cordata*), on the other hand is almost certainly native to Britain. Its site is suburban today, but only because the City has spread into the countryside. Field Eryngo might be native there too, or it might not. It has certainly been known on limestone rocks and open spaces in the City for 350 years and is fertile in and around Plymouth as nowhere else in Britain. You can find it today in the flowerbeds on Plymouth Hoe, after 'twitching' the Plymouth Thistle on its crumbly cliff edge. Happily the City authorities have agreed to take these plants under their wing. The Plymouth Thistle is protected under an enlightened local by-law made in the 1930s;

London Rocket (*Sisymbrium irio*) owes its English name to a population explosion after the Great Fire of London in 1666. [Bob Gibbons]

Right: Plymouth Pear (*Pyrus cordata*) is probably a native tree, surviving in a farmed and now partly suburban landscape. [Peter Wakely/English Nature]

Far right: The scented flowers of Italian Catchfly (*Silene italica*) open only as the evening approaches. [Bob Gibbons]

and the Eryngo is granted an unusual degree of tolerance even when its prickly heads obtrude among the geraniums and petunias.

Once established, naturalized plants can be extremely persistent. The wild colony of Asarabacca (*Asarum europaeum*) on a hedgebank near Redlynch, Wilts, has been known since 1782, and is one of the county's botanical show-pieces. Here, quite exceptionally, the plant produces seeds and seedlings. It has always been a favourite among plant hunters, with its canopy of shiny Cyclamen-leaves and brown, cup-shaped flowers lurking beneath in the semi-darkness where they are pollinated by wood-lice and other creeping things. The plant is a relative of tropical gingers, and was used by herbalists as a rather drastic pick-me-up – according to one authority, it 'purges violently, upwards and downwards'. William Turner knew it 'onely in gardines that I wotte of', and in some places it may not so much have escaped from as survived the garden, like a ruined stone arch or well, or a broken statue half-hidden in the weeds.

A rare tale of weedy persistence comes from the unprepossessing suburbs of Dartford, Kent, where the Italian Catchfly (*Silene italica*) has been hanging on grimly for at least 135 years. It might have been there much longer, since Italian Catchflies look rather like native Nottingham Catchflies, differing mainly in their longer flower-stalks and habit of opening their flowers in the evening. Nor is it always found at its pristine best in Britain. J.E. Lousley[106] noted that 'it is usually seen by botanists with the flowers closed – its dingy appearance being enhanced by the collection of road dust on the sticky hairs with which it is covered. I [only] once saw the Italian Catchfly in perfect condition when I took a friend to see it late one evening, and the extent of the transformation was a great surprise.' The classic places to see it, by torchlight if necessary, were disused chalk-pits at Greenhithe – the hottest, most arid places in town. Unfortunately in that part of the world pits tend to become useful places for tipping rubbish. The last surviving site of Italian Catchfly was filled in during the 1980s.* The site is now 'dull and derelict' suburb – houses, roads and wasteland. Yet the Italian Catchfly did not give up. It survives on a nearby roadbank, more dust-blown than ever, in dramatically changed surroundings. About 100 plants were counted there in 1987. Rosemary FitzGerald has found intriguing evidence that the history of this Italian Catchfly colony is even more convoluted than we knew, for it has probably exchanged habitats not once but twice. Close by she found the remains of a flint wall, part of the former walled garden of a house that once belonged to Mrs Beaton, the noted cook. The house itself was long ago demolished and swallowed up by a chalk pit. But might the Italian Catchfly have originated there as a garden plant, followed the quarrymen into the chalk-pit, and finally put down roots on a suburban road bank? There is evidence of it escaping from gardens in Dorset and Sussex. Though not a showy plant, it might have been grown for its pleasant evening scent, or by botanists. Its setting at Dartford has grown progressively downmarket from the pleasant leafy place Lousley knew – but the Catchfly's ability to adapt to all circumstances is impressive.

Modern floras agree that the enterprising grass, *Cynodon dactylon* is probably introduced in Britain, though it is still in the Red Data Book. This is one of the world's most successful grasses, whose fans of dagger-shaped leaves at the end of tough stolons are 'instantly reminiscent of countless Mediterranean lawns'.[58] In South Africa it is called 'Kweek', in India 'Doob', in America and Britain, 'Bermuda Grass', and in books, 'Creeping Dog's-tooth'. The 'teeth' are the flowers, set in rows like the jawbones of dogs, splayed out like an open hand. Though rare in Britain, the grass is very much at home near the car-park on Minehead golf course as well as some other maritime lawns like the seafront at Weston-super-Mare, or the lawn at Kew, or even by a bus stop in Reigate. But appearances can be deceptive, and the plants may have been there long before the bus shelter or the golf club. *Cynodon* was first recorded in Britain as long ago as 1685, when John Ray found it growing 'plentifully' between Penzance and Marazion on what was then a wild smuggler's coast. It is still there, forming, among other things, a fine xerophytic lawn in Lannoweth Terrace, Penzance, where it grows among native vegetation. The grass has been known around Poole Harbour since 1780, and although most of its surviving sites are now dismal subur-

* The infilling of the Greenhithe pits was not without botanical drama. A field excursion on 27 July 1975 reported one pit being filled with industrial waste and producing aliens like Madrid Brome, Wormwood, Tobacco-flower and Duke of Argyll's Tea-plant. The party moved on to another vast disused quarry 'which was about to be filled in with some of London's refuse. We saw, probably for the last time in this pit, the grass *Nardurus maritimus*, while in a woodland area were scores of Green-flowered Helleborine (*Epipactis phyllanthes*), Twayblade and several hawkweeds' (*Watsonia*, Field reports, Jan 1977).

ban road verges, its original habitat was 'sandy turf near the shore'. All in all, *Cynodon* clings to our warmest shores with brave persistence, and the main reason for doubting its nativeness is the lack of any fossil history plus its readiness to take root anywhere sufficiently free of frosts and herbicides.

Cynodon's seafront site at Weston-super-Mare has one of the strangest plant communities in Britain. At first sight it looks like ordinary close-cropped grass, with ornamental trees and shrubs, fronting a row of Victorian guest-houses. A closer inspection from the botanist's usual kneeling posture will reveal that parts of this lawn are anything but ordinary. Here *Cynodon* is part of a rich flora of intermingled naturalized and native flowers like Suffocated Clover (*Trifolium suffocatum*), discovered a century ago and recently refound 'among the deck chairs'. In at least one place there are patches of a rare introduced clover, *Trifolium tomentosum*, with spectacularly woolly heads. There is an abundance of Clustered Clover (*Trifolium glomeratum*) and Smooth Rupturewort (*Herniaria glabra*) 'in patches worn bare by local schoolboys' football goals'.[58] Most remarkably of all, some rough grass by iron railings in the town's central park contains the very rare *Equisetum ramossissimum*, recently dubbed the Branched or Ramose Horsetail. This primitive plant is supposed to be another accidental introduction and so is rarely illustrated in the floras. It was discovered here about 1963, but was considered to be an odd form of the Field Horsetail (*Equisetum arvense*). Only in 1986 was its real identity established. Where did it come from? The area was occupied by American troops returning from Europe in 1945, and they seem the plant's most likely mode of origin – there is even a scientific word for plants carried about by armies: polemochores. At its only other British site, in long grass by a straightened river near Boston, Lincs, 'Branched Horsetail' was probably introduced with imported ballast, offloaded from ships to consolidate the sea wall.

The flower with perhaps the most resonating historical association of all is Woad (*Isatis tinctoria*), the plant which famously supplied the blue juice that the ancient Britons daubed themselves with, according to Caesar. You need a lot of Woad to make a blob of dye, and the amount of Woad growing wild today would barely daub a brace of Britons. They must have had fields full of the stuff. Like oil-seed rape today, some of this cultivated Woad would have escaped, and in a few warm sheltered cliffs free from competition it went native. After farmers ceased to grow it as cheaper indigo dyes became available, Woad stopped escaping. Today, it survives in only two wild places: on a cliff of red marl above the River Severn at Tewkesbury – which, as a wool-town, had a history of Woad-growing – and in an old pit in suburban Guildford, also a centre of the downs wool trade. It has lived in both sites for the past 200 years, but there are signs that it may not survive there for much longer. At Tewkesbury, Woad is at risk from Sycamore which now forms dense woodland over what was once a warm sunny slope.

I once risked my neck to photograph the one Woad-plant I thought I saw, yellow and frothy against the red earth, almost at the top of the cliff. I hauled myself up using the sycamores as handles until, caked in mud and with no idea how to get down again, I reached the spot and saw that the Woad had somehow turned into a small, mis-shapen Rape-plant. But perhaps, after all, we do not need to worry about Woad. It is coming back into fashion as a game-crop, and turned up recently on the Tewkesbury by-pass, as mysteriously as the Mona Lisa's smile.

Is it a native species?

In 1985, the late, great Irish botanist, David Webb, published a provocative paper in *Watsonia* entitled: 'What are the criteria for presuming native status?'[200] Since, in terms of conservation value, there is a world of difference between a native and a non-native species, it is surprising how little has been published on how one tells the difference. Webb was caustic about how such decisions are often made, 'based all too often on inappropriate criteria, on irrelevant emotions such as local patriotism, or misinterpretation of fossil data, or on an uncritical acceptance of earlier opinions'. For example, 'CTW'[30] inferred that the Field Eryngo was a native species on the grounds that it had been established at Plymouth for 300 years. But on such a basis Medlar, Walnut or even Sycamore would have even better claims to be native plants. Webb produced his own list of probable alien species, masquerading in the floras as native, which must have surprised many of his readers for they included such everyday plants as Red Dead-nettle and Annual Meadow-grass, as well as Red Data Book species like Corncockle, Fingered Speedwell and Field Cow-wheat. By contrast, he found few examples of the opposite tendency (though he thought that Rhododendron might be much maligned in this respect). Our native flora, in Webb's view, is smaller than we think.

Webb's main contribution, apart from encouraging botanists to think about the matter, was to set out what native plants do and don't do. The most conclusive evidence of native origin is in the fossil record between the last glaciation and the first extensive agriculture in the Neolithic period (though even that is not absolutely conclusive – the late glacial flora included Sanfoin, for instance, but it died out only to be reintroduced as a fodder crop in the seventeenth century). Pollen analysis from peat deposits has produced a long list of such species, though most of them are plants which no one has ever supposed to be anything but *carte blanche* natives. Absence from the fossil record, however, does not prove that a plant is not native – its pollen may be rare, or impossible to identify. More reliable evidence can be found in a plant's behaviour. Native plants do not, as a rule, spread suddenly and rapidly all over the place, like Canadian Pondweed or Oxford

Ragwort did, nor do they normally shrink to vanishing point just as suddenly, as did Cornflower or London Rocket. A native plant should be more or less in balance with its wild environment. Webb was suspicious of any plant which did not seem to have a natural habitat (like Red Dead-nettle), or whose distribution is unlike that of any other native species. Another indication of exotic origin is the ease with which a plant can become naturalized. The Fly Honeysuckle (*Lonicera xylosteum*) is traditionally considered native near Arundel, W Sussex, where it has been known since 1801. But Fly Honeysuckle is a bird-sown plant which 'escapes' from gardens very readily. Moreover it has been in cultivation since at least the seventeenth century. Though you could make a case for a native origin on the grounds that its Sussex sites are old woods on the chalk similar to those in France where it *is* native, Webb was unconvinced.

Native species should, in theory, show a degree of genetic diversity. Though there was little available evidence at the time, recent DNA analysis has shed further light on species which, until recently, were regarded as only doubtfully native. The Plymouth Pear, for example, has different genotypes in Devon and Cornwall, which strengthens its claim to be an ancient native tree in the South-west. Meadow Clary shows a cline of genetic variation across the country which is fairly convincing evidence of nativeness, and contradicts the story that it was introduced from cultivation by the Romans. Native species should also be capable of reproducing by seed, and suspicion alights on plants that subsist only vegetatively or occur only in dense patches. It seems unlikely that any plant could survive 12 000 years without some exchange of genes, with the possible exception of apomictic plants which produce seed wholly female in origin, without fertilization. Webb's criteria may not always provide a conclusive answer to which species are native to Britain and which are not, but together they point to qualities we should expect to find in a genuine native plant. Of course it all requires a cool head. In the remainder of this chapter I introduce some rare plants which have been or are regarded as native in Britain but about which there are considerable doubts if one looks at them in an unbiased Webbian light.

In 'the grey zone'

Matthiola incana or Hoary Stock does not look like a native wild flower, and in most of the places where it has gone wild, it isn't one. The exquisitely scented purple or white flowers have been cultivated in gardens for centuries as ancestors of gilliflowers and stocks. Even plants found in wild places on sea cliffs and shingle banks often reveal, on closer inspection, the large or double flowers, the Neopolitan ice-cream colours, or the inflated pods of garden varieties. Nor does their location near resorts, like Brighton,

Ramsgate or Bournemouth, allow much doubt about where the plants escaped from. Hoary Stock was very popular in seaside gardens (see picture on p. 164). However, botanical opinion is more divided about the more remote cliff localities at Nash Point, Glamorgan, and the chalk cliffs of the Isle of Wight, where, as Geoffrey Grigson noted, the Stock has 'taken refuge in the warmth like an old lady'.[75] On these sheer cliffs, the flowers are a more constant purple-pink – forming a vivid contrast to the white chalk, like plum jam on a cream scone – and the plants have the consistent character of wild-type Hoary Stock. On one cliff on the Isle of Wight, it is accompanied by the only known wild-type Sweet Alyssum (*Lobularia maritima*) in Britain, growing in habitat typical of both species in their native France. Could they be native here at least? The *New Flora of the British Isles*[180] regards it as possibly native on sea-cliffs, a view which is shared by local naturalists. On the other hand, an authority on the Cruciferae, Tim Rich, points out that a relatively rapid reversion to a wild appearance is characteristic of the crucifers, the best-known example being the Wild Cabbage, a descendent of ancient vegetable plots.[162] Hence, the line between 'wild' and 'gone wild' is even more woolly in the Cruciferae than for most other families. But, whatever their origin, the well-established stocks and alyssums on the Isle of Wight are a unique part of the local flora, and are of *scientific* interest. The term Site of Special Scientific Interest could have been made for them.

The Purple Coltsfoot, *Homogyne alpina*, is one of our most puzzling mountain plants (see picture on p. 164). It was 'discovered' by Don around 1813, but not seen again until A.A. Slack refound it in 1951. *Homogyne* is confined to a single broad cliff ledge 'somewhere in Glen Clova', where its shiny green, kidney-shaped leaves cover a few square metres of turf. It flowers rather reluctantly, producing tufts of purplish petals pleasantly described by John Fisher as 'upstanding in a bunch but in some disarray, like the hairs of a misused paintbrush'.[56]

I once clambered up to the *Homogyne* ledge – something we are asked not to do nowadays – and took a look for myself. It would be hard to imagine a more convincingly natural habitat, with a varied alpine flora and a spine-tingling view all round (a stomach-churning one, too, when you look down). However, this plant may not be all it seems to be. It was one of the later discoveries of George Don, who knew Glen Clova almost as well as he knew his garden nursery. Don says he found it 'on rocks by the side of rivulets in the high mountains of Clova, as on a rock called Garry-barns'. So in more than one place, apparently, though, unfortunately, we no longer know which particular rock was called Garry-barns. This particular find of Don's has been questioned in the context of several other plants recorded only by him, like Alpine Buttercup (*Ranunculus alpestris*) on '3 or 4 rocks on the mountains of Clova', and *Potentilla tridentata* from 'the hill of Werron'. Slack's discovery seemed to vindicate Don (a second record of

Homogyne, from the Outer Hebrides in 1955 is now known to be fraudulent) and establish *Homogyne* as a native British plant.[158] There are indications, however, that *Homogyne* was planted, most likely by Don himself. Unlike most native British alpines, it does not occur in Scandinavia, the nearest wild plants being in central Europe and the Pyrenees. There are signs that it is not well adapted to Scottish conditions: by contrast with its rampant habit in the Alps, our *Homogyne* flowers sparingly, does not set viable seed, and manifestly shows no capacity to move off its ledge. It looks very much like a clone. No native mountain plant is quite so undemonstrative, and none is quite as rare either. It certainly is possible to introduce exotic plants to high Scottish mountains, as the establishment of the Pyrenean Columbine (*Aquilegia pyrenaica*) in Caenlochan Glen has shown, though they never do very well. With the tenacity of a coltsfoot, *Homogyne* seems able to hang on, but not to spread. Surely it could not possibly have survived in this way for thousands of years. One can imagine Don busy with his trowel on broad remote ledges in Glen Clova, creating a kind of hanging garden, a fashionably creative embellishment of nature, his own little bit of Glen Clova which no one else was likely to disturb. If so, *Homogyne* does credit to his skills as a nurseryman.

Could the beautiful Spring Snowflake (*Leucojum vernum*) be a native plant? Modern floras cautiously accept that it is 'possibly native' in two places in the west country, although the Snowflake has now been banished from the Red-list. The 'native' sites are both remote valleys and their Snowflakes are long established. In March 1997 Rosemary FitzGerald and I visited both places in order to make up our own minds. The Somerset one, near Stogumber, is a secluded glen where a stream cuts through a bank of red clay. The Snowflake grows in large patches under alders and willows on rich spongy silt (see picture on p. 165). The vegetation is impeccably natural, indeed quite distinguished, and includes both species of golden saxifrage and a notable abundance of Wood Speedwell (*Veronica montana*), one of the trademark species of ancient woodland. Perhaps the sprinkling of Snowdrops introduces a mote of doubt, but some regard even Snowdrops as respectable plants in this part of the world. This site was discovered in about 1910 by a Miss M.A. Hellard, and has been much visited since. Today it is celebrated by an annual 'Snowflake Party' thrown by the proprietor. Spring Snowflake is a difficult plant to time right. On our visit, a chilly day in late March, it was already nearly over. Unlike Snowdrops, which retain their flowers for ages, Spring Snowflakes quickly balloon into an inflated cup, like a bridal gown, and then sag into spent, crumpled petals, like party decorations the morning after. The first flowers have an almost alpine freshness on their short stems and still unfolding leaves, but the last are lost in the usual tight leafy mass of a bulb plant. Frankly it was a bit disappointing. We were told that it had increased slightly during the past few years, despite erosion and the occasional raid by a rogue gardener.

A mile upstream lies the clue to its origins. Above a steep stream bank is an old wall that once enclosed a garden. The slope below has the appearance of a garden gone wild, with an abundance of daffodils, two kinds of snowdrops, and, surprise, surprise, another large patch of Spring Snowflake. It is not difficult to imagine bulbs entering the stream through bank erosion, or the scrapings of a pheasant or badger, and washing down in the current to be washed up later with a convenient bedding of debris and river silt. After all, this valley is in no way unusual – in the Quantocks of Somerset, wooded valleys and fast flowing streams occur in plenty, and if the Snowflake really is a native species, surely it would have survived elsewhere? And yet it is confined to a mile or so of stream below an old garden. We reluctantly decided it was an introduction, though a well-established one.

The Dorset site has more convincing credentials. There the flowers grow among old coppiced alders by a stream which cuts through soft Greensand. The site seems to have changed remarkably little since the Spring Snowflake was discovered here by a local botanist, Mansell-Pleydell, in 1866.[118] Then, as now, you pass Snowflakes growing in tufts and patches attractively mixed with golden saxifrage over about quarter of a mile of streambank. Then, quite suddenly, as Mansell describes, you reach a band of impervious clay, now much trodden by cattle, and once you hit the clay there are no more Snowflakes: it is an unbridgeable barrier even after 130

Above: The wild form Hoary Stock (*Matthiola incana*) growing on the chalk cliffs of the Isle of Wight.
[Bob Gibbons]

Left: Purple Coltsfoot (*Homogyne alpina*). Common in the Alps but known in Britain from a single cliff ledge.
[Bob Gibbons]

Spring Snowflake (*Leucojum vernum*) photographed at its traditionally native Somerset site. [Peter Wakely/English Nature]

years. The only nearby settlement is a couple of ruined labourer's cottages on a hill and a mansion house with a large garden about a kilometre away. It has been shown that these Dorset plants differ from the type normally grown in gardens, and are similar to unquestionably wild plants in France with greenish marks at the tips of their petals.[9] Our conclusion was that the Dorset population could well be native. If so, the Spring Snowflake would rank as one of our rarest plants, with a total of about 2000 bulbs covering an area no larger than a tennis court.

It would be nice to claim Hartwort (*Tordylium maximum*) as a native British species, for it is a distinctive looking umbellifer, whose long calyx teeth give its bunched flowers a spiky look. Even more unmistakable are its large, hairy, heart-shaped seeds. Stace regards it tolerantly as 'possibly native', and the conservation world has embraced the plant to its bosom and Red-listed it. Unfortunately, Hartwort's credentials seem rather weak. Its native range lies far to the south, in the Iberian peninsula, stretching eastwards to Turkey and the Crimea. Its British habitats lay on waste ground or rough scrubby grassland near the Thames. In the seventeenth century, Ray's helper, the apothecary Samuel Doody found it in several places 'about Isleworth', where it persisted for at least 150 years, and also turned up at Twickenham.[97] In recent times Hartwort has been confined to the Tilbury area in Essex, where it was first discovered in 1875. Originally it grew 'in considerable abundance' by the banks of ditches. By 1968, however, the Hartwort had almost disappeared through a combination of scrub invasion, regular mowing and general tidying up.[20] As a monocarpic species that flowers only once and then dies, Hartwort thrives on untidiness and soil disturbance. Fortunately seed from Tilbury has been grown on at Kew and Cambridge, and, thanks to restocking, Hartwort still occurs in its original site. All the evidence suggests a naturalized plant, but before consigning its native status to *auld lang syne* it is as well to remember that the all-but-vanished natural alluvial clays and gravel ridges of the Thames estuary were once quite rich in rare plants. The Berry Catchfly (*Cucubalus baccifer*), for example, once grew 'in considerable abundance' on the Isle of Dogs until it was destroyed by dock development.[97] In the context of the present-day suburban sprawl, perhaps it should not matter very much whether Hartwort is native or not. At least it predates most of its surroundings, and deserves protection on that account alone.

Authorities differ about the status of Labrador Tea (*Ledum palustre*), that remarkable scented shrub of peat-mosses. Traditionally it is regarded as 'possibly native' at the 'Bridge of Allan', Stirling, the botanists' soubriquet for Flanders Moss, but naturalized everywhere else. This view does not really stand up. In nearly *all* its stations Labrador Tea grows about as far away from houses and gardens as it is possible to be. I remember being taken to see it at Flanders Moss (as it happened by Mark Young, the illustrious author of *The Natural History of Moths* in this series). I had half expected to be shown a few bushes in some peripheral ditch or about the ruins of an abandoned croft. But no, we squelched on over the moss, through cotton-grass and heather to wetter ground with hummocks and pools of Sphagnum moss and swarms of biting insects. And there it was: 'strange hummocks on a remote moor', as the late A.R. Clapham once described it. Perhaps they are a common sight in Labrador, but here, on the flat wet mire under a big Scottish sky they looked interesting and unlikely. Its stations in England sound very similar. On Scaleby Moss, Cumbria, where it was found in 1948, some 20 bushes survive, despite periodic fires.[78] High up on Bleaklow, Derbyshire, there are a few more, doing a good job of stabilizing the loose peat. There are also Labrador Tea bushes on the remote moors of Yorkshire and Lancashire. (We can ignore records from southern England where it had clearly escaped from a nursery.)

There are clues to the origin of our Labrador Tea-plants, but they point in opposite directions (see picture on p. 168). The British plants are all subspecies *groenlandicum*, which is indigenous to Western Greenland and North America (including Labrador), and not, as you would expect, subspecies *palustre* of nearby Scandinavia. Recent survey work suggests that the English colonies, at least, are each a clone established from a single bush. This decidedly points to an introduction. On the other hand, Hugh McAllister found that the chromosome number of British plants is 52, double that of plants in Greenland and Canada.[78] This suggests they have been genetically isolated in Britain for a long time. The colonies, though small, look firmly established. How did they reach these remote places? Human agency seems very unlikely. The most plausible theory is that they were brought there by migrant birds, roosting high on the peat moors on their passage south. But I have yet to hear which birds migrate to Britain from Labrador or western Greenland. Geese? Wheatears? There is, of course, a huge difference in conservation value between introduction by natural means and by human agency. All in all, it is a rather mysterious plant.

Trees are often overlooked by county floras. The distribution of what is arguably our most magnificent native tree, Black Poplar (*Populus nigra* ssp. betulifolia), was poorly known until the 1980s, partly through confusion with planted hybrid Poplars, partly because few field botanists, except the late Edgar Milne-Redhead, bothered to search for it. The same was true of two long-living native trees, the Large and Small-leaved Limes, at least until Donald Pigott produced his two classic monographs on the native limes.

Foresters knew them, palaeo-botanists reconstructed their role in the Stone Age landscape, but field botanists tended to pass them by, looking down not up, recording the scarce flowers among the roots without noticing the tree. Large-leaved Limes hundreds of years old have remained hidden and unknown until very recently – and some of them were in nature reserves! By the same token, long-naturalized trees and shrubs are seldom given the same attention in floras as naturalized flowers. Sweet Chestnut (*Castanea sativa*) and even Box (*Buxus sempervirens*) are often regarded as not quite respectable wild trees, though one of the oldest trees in the country is a Sweet Chestnut and ancient placenames commemorating Box are numerous. Walnut (*Juglans regia*) is generally regarded as a garden tree, though in the warm chalk coombes on the Chequers estate it regularly ripens and produces seedlings in the wild. Holm Oak (*Quercus ilex*) has a prominent role in the natural landscape of the Isle of Wight and is even accompanied by some dependent species, like the big mushroom *Amanita ovoidea*. Among all these oversized botanical Cinderellas, there is one in particular that deserves more attention than it normally receives. This is the Medlar (*Mespilus germanica*), a small, branchy tree with oblong leaves, white flowers similar to apple blossom and strange yellow-brown fruits that soften only when ripe. Though it is unlikely to be native in Britain – for the nearest indigenous Medlars are in the Balkans – it has grown in wild places for a very long time. Medlar was once cultivated in gardens and orchards for its beauty and unusual, piquant fruit – usually just a few trees in a corner, near the hedge. Like Victorian plums, Medlar fruit was notoriously laxative (to Ray, the flesh of Medlars even looked like human excrement, though he insisted this was coincidence). Both Chaucer and Shakespeare made allusive jokes about it, which could not have been understood if the Medlar had not itself been a standing, or rather, a sitting, joke among our ancestors. It has a deeply Chaucerian old name, 'open-ers' (pronounced open-arse) – Medlar was a politer, French-derived name that came later – that suggests it was grown by the Saxons, and it might well have come to Britain on Roman galleys.

If the Medlar had stayed where it was, in the corner of King Alfred's cider orchard, it would be no more than a reminder of the funny things people ate in the olden days. However it was already naturalized at the time of the first English herbals. Gerard, for example, found it 'often-times in hedges among briars and brambles'. Hooker, writing in the early nineteenth century, noted that it grew convincingly wild in 'hedges and thickets', especially in Kent and Sussex. Wolley-Dod, author of a celebrated *Flora of Sussex*,[209] went further and regarded it as fully native in East Sussex. Unless it has been grossly under-recorded – which is possible – the Medlar has declined seriously since the sixteenth century. For example, in the whole of Hampshire, only four convincingly wild Medlar trees were recorded this century: one, at Pains Hill, Lockerley, was cut down in 1924; a copse on the bank of the River Hamble once had a 'well-grown bush flowering and fruiting freely amongst native trees and shrubs', though by 1988 it had grown moribund; and there are or were two more trees in 'the hedge of a lane' near Flexford Mill Cottage, Sway.[17] The best known wild Medlar is a hedgerow tree at Redhill, Surrey, first recorded by the philosopher John Stuart Mill in 1831; it was in wild country then, but is now surrounded by suburban houses and roads.

Above: The Medlar (*Mespilus germanica*) has grown wild in hedgerows and woods for at least four centuries, yet is largely passed over by naturalists. [Bob Gibbons]

Left: Labrador Tea (*Ledum palustre* ssp. *groenlandicum*), a mysterious low shrub of remote peat mosses and moors. [Bob Gibbons]

Wild Medlars differ from those still in cultivation, having more thorns and smaller fruits, implying that they have led a separate existence in the wild for a long time.[152] Individual trees have been known for a century or more, and often they have deepened their hold on a hedgerow by vigorous suckering. This may be one explanation for the relative abundance of Medlar trees in hedgerows in the Battle-Ashburnham district of Sussex. Medlars were also sometimes cultivated by grafting onto wild hawthorn. The Medlar would grow up in this pick-a-back way and so establish itself in the hedge. The practice was finnicky and needed skill, and perhaps survived only as long as there were local squires with a taste for Medlar jelly who employed a lot of gardeners and hedgers.

New colonists? A tale of Tongue Orchids

Orchids, with their light spore-like seeds, ought to be among the more efficient colonizers in the British flora. Within limits, some species probably are. The massed blooms of spotted and marsh orchids on industrial waste tips suggests that their seed is transported by air currents. Even some of the scarcer species seem to share this ability, like the occasional appearances of Lady Orchid well outside its normal range. In at least two cases, wind-

Tongue Orchids, left *Serapias lingua* and right *Serapias cordigera*, possible colonizers of southern Britain in the advent of global warming. [Bob Gibbons]

blown seed may well have arrived from neighbouring countries. That would certainly account for the colony of Military Orchids in Suffolk, which look more like German plants than their congeners in Buckinghamshire. Similarly, the colony of Dense-flowered Orchid on the northern tip of the Isle of Man probably arrived there in a favourable wind from the Irish Republic. If so, it is surprising that new *species* of orchids are not reported more often, since there are many non-British species just on the other side of the Channel. The few European orchids reported from Britain before 1976 were 'casuals' of uncertain origin, like the plant of Short-spurred Fragrant Orchid (*Gymnadenia odoratissima*) found on limestone rocks on the Durham coast in 1912. More recently, a mysterious succession of new species have turned up in various places near the south coast and in the Channel Islands. First was a single plant of 'Bertoloni's Bee Orchid', *Ophrys bertolonii*, found by R.E. Webster in April 1976 on a dry, stony Dorset hillside facing the sea in an area noted for rare plants. A few weeks later the specimen had vanished, perhaps grazed by a rabbit. On the strength of that, this striking pink, blue and black insect mimic achieved an entry in Britain's Red Data Book (1977),[142] but word got around in botanical circles that someone had scattered seeds 'of various Mediterranean species of orchid' in that part of the county. An alternative story was that the orchid had been planted. However

D.M. Turner Ettlinger, shown the site by Webster the following year, found no signs of disturbance.[194] A small non-flowering rosette of an *Ophrys* was still there, but it may have been Early Spider Orchid which is common in the area. By 1978 it had gone. *Ophrys bertolonii* is not a likely colonist, being confined to the Mediterranean coast, and local even there. But in the light of what follows, people may have jumped to conclusions too soon.

The Tongue-orchids (*Serapias*) are among the most exotic-looking plants in Europe, with their lurid colours and broad protruding lip. It does not need much imagination to see them pulling a face at you. Three species are found on the opposite coast of the Channel, most commonly in Brittany. The first Tongue-orchid to be found in Britain was, however, a Mediterranean species, *Serapias neglecta*, in 'a cornfield' on the Isle of Wight in 1918, where it was promptly dug up. It may have arrived in imported grain. A much more convincing example of apparently natural colonization was the Small-flowered Tongue-orchid, *Serapias parviflora*, found in May 1989 by Paul Cobbing while erecting a fence near the sea in south-east Cornwall. The spot is a remote one, in short, rabbit-grazed natural grassland and gorse scrub. Three flowering spikes appeared the following year, and five each in 1991 and 1992 (though two of them were unfortunately squashed by photographers).[117] Since then no more plants have been seen, but the site had by then become flattened and compacted by their kneeling admirers. Seedlings successfully germinated at Kew were planted at the site in 1993, but did not survive.

To the dismay of local botanists, *Serapias parviflora* was dismissed as another 'probable introduction'. However carefully done, planting is nearly always detectable; and those who know the site are convinced that the orchid established naturally from wind-blown seed. The species has, after all, been expanding its range in Europe, and has colonized parts of Brittany in the past twenty years. Other suggested alternatives – a migrating bird, or stuck to a horse and cart delivering military supplies from Plymouth – seem unlikely.

I myself had a small part to play in the discovery of a second species of *Serapias*, on Guernsey in May 1992. A party of us were spending a holiday on Sark, but one member of it, John Finnie, had decided to stay behind for a night in order to attend Mass. Sensibly he decided to spend the delay botanizing, and took a bus to the western side of the island to see the wet meadows where the Loose-flowered Orchid (*Orchis laxiflora*) grows. It was in this area, on dry, natural grassland at the edge of a golf course, that he found a different and most unexpected orchid. Making a very passable sketch of it on the back of his bus timetable, John returned to St Peter Port, and the next day rejoined us on Sark. It was at this point that I enjoyed my five seconds-worth of botanical glory when John showed me his sketch, which I identified as a *Serapias*, possibly *Serapias lingua*. On our return, a week later, some of the party revisited the site, where the orchid was still in flower, and photographed it. Later, David McClintock confirmed that the species was indeed Tongue Orchid, *Serapias lingua*. Like *Serapias parviflora*, it too is a Mediterranean orchid but also occurs on the Atlantic coast of France as far north as Brittany, and so its natural establishment in Guernsey seems not at all unlikely. As far as I know, the plant has not reappeared, but it was a large, healthy specimen, and, again, it was found in an out-of-the-way spot with no signs of planting. In 1998, *Serapias lingua* turned up in a grass ley in Devon.

In June 1996, a third *Serapias* appeared, this time in an old chalk pit in Kent managed as a nature reserve by the Kent Wildlife Trust. This was *S. cordigera*, the Heart-flowered Tongue-orchid, perhaps the most spectacular of the three, with its fat crimson furry labellum, exactly like a tongue after an indulgent night out. Two flowering specimens were found on a south-facing bank below a cliff, almost within sight of the French coast. Moreover it was accompanied by a fine collection of native orchids.[160] Again, the plants seem to have established naturally, and again there is an obvious source of seed on the Finistère coast of Brittany, where it is widespread on grassland, sandy heaths and dunes, and by the banks of streams.

These Tongue-orchids look very un-British, and akin to the animal-like orchids of the tropics, with their great stuck-out tongues the colour of raw meat. Given their potential significance as indicators of global warming, it is surprising that this extraordinary sequence of arrivals has not been more widely reported. Possibly they all incur a suspicion of botanical fraud in the wake of tales about *Ophrys bertolonii*. These Tongue-orchids represent something new or overlooked in the British flora, neither settled 'native' species, nor 'introductions', but potential colonists, perhaps a promise of what the new century may hold in store. They are dramatic, and perhaps rather appropriate, harbingers of change as they pull faces at us from their landing places.

Island living: the strange history of the Wild Paeony

The islet of Steep Holm lies in the Bristol Channel, roughly halfway between Weston-super-Mare and Barry Island. Though it is surrounded by cliff and measures only 700 metres from one end to the other, Steep Holm has yet seen a remarkable amount of human activity. Among the rocks and weeds lie what is left of a medieval priory, Victorian gun batteries, various roofless huts and sundry World War bric-a-brac. For many years after the last war, the island looked like a bomb site, strewn with rubble, iron girders, Nissen huts, broken glass and steel cable. It has since been cleaned up and is now looked after by the Kenneth Allsop Trust.

With all this disturbance, you would expect to find a few naturalized weeds and garden escapes. What is remarkable about Steep Holm is that the escapes have taken over altogether. The interior of the island is a mass of neck-high Alexanders (*Smyrnium olusatrum*), euphemistically called a meadow. It was formerly grown in gardens as a pot-herb and for Lenten 'greens', and was probably once cultivated in the priory garden. Among other herbs still found on sheltered sunny spots on the island are Henbane (once used as a sedative), Caper Spurge (a laxative), Coriander (spice), Greater Celandine (eyedrops) and monstrous Wild Leeks (an anti-scorbutic) with heads like frizzy purple tennis balls. These Mediterranean plants have survived on Steep Holm for a very long time; the island's Alexanders merited a mention in the *New Herball* of William Turner, written in the middle of the sixteenth century. Their probable origin lies in the early Middle Ages, when a community of monks established a priory on the island between 1166 and 1260. Such a colony would have to be self-sufficient, and no doubt the monks' garden of physic was well stocked. That some of their herbs should have survived so long after their departure may be due to the mild, almost frost-free climate of Steep Holm, but perhaps also because the island's natural flora is so impoverished that plant competition is significantly lower. Even the island's tree, the Sycamore, is an invader.

Was Steep Holm's most famous flower, the Wild Paeony (*Paeonia mascula*) also brought there by the monks? If so, it was probably not the plush crimson flowers they valued so much but the big black seeds. Paeony seeds, though poisonous, could be used with care as a 'hot' spice. They could also be worn on a necklace to charm away evil spirits. Paeony could help those oppressed by nightmares. The root, chopped up fine and brewed in water and wine was useful in the treatment of rheumatism. One can see why they might have grown it. The only reason for doubting it is that, although the island has attracted visitors for hundreds of years, none of them noticed the Wild Paeony until 1803. 'Did some romantic mystifier plant *Paeonia mascula* on the island deliberately, as a hoax to catch all the botanists of England?' wondered the suspicious Geoffrey Grigson,[75] aware that Gerard had once planted a Paeony and tried to pass it off as a native. Or did the visitors miss it because they came in summer, after the Wild Paeony had finished flowering? Probably the latter. Local fishermen assured its discoverer, Francis Bowcher Wright, that the plant had been there for at least sixty or seventy years – as long as any of them remembered.

Wild Paeony (*Paeonia mascula*) has grown on Steep Holm since at least 1803, and was introduced to Flat Holm in the 1980s. [Bob Gibbons]

The contrast between the exotic flower and its 'abrupt and high' island setting inspired at least one poet, William

Lisle Bowles (1762–1850), sonneteer, vicar and mentor of Coleridge.

> And desolate, and cold, and bleak, uplifts
> Its barren brow – barren, but on its steep
> One native flower is seen, the peony;
> One flower, which smiles in sunshine or in storm,
> There sits companionless, but yet not sad.

The flower, you see, represented Virtue, smiling, hidden, on her 'Rock of Care'. But in truth the Paeony has not had much to smile about. Its status has always been somewhat precarious, clinging to the top of the cliff at the eastern end of the island, especially as its admirers were often keen to return home with a souvenir root or two, or a pocket full of seeds. In time, attitudes became more protectionist. In 1907, a party was 'pledged before landing not to touch it, *but to think of it*'.[159] Later parties were only allowed to land by permission of the tenant, Harry Cox, who exercised his proprietorial rights over the plant. On one occasion, having allowed visitors to view the Paeony, he 'afterwards alleged that someone had picked a bloom and made a tremendous fuss before escorting the subdued party back to the boat'. On the other hand, he would sometimes provide seeds or even whole plants to more favoured guests.

By the 1950s, there were only about a dozen plants left. For a few years, Harry Savory of the Steep Holm Trust made the Paeony his special concern, and cleared away encroaching bushes from around the main patch, known as 'Peony Glen'. As a result of this, the plant put on a brave annual show of 50 or so flowers in the mid-1960s. But disaster struck in 1969 when the plants became infected by mildew and their roots began to rot. A man from the Min of Ag was brought over to inspect them. It seemed to him that droppings from starlings roosting in the sycamores had contaminated the vegetation. But almost certainly the root cause was increased shading, producing coolness and damp in place of the full strength sunshine craved by this Mediterranean flower.

From this point onwards, the story of Steep Holm's Paeony turns full circle back to the days when it may have bloomed in beds in the priory garden. Fearing the possible loss of the wild colony, the Trust decided to dig up the healthiest remaining plant and transferred it to a drier and more accessible position by the visitor centre. Since then, several more plants have been cultivated by the Trust's Warden and planted out in a walled enclosure, within cylindrical wire cages. Visitors can buy Paeony seeds, harvested from these captive plants, at the Trust's shop, five per packet. It is all well-meaning and visitor-friendly, but the experience is not unlike a visit to a garden centre. Meanwhile, the state of the wild colony is reported to be more precarious than ever: in 1989, Ian Taylor and Rosemary FitzGerald could find only a single plant. Thanks to a regular diet of gull and starling droppings, the island's vegetation has grown much thicker. What is now called the Sycamore Wood shades what was once the sunny Paeony Glen. Although attempts had been made to cut back some of the growth, the investigators considered they had been 'rather haphazard and far from successful'. A crash programme of felling and scything was recommended, followed by the planting of locally cultivated plants onto rocky ground nearby. If this work succeeds, the Paeony may once again become the Wild Paeony. Let us hope it does.

Fortunately the Paeonies in their wire cages or bespattered beneath the starlings' roost are not the only ones to have survived. At least three nineteenth-century visitors transplanted the Paeony to gardens and other sites that took their fancy. One such site was Blaise Castle where Paeonies flowered for many years, until eventually being 'exterminated by the public' according to J.W. White.[203] Another was at a place called The Rocks, near Bath, where they were seen by C.C. Babington. Interestingly, their planted origins were eventually forgotten, and both colonies came to be regarded as native. They have long since vanished, but a third substantial colony of Steep Holm Paeonies survives in a wild garden near Weston-super-Mare, where they are said to have been planted by the Bishop of Bath and Wells some time between 1824 and 1835. If so, these plants date back almost to the time of discovery of the Wild Paeony, and over 150 years they have become thoroughly naturalized. Like their island siblings, this colony suffers from over-shading and competition from brambles and ivy, and until recently seldom flowered. However the owners appreciate their value, and have maintained the Paeony patch by clearing brambles and ivy. Some 26 plants were seen by Taylor in 1989, plus a further 4 transplanted from 'vulnerable locations near the path' to a site nearer to the house.

Planted flowers are normally part of the garden heritage and have little or no nature conservation significance. In the context of the story of the Paeonies of Steep Holm, however, this naturalized colony is a God-send, as Ian Taylor (pers. comm.) describes:

> The woods in which *Paeonia* is naturalised are now managed rather like a wild garden ... This population is of considerable importance: it has a recorded history almost as old as the Steep Holm colony and the plants certainly represent a gene pool which has been subjected to the selective forces of the British climate for a long period. Also, considering the precarious nature of the Steep Holm population, this site may provide an important back-up stock of original material.

A last twist in the tale, then. While the island colony seems to be almost extinct, the genetic stock which may prove its salvation survives, thanks to the much-reviled Victorian collectors!

Chapter 9

Lost flowers

Nationwide extinction is a very rare event among our native flora. Although many species are Red-listed as endangered or even 'critically endangered', less than one per cent of our flora has actually died out since botanical records began. Small populations of plants can be surprisingly stable, and some rare species may have been rare throughout most of their existence. Native British wild flowers are almost by definition, hardy, tenacious plants which either survived the Ice Age or colonized the land afterwards, and have held on ever since. Of course many of them have become much less common in recent years through habitat loss, and the rarer you are, the more vulnerable you become. Very rare flowers are *always* potentially at risk. But native plants are good at surviving. Though *local* extinctions happen all the time, *national* extinctions are unusual. Some flowers that died out were not native in the first place. For a plant to die out nationally some form of wholesale disaster must have overtaken it. The world has lost very few British plants. All but two of our score or so of extinct flowers can still be found on the European mainland, in some cases on the opposite shore of the Channel. The two that are globally extinct in the wild were endemic species. One of them, the Interrupted Brome, *Bromus interruptus*, survives in cultivation. It probably evolved by way of genetic mutation. The other, Broad-leaved Centaury, *Centaurium latifolium*, died out before genetics were discovered, and may not have been a fully formed species at all. There are also a few apparently extinct microspecies, like the hawkweed, *Hieracium subramosum*, known only from a single collection on the Fife coast in 1876, or the endemic 'Pugsley Hawkweed', *H. hethlandiae*, whose site was destroyed by the Sullom Voe oil terminal, but which survives in cultivation.

How is extinction observed? It is hard to prove a negative. Since you are rarely present at the actual moment when the last sterile flower fades away, extinction is usually concluded after the event, sometimes long afterwards.

On occasion extinction takes everybody by surprise, for instance when someone at the Biological Record Centre noticed there were no recent records of Thorow-wax (*Bupleurum rotundifolium*) or Swine's Succory (*Arnoseris minima*). Sometimes a plant disappeared on a supposedly safe site, like the Stinking Hawk's-beard (*Crepis foetida*) at Dungeness. The actual cause of extinction may have little or nothing to do with the reasons which caused the plant to decline. It was just rotten luck that, for example, some forester chose one of the last stands of the Summer Lady's-tresses to plant some more Christmas trees, or that Davall's Sedge (*Carex davalliana*) happened to occupy prime building land with a fine view of Bath, or that the lonely sand-pit where the last Narrow-leaved Cudweeds (*Filago gallica*) grew became a convenient place for tipping noxious waste from a mushroom farm! It is interesting that no native plant has died out completely since the conservation industry gathered steam in the 1980s; the rarest species receive quite a lot of attention nowadays. I suspect that the next extinct plant, if there is one, will come from the ranks of the rare but still relatively widespread species. Even relatively common plants could conceivably become threatened through climate change (like snow-patch plants in Scotland), or from disease (like Wych Elm).

Curiously enough, we have fewer 'lost plants' today than 25 years ago.[140] No fewer than seven species were rediscovered in the wild after being declared extinct (*Orchis militaris*, *Schoenus ferrugineus*, *Senecio paludosus, Atriplex pedunculata, Polygonum maritimum, Spergularia bocconii* and *Bupleurum falcatum*); most famously the Military Orchid which went missing for 40 years until J.E. Lousley refound it at his famous picnic (see Chapter 11). Some of these refound species probably shift their ground from time to time, like Sea Knotgrass (*Polygonum maritimum*), or occur in poorly surveyed areas, like Brown Bog-rush (*Schoenus ferrugineus*, and see Chapter 13). At least one of them, Fen Ragwort, probably germinated from buried seed, brought to the light by ditch maintenance works. Plants like Cottonweed (*Otanthus maritimus*), whose seed is transported by ocean currents, or Stinking Hawks-beard, whose light parachute-seeds could perhaps float over from Normandy on a southerly thermal, might re-establish themselves one day without any help from us. When, therefore, we say a species is nationally extinct, we really mean it is nationally extinct *as far as we know*. Judging from the past half-century, the future will hold plenty of surprises.

National extinction is a fairly rare event; it is not, therefore, a *likely* fate. As far as we know, we have the same native flora today as existed at the time of King Canute, and, probably, of Boadicea. The pollen and other preserved material from Saxon and Roman times consists almost entirely of the same species we find today. A few exotic flowers like Alpine Poppy (*Papaver alpinum*) or Polar Willow (*Salix polaris*) occur in late Ice Age deposits, but they probably died out naturally many thousands of years ago as the climate grew warmer. Our island flora is remarkably stable: as we have seen, even the flowers of the late Ice Age are still with us in the coldest or barest places in Scotland, Wales and northern England. There is one major exception, and this is the flora of arable crop fields, which as a class are much more extinction-prone than other plants since their habitat is entirely at the mercy of the farmer, and modern agricultural methods leave no room for rare weeds. However, flowers that are confined to crop fields or ephemeral patches of open waste land are not likely to be strictly native in the first place – though some of them may have been with us for as long as crops have been grown. Crop weeds are within the human domain – what they do depends on what humans do. Some crop weeds are as near extinct as makes no difference. In the distant past, cornfields near Swaffham, Norfolk, were gay with wild Larkspurs (*Consolida ajacis*) and Violet Horned-poppies (*Roemeria hybrida*). Today your best chance of finding the Larkspur is to search gardens or allotments on sandy soil, while the Violet Horned-poppy is extinct even as a 'casual'; it was last seen, as far as I know, on a chicken-run near Lakenheath in 1957.[179] The well-known Corncockle grows mainly where kind people have sown it (see Chapter 8). The Bur-parsleys (*Caucalis platycarpos* and *Turgenia latifolia*) so familiar to Johnson or Ray, have effectively departed the scene, as has Branched Broomrape (*Orobanche ramosa*), which died out when farmers stopped growing hemp. However their capricious appearances and disappearances are in a different category from the lost traditionally native wild flowers of natural habitats to which I devote this chapter.

Boyd's Pearlwort (*Sagina boydii*), drawn here by Stella Ross-Craig, is still available from alpine nurseries but has been found only once in the wild.

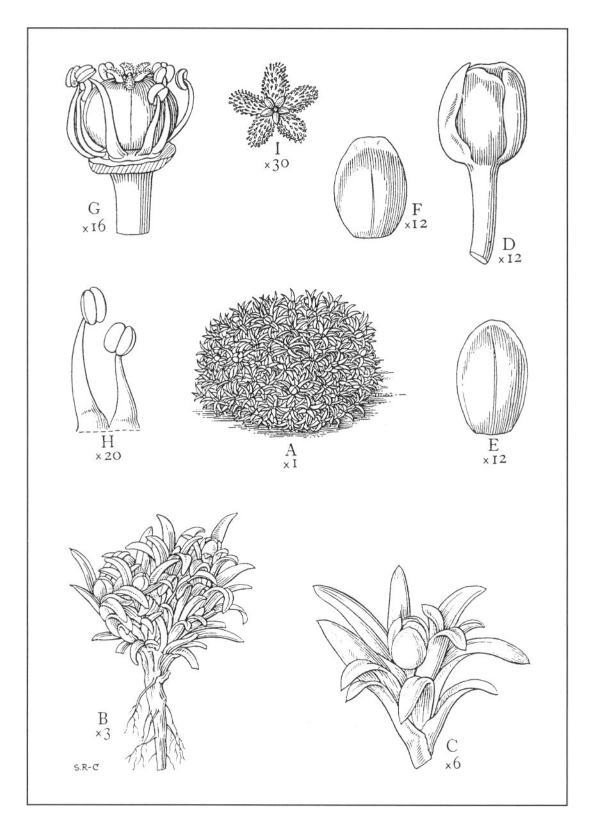

One plant which is almost impossible to categorize, is the mysterious Boyd's Pearlwort, *Sagina boydii*. This little cushion plant, more like a moss than a flower, was collected in 1878 by the eponymous W.B. Boyd, supposedly from Ben Avon in the Cairngorms, and described as new to science by F. Buchanan White. It has never been seen wild again, but survives in cultivation. Boyd had certainly found it. The question was where. Druce, who interrogated the forgetful Boyd on this point, related that 'he does not remember gathering it, but found it in his potting shed among other specimens he brought back from Braemar, and his impression is that he had gathered it on Ben A'an'.[45] Unfortunately, there were also some living plants from Switzerland lying about in Boyd's potting shed, and so there is a chance that they might have come from that quarter, though 'in later years his memory rather crystallised upon Ben A'an'. Ben Avon is a huge and remote hill, and Boyd's Pearlwort a tiny and insignificant plant. If Boyd did find it thereabouts, then the plant is almost certainly still there. If not, it is probably not a British plant at all. Either way, we can eliminate it from our list of extinct British flowers.

The rediscovery of an 'extinct' plant

National extinction is not the prerogative of the beautiful. Few other than the most devoted of botanists would have noted the passing of the Stalked Sea-purslane or *Atriplex* (formerly *Halimione*) *pedunculata*. You can find its portrait on Plate 72 of Keble Martin's *Concise British Flora*, a pallid little thing, not unlike the common Sea-purslane (*Atriplex portulacoides*) apart from its stalked fruit-heads. Keble noted that it was 'very rare ... in salt marshes, E England'. So rare, in fact, that it had not been seen at all since the 1930s, which might explain why his specimen looks as if it had been kept in a drawer.

Since the Stalked Sea-purslane was obviously extinct, nobody bothered to search for it. By 1987 perhaps no one alive had ever seen one in Britain. Its rediscovery by a sea-wall near Shoeburyness, Essex therefore came as a great surprise, especially as it had never been an Essex plant: the last known colonies were in East Kent, The Wash and Suffolk. On 29 September 1987, Simon Leach was near the end of a long hot day in the field, surveying coastal grazing marshland for the NCC with a colleague, Shaun Wolfe-Murphy. 'Both of us were tired and *very* thirsty,' he told me. 'The farmer told us we were close to the driest spot in Britain.' They reached the last ditch, a promising looking one, full of Beaked Tasselweed (*Ruppia maritima*), with the very local Sea Barley (*Hordeum marinum*) beckoning from the far bank. But it was just another ditch, one of dozens they had already looked at that summer.

> Nothing – but *nothing* – could have prepared us for what we saw. Two minutes later I was crouching in this tiny patch of salt-marsh hyperventilating, while all about me were hundreds of plants that looked like nothing I had ever seen before. With their grey leaves and 'mealy' appearance they had the 'jizz' of a small orache or sea-purslane, but the fruits – with pods like a Shepherd's-purse – were on long *stalks*. What on earth was it? After a while I recalled a little-known plant called Stalked Sea-purslane. Just the *name* was enough to convince me. Once Shaun had calmed me down a bit, we took out the *Excursion Flora*. It keyed out right. I phoned Lynne Farrell (then head of the rare plants section in the NCC). 'Are you *sure*?' she asked, hardly able to believe it.

Three days later, Lynne visited the site. It was indeed *Atriplex peduncu-*

lata, and they counted 1,714 of them over about a hundred square metres.

Had the plant been there all the time, or could it have arrived recently, unannounced, perhaps in the crops of Brent Geese from Denmark? The discovery inspired people to search its old sites, but it has not yet appeared anywhere else. As Simon put it, 'we suspect this is a plant that can't be found if you deliberately look for it. You have to *stumble* on it'. Unfortunately the original colony has since dwindled to nothing, which tends to support the recent-arrival conjecture. The Stalked Sea-purslane survives only in two places where seed was deliberately sown to try to boost its numbers – and in one of those the introduced plants were almost immediately wiped out when well-meaning ornithologists decided to build a scrape for waders at that very spot! We shall have to wait and see whether it survives or becomes extinct again.

A 'doomwatch' of Britain's extinct wild flowers

The cross sign † precedes the year of the last record.

Cotton Deergrass, *Trichophorum alpinum* (= *Scirpus hudsonianus*) † *c.* 1813

If 'Cotton Deergrass' was really a native plant, a practised survivor of climatic change and the coming and going of habitats, then it was a uniquely unlucky one. No sooner was it discovered than it died out. It was also the first recorded wild flower extinction in Britain.

Cotton Deergrass is a book name. At the time of its discovery it was thought to be a small kind of cotton-grass, and named *Eriophorum alpinum*. Though it does look a little like one, with its attractive white, bristly heads, it is clearly a deergrass, with tufts of erect stems rising from a creeping rhizome rooted in moist peat. Later it was renamed *Scirpus hudsonianus*, named after Hudson's Bay, where it is common. Though a widespread plant in northern regions, Cotton Deergrass was confined in Britain to a single site, Restenneth Moss, near Forfar in Angus, where it was first found in 1791. Not, I suspect, altogether by coincidence, the site lay only a few miles from Dovehillock, the home of George Don, the noted alpine gardener. That year he was visited by a young medical student and keen botanist, Robert Brown, and the two went off botanizing together. One of the plants they found – out of a most remarkable list Brown later compiled under his modest cognomen of 'Jupiter Botanicus' – was the new species. The discovery was credited to them both, and a specimen sent to the Linnaean Society by Don's friend John Mackay. Sir J.E. Smith considered it an elegant plant with its 'extreme white', delicately textured seedheads, and a fine portrait by Sowerby graced the *English Botany* of 1796. No one seems to have doubted that it was a native plant. Why not, indeed? The botanical exploration of Scotland north of the Tay had barely begun.

At least one more botanist came to pay his respects to the new 'cotton-grass', courtesy of George Don, and he has left us with the most substantial account of it that we have. It was by appointment, at 6 o'clock in the morning, that Dr Patrick Neill of Edinburgh met Don at an inn in Forfar and was conducted to the moss near Restenneth Priory – about a four-mile round walk. There he 'had the great satisfaction of procuring a living patch of *Eriophorum alpinum*, and a number of fine specimens for drying'. He goes on to say that Restenneth Moss had been partially drained 'for the sake of a rich deposit of marl' (lime-rich clay) but at one end there was still some wild marshy ground with Fen-sedge and Cotton-grass, and of course, the rare *Eriophorum alpinum*, which grew on the drier or firmer parts of the Moss. Don remarked to him that in a few years the plant would disappear, and disappear it did. At that time marl was much in demand for farming improvements in the area, and to reach it the bog has first to be drained and then stripped of surface peat. The dredging of marl left a big hole in the ground that usually filled with water. Exactly when *Eriophorum alpinum* disappeared is uncertain – possibly by 1804. There are specimens of later date, but they may have been from gardens (for Patrick Neill had procured a *living* patch). At any rate, it had certainly gone by 1813.

Scirpus hudsonianus

Cotton Deer-grass (*Trichophorum alpinum*, captioned *Scirpus hudsonianus*). Less than twenty years separated its discovery from its extinction in Britain.

Surely there is something fishy about all this. Granted that it is biogeographically possible for 'Eriophorum alpinum' to occur in Britain, for it is a circumboreal plant found in North America, Siberia and Northern Europe, we also know that Don kept a well-stocked plant nursery at Dovehillock, and his name is associated with quite a number of dubious plant finds in Scotland. It was others, not Don, who claimed that the 'Cotton Deer-grass' was truly wild and published the discovery. Bad luck, too, that it grew in a place that Don knew was doomed, and not in a single one of the many other peat bogs in the area, some of which have survived. Much as one would like to believe in this attractive little plant, there are too many coincidences to inspire much confidence in a natural origin. Surely it was planted.

Northern or Arctic Bramble, *Rubus arcticus* † *c.* 1841

Rubus arcticus, a pretty pink-flowered mountain bramble, is a plant of the far north, especially the Scandinavian arctic, where they make a delicious red liqueur from its berries. In Britain, it is one of the 'doubtful' species, a ghostly plant whose existence rests on a few old specimens in herbariums, and the largely hearsay testimony of long-dead botanists. Yet there must be many who, when walking on the high hills of Atholl or Mull, dream of rediscovering it.

The first indication that *Rubus arcticus* occurred in Scotland came from the Revd Dr Patrick Walker, who found it 'in rocky mountainous parts of the Isle of Mull' in 1768. It is said to have been refound in the nineteenth century by J.T. Boswell Syme, specifically on Mull's Ben More. There were specimens labelled from the island in James Dickson's fascicle of pressed plants dated 1802; and the one still preserved in the Natural History Museum has been authenticated as *Rubus arcticus*. However, Mull is an unusually well-explored Hebridean island, and there has been no sign of *Rubus arcticus* during the past 150 years. The authors of the recent environmental survey of the island consider that 'in spite of these early records we must continue to regard the locality of this species as dubious, at least until its status in the British flora is clarified'.[90]

Another possible locality is on Beinn-y-Ghloe, above the headwaters of Glen Tilt in Atholl. James Sowerby was given a specimen from a Richard Cotton who had gathered it from near the summit. There are also unauthenticated records from Ben Lomond and Ben Lawers. Professor Hope considered it to be a Scottish species on the basis of specimens supplied to him by Adam Freer (though from where we do not know, since Hope's herbarium has perished). Finally the Botanical Society of Edinburgh reported a possible specimen, 'too imperfect to decide', collected by J. Robertson, from the head of Glen Tilt in 1841. If all these 'may be's' add up to a 'must be', then *Rubus arcticus* occurred wild in Britain in several widely separated places between 1768 and 1841.

Druce suggested that its finders had mistaken the dwarf alpine form of Water Avens (*Geum rivale*) for the Northern Bramble,[45] but it is unlikely that people of the calibre of Walker, Syme and Dickson would all have made the same mistake. Perhaps the plant was an irregular natural introduction, sprung from the droppings of some passing Fieldfare from Scandinavia. Less likely, *Rubus arcticus* might have been planted, not necessarily by its finders, but by one of the many nurserymen exploring the virgin hills of Scotland at the time. Or just possibly the plant is with us still, but only sporadically or in vegetative condition. If so, one day some lucky botanist will refind it.

Davall's Sedge, *Carex davalliana* † *c.* 1845

The suspicions I, perhaps ungallantly, raised about Don's 'Eriophorum alpinum' might at first sight seem to apply also to the mysterious sedge *Carex davalliana*. It too was confined to a single, apparently unexceptional, locality, and it too was destined for early oblivion. But the sedge has more plausible claims to be a native plant, firstly in the absence of a known commercial nurseryman to introduce it, and, more significantly in the broader context of its site at Lansdown, a south-facing limestone bank above the River Avon, near Bath. If one was hoping to find this European sedge in Britain, a warm, sunny position around a spring below a limestone hillside would be the very place. Furthermore, the Avon valley is rich in rare sun-loving native flowers, at least one of which, the extinct Hairy Spurge (see below, p. 190), was also confined to hillsides near Bath.

'Davall's Sedge' was first found in 1809 by a Mr Groult 'in a boggy place on the slope of a hill' crowned with firs. Groult showed the plant to Ernest Forster (he of *Luzula forsteri* fame), who identified it and, as was then the custom, sent specimens to Sir James Edward Smith in Piccadilly. Its portrait appeared in *English Botany* the same year, and the pressed plants eventually made their way into the British Museum herbarium where they still reside. The living sedge was much more vulnerable. The Lansdown property on which it grew was sold for housing and the land was drained and developed. When T.B. Flower visited the site in 1852, all traces of the bog, and the rare sedge it contained, had been destroyed. The site is now a suburb. The only addition to the story is two records of this species from Yorkshire, which may have been garden plants or mistakes.

In Europe, *Carex davalliana* is well known to be a species with peculiarly exacting requirements, so much so that it has lent its name to an entire plant community, the *Caricion davallianae* – which it often dominates. It grows in mossy, lime-rich fens gathered around seepages and springs.[25] It is vulnerable to drainage, and is becoming rare in lowland areas as more and more wet meadows are drained to grow crops. In Britain it seems possible that Davall's Sedge was a very rare native plant at the limit of its range, and that the Lansdown locality was its very last site. Several other British sedges, like *Carex depauperata* and *Carex norvegica*, are almost as rare with tiny, circumscribed populations. Unfortunately sedge pollen is notoriously difficult to assign to a particular species, and so there is no confirmation from pollen analysis that Davall's Sedge was here in the post-glacial period. Perhaps the only reason it was found at all was that the kind of educated people who congregated at Bath – physicians, antiquarians, art lovers, etc – were sometimes interested in botany, and so the woods and hillsides near the town were well explored. But if *Carex davalliana* really was a native plant, it is surprising that it has not turned up elsewhere. After all, we *do* have '*Caricion davallianae*' vegetation in Britain, for example in Upper Teesdale.[169] Unfortunately we seem to have the plant community without its defining species.

Broad-leaved Centaury, *Centaurium latifolium* † 1872

Unlike most of our lost flowers, this one is totally extinct, for it has been found nowhere else in the world. The question is whether it was really a genetic species at all. It was named by Druce, a champion 'splitter', mainly on the grounds of its distinctive ovate leaves and also on minor floral characters. Druce based his description on dried herbarium plants, not living material. He did not lack dead material however: there are so many herbarium sheets of *Centaurium latifolium* that over-collecting has been suggested as the cause of its downfall.

The first published record of the Broad-leaved Centaury as a distinct plant, though not yet as a species, is in 1803 (the earliest herbarium specimens predate this by a few years) – *'In arenosis maritimus prope Liverpool'*, that is, on sand by the sea near Liverpool, found by D. Bostock and D. Shepherd. The last is 'a few plants near Freshfield railway station on some sandy ground now enclosed' in about 1872. Did enclosure in some way eliminate the plant? Asparagus beds were being dug in the Freshfield dune-slacks at about this time, and there was also much disturbance from military activity. But whether collectors or asparaguses or soldiers were implicated in the extinction of one of our few endemic flowers is unknown.

As a globally extinct British plant, Broad-leaved Centaury deserves more attention that it has received. It might have been a genetic mutation, like *Bromus interruptus* (see p. 206), which produced a distinctive local population that bred true for a while. In 1976, Francis Ubsdell found a considerable amount of genetic interaction between the two species of centauries on the Lancashire dunes, Common Centaury, *Centaurium erythraea* and Sea Centaury, *C. littorale*.[195] Not only are these species themselves very variable, but there were also apparent hybrids, and some of these hybrids were fertile. An analysis of this swarm of centauries suggested that a new strain had formed through hybridization which had become genetically isolated from its parents through a change in the chromosome number, and was able to compete successfully with them. This is close to becoming a new species. In these circumstances, it is likely that short-lived genetic races are thrown up from time to time among the centauries of the Lancashire dunes, and 'Broad-leaved Centaury' might have been one of them. Perhaps, then, it was not a clearly separated species so much as a species in the making, an evolutionary *possibility*, that either failed after a time and was 'absorbed' back into the gene bank, or was destroyed.

Marsh Fleawort, *Tephroseris palustris (= Senecio palustris* or *S. congestus)* † 1899

The drainage of the Fens in the nineteenth century caused the temporary extinction of the Bittern, the permanent extinction of a great many invertebrates (most famously the Large Copper butterfly) – and just one flower, the Marsh Fleawort, a flower of pond margins and fen ditches. It was an attractive plant, with bunched masses of large yellow 'daisies' born on leafy stems up to a metre tall. It differed from its fenland 'twin', the Fen Ragwort, in its woolly stems and leaves, shorter, stouter habit, and, usually, earlier flowering. Our plant was subspecies *congesta*. It made its botanical debut early as 'Hoary Fleabane' in How's *Phytologia* of 1650, where it is recorded by the noted Irish botanist, Mr Heaton 'a stones cast from the East end of Shirley Pool ... in Yorkshire'. Shirley Pool still exists, but unfortunately minus its Fleabanes. John Ray also recorded 'Hoary Fleabane' during a hurried tour of Wales in 1662 – it was about the only plant he did record on that occasion, and grew 'near Aberavon, on the sandy meadows by the sea-side' – which seems to be somewhere between Carmarthen and Gwent. In the eighteenth and early nineteenth century, Marsh Fleawort seems to have been the commoner and more widespread of the two fenland 'ragworts'. It occurred in ditches near the coast from the Humber to the Suffolk coast, but most abundantly in the Norfolk Broads where the local marshmen knew it as 'Trumpets'. There was also an isolated site in West Sussex. Sir Joseph Banks (1789) was the first to notice that in some years the 'Trumpets' blew in vast abundance, but in others they were hard to find. This fluctuation may have been linked to ditch maintenance. Marsh Fleawort is a short-lived plant which needs open muddy spaces to establish its first-year leaf rosettes. Interestingly, it is an early colonizer of the bare mud of newly created polders in Holland. It is therefore an opportunist, and depends on periodic disturbance to maintain its numbers. But unlike some of its erstwhile neighbours, like Fen Violet or Fen Ragwort, the seeds of Marsh Fleawort are evidently short-lived. If so, then the flower would be very vulnerable to changes in land-use, including neglect. It was long in decline, disappearing from Sussex as early as 1725, from the fens of Cambridgeshire, Lincolnshire, and the Humber area by the early part of the nineteenth century, and from its bastion in the Norfolk Broads by the 1890s. The last records, both from Norfolk, are from Fleggburgh in 1898 and Dersingham in 1899. Drainage and agricultural change eliminated the Fleawort from many of its old sites, but it may be that neglect was the root cause of its extinction in the Broads. Charles Rothschild, who knew the Broads well, remarked that the level of maintenance in Broads marshland was lower at the turn of the century than it had been thirty years earlier, when the demand for marsh hay was greater. Possibly the decline of the Norfolk marshman, with his ditching tools and scythe, spelt doom for the cheerful sulphur yellow daisies he knew as Trumpets.

Black Pea, *Lathyrus niger* † *c.* 1900

The Black Pea is a straggling weak-stemmed vetch, with clusters of pretty magenta ('livid') flowers but without the tendrils or winged stems of so many common vetches. At one time the origin of a plant was judged by the wildness of its surroundings. At least two of the localities of Black Vetch –

the Den of Airlie in Angus and Killiecrankie in Perth – scored highly in this respect. Sir W.J. Hooker included it as a presumed native in his seminal *British Flora* of 1821. Having taken root there, the Pea flowered regularly in successive floras – Bentham and Hooker, CTW, Keble Martin, right up to the mid-1960s, when people began having second thoughts.

These belated doubts were due to the nature of records of *Lathyrus niger* and its occurrence elsewhere as a known garden escape. All but two of the 'native' Scottish sites consist of just a single record: it appeared, someone saw it, then it was gone. Of the two exceptions, it was first found at the Den of Airlie in the 1820s, on rocks near the river, and was last seen in 1844. At Killiecrankie, it appeared about 1850, and had disappeared by about 1900.

If Black Pea were a native plant, one would expect to find it in Sussex or Kent, not mid-Scotland, for it is a species of central and southern Europe. Its finder at the Den of Airlie was Thomas Drummond, a local nurseryman, who had taken over Don's old garden at Dovehillock. Above the Den stands the mansion house of the Ogilvies, with its extensive wild garden, and it does not need much imagination to see how a plant like Black Pea might climb over the garden wall or be established by a gardener at a favourite spot by the river. The Killiecrankie plants are also now thought to have originated from neighbouring gardens. There is little doubt now that Black Pea was no more than a garden escape – one of a large number of such plants in Scotland, and less successful than some.

Alpine Butterwort (*Pinguicula alpina*) grew on in a Ross-shire bog for nearly a century. It was last seen there in 1919. [Bob Gibbons]

Alpine Butterwort, *Pinguicula alpina* † 1919

Many of our extinct plants are ones that most people would step over or on without a second glance. *Otanthus* is an obvious exception, and Alpine Butterwort is another. It is very attractive, with crisper, neater leaves than the common Butterwort, and pert white flowers with yellow throats and short tapering spurs. It is a familiar plant to walkers in Scandinavia or the Alps, but in Britain, with the exception of a doubtful old record from Skye, Alpine Butterwort is known from only one out-of-the-way place, at Avoch on the Black Isle of Cromarty. Here it survived for just under a century.

Its story has been pieced together by Ursula Duncan.[48] In 1831, the Revd George Gordon picked an unknown carnivorous plant growing in great abundance in 'the bogs of Auchterflow' on the Rosehaugh estate in the parish of Avoch. At first it was thought to be a form of Pale Butterwort

(*Pinguicula lusitanica*), but its true identity was determined by Dr Hewitt Watson, and the species was added to the British list in 1836. Collectors converged on the Black Isle, digging up plants for their private herbaria, and perhaps a few more for their gardens. Ursula Duncan examined no fewer than 38 herbarium sheets from Avoch, representing an intensity of collecting she rightly castigates as 'scandalous'. The labels on these sheets vary – 'Rosehaugh Wood', 'the moor behind Rosehaugh house', 'bog of Shanggan' or 'between Munlochy Bay and Invergordon', but they all seem to be different ways of referring to the same place. By 1848 the Alpine Butterwort was already under threat, though not it seems from collecting so much as from drainage and agricultural improvements in the area. As a compromise, the proprietor, a Mr Fletcher, raised a wall around the plot of ground where the Alpine Butterwort grew. A description from 1882 mentions the 'walled-in enclosure of bog amid cornfields under high cultivation', which sounds ominous – for bogs in the middle of cornfields tend not to thrive. Soon seedling conifers were spreading over the site, which suggests it was drying out. Druce reported the subsequent extinction of the Alpine Butterwort by 1919, attributing the cause 'to seedling conifers drying the bog in which it formerly grew'.[44] It seems to have been an early example of the futility of putting walls or fences round endangered species.

If the Alpine Butterwort really was a native species, its confinement to that particular spot, apparently not in any other way exceptional, is hard to explain. Could seed have been blown, or carried across, from Norway? Or, more likely, did an unknown eighteenth-century proprietor indulge in a little 'wild gardening'? – quite a number of gardeners at Forfar or Edinburgh would have been able to supply it. The floras continue to regard it as probably native, but it seems to me unlikely that a native bog plant would have survived at one rather ordinary place on the Black Isle but nowhere else.

Jagged Chickweed, *Holosteum umbellatum* † 1930

The jaggedness of this small, undeniably dim annual flower lies in the petals, which bear irregular notches, like saw blades. Some books also call it 'Umbellate Chickweed', from the characteristic corona of tiny white or pinkish flowers, radiating from a point. I have to say I doubt whether anyone used either name. People who found plants like this used scientific names. *Holosteum*, then, was a rare flower associated with roof thatch and crumbling walls in East Anglia and Surrey. Surprisingly, Gerard seems to have known it, but the first particularized record is on the city walls of Norwich in 1765, where it persisted until 1887. Another site was on a wall 'leading from May Water Lane to Southgate Street' in Bury St Edmunds, and also on the thatch of houses near the railway station. Elsewhere in Suffolk it occurred for a while on walls at Eye, and 'sparingly' on the ruins of Hoxne

Abbey. In Surrey, where it was first found in 1905 and lingered until about 1930, it grew on the ruins of Newark Priory. Evidently *Holosteum* had a predilection for ancient stones.

So far this does not look like the record of a native plant. But *Holosteum* is more interesting than that. These walls and roofs were the places where botanists spotted it, but, though no one knew it at the time, its aboriginal habitat had been on the heaths and lake margins of the Breckland in Norfolk and Suffolk. The great A.G. Tansley suspected as much. He knew *Holosteum* from the dry heaths of northern Germany, which closely resemble the Breck, and theorized that its peculiar habitats in Britain resulted from transference from a wild place to artificial ones: 'it got from the heaths into the rye-fields, perhaps, and so on to the roofs with the straw'.[190] Tansley was to be proved right. Confirmation of *Holosteum*'s real status came later, when quantities of its preserved seed was found in the bottom sediments of the Breckland ponds, known as meres. It must have been common there once, perhaps when the climate was warmer and more continental. In more recent times *Holosteum* has persisted as a weed on walls and thatch.

We are left with the question of why it died out there too. Evidently it had become very scarce by the 1860s. Were the old crumbly walls made with local stone and local mortar being pulled down and replaced with wire fences? Were the ruined abbeys tidied up, and given a squirt of weedkiller? And did its survival on thatch depend on fields of rye that have long since been replaced with barley? Or perhaps the extinction of *Holosteum* was caused by climatic change. It has become much scarcer in northern Europe in recent times, and in the North Limburg area of Holland is confined mainly to churchyards and cemeteries, where it flowers in the spring with other annual chickweeds. It all seems very mysterious, and I am left with the feeling that it may yet be flowering unseen, somewhere in the Breckland district.

Cottonweed, *Otanthus maritimus* † 1936

The beauty of this fluffy seaside flower, with its cups of yellow set on fleecy stems, may be the reason why nearly everybody has preferred its scientific name to the rather demeaning English one. In the last century it was called *Diotis*; in our own, *Otanthus* – and *Diotis* was less rare than *Otanthus*. *Diotis*, *Otanthus* or Cottonweed, this is one of our remarkable flowers. It inhabits the warmer fringes of the Atlantic Ocean and establishes itself just above the tide line on soft sandy shores and offshore sandbanks. With the help of warm coastal drift, *Otanthus* has at one time or another colonized Jersey, the Isles of Scilly and some of the wilder, milder shores of England, Wales and Ireland. But few of these colonies seem to have persisted very long, apart from one near Wexford on the south coast of Ireland. During the twentieth century, *Otanthus* has been no more than a temporary colonist in

England. One can imagine its light waterproof seeds, perhaps from flowers growing under a hot sun far to the south, drifting in the current and occasionally washing ashore in the sediment. Every now and then it is able to root and flower for a while, before being destroyed by frosts, or storms, or natural succession. Possibly our weather is now too cold and wet for it, though *Otanthus* seems quite at home not far away on the coast of Normandy (the D-day landings would have run over it). But this is a flower of natural shorelines, of shifting dune strands facing warm incoming currents. Many of its former English sites are now unsuitable – either built over, or contained by sea walls, or teeming with holiday makers, and it may be that there is simply too little wild shoreline left for a good chance of stray seed establishing itself.

Cottonweed (*Otanthus maritimus*) comes and goes with ocean currents, but has not been seen in Britain since the 1930s. [Bob Gibbons]

It seems that *Otanthus* was formerly much less rare in Britain. Gerard knew of it from a record at Mersea Island in Essex. Another early botanist, John Goodyer, found a single plant 'on the seashore on the south parte of the Iland of Haylinge' in 1621. Thomas Johnson was sent a specimen in about 1629 by 'my worshipful friend Mr Thomas Glynne, who gathered it upon the sea coast of Wales'. Later, during one of his journeys as 'the botanical Mercury', he was shown the plant by the beach within a mile of Glynne's house at Glynllifon in Caernarvonshire. John Ray was equally fortunate. He found it 'plentifully on the Sand near Abermeney-Ferry, in the Isle of Anglesey' in 1662, and again on the beach at Penzance, growing with the rare Purple Spurge (*Euphorbia peplis*, see p. 195), another ocean wanderer from warmer climes. Half a century on, Samuel Brewer's diary mentions finding *Otanthus* 'in great plenty for a mile together' at Llanfaelog, also on Anglesey, in September 1727 (it was still there in 1773 when Sir Joseph Banks rediscovered it). On the other hand, Brewer was unable to refind Ray's plants at Abermenai.

Apart from these early records, *Otanthus* has been found in at least seven English and Welsh counties as well as the Isles of Scilly and the Channel Islands. Its appearances and disappearances are as follows:

Cornwall Ray's site at Penzance was thought 'long extinct', until a plant turned up again for a season in 1927. A single plant was recorded at Praa Sands, a few miles east of Penzance, in 1881, and another in the same area in 1933. Keble Martin says *Otanthus* was rediscovered in Cornwall in 1915–16,[95] though the Cornish flora does not confirm this.

Devon Recorded 1801 'On the Devonshire Coast'.[95] It was found on both the north and south coasts at Bideford and Babbacombe [Torbay] in 1855–6.

Dorset In the last century it occurred on Brownsea Island in Poole Harbour. There is also an eighteenth-century herbarium specimen from Burton Bradstock, near Bridport, and a painting dated *c.* 1890 which implies its presence near Weymouth.[118]

Hampshire Long gone from Goodyer's Hayling Island. In 1879 plants were found on a sand bank at Mudeford by Christchurch Harbour. They were soon washed away by the tide, but it re-established nearby until 1891.[17]

Kent Long extinct. Recorded from Sheppey in Hudson's eighteenth-century *Flora Anglica.*

Suffolk In the nineteenth century *Otanthus* grew on a number of sand and fixed shingle beaches along the coast between Lowestoft and Orford, the last record being a herbarium specimen collected in 1880. Francis Simpson considers that this coast is now 'too developed and over-run for it to grow and survive'.[179]

Anglesey Long gone from Ray's site at Abermenai, but a single plant was found at Brewer's site at Llanfaelog in 1894.

Alderney and Guernsey Last seen on Alderney in 1838. The sole record from Guernsey is probably a mistake.

In our own century, the only well-established colonies of *Otanthus* were on St Martin's in the Isles of Scilly and on Jersey. At St Martin's, a large patch was discovered in 1877 on sand dunes at the north-east corner of the island. By the mid-1920s, when E.J. Salisbury paid it his respects and took a series of lantern slides of it for his popular lectures, there were several hundred plants scattered over some thirty yards. But by the time J.E. Lousley visited the site in 1936, there were only a few weak, non-flowering stems left. His photograph reproduced in *Flora of the Isles of Scilly* suggests that 'the Otanthus patch' had become over-run and shaded by taller vegetation. Two years later there was no trace of it left. Lousley speculated that two unusually severe winters in succession had killed off the last plants, although he went on to suggest that 'the end may have been hastened by a grazing animal tethered on the spot, or by the dumping of a load of seaweed'.[107]

On Jersey, *Otanthus* initially flourished and then declined over a similar period of about 60 years. It was discovered there in 1835, and in the middle years of the last century grew in some abundance on the sands of St Ouen's Bay on the western shore of the island. It had become much rarer by the 1870s, and disturbance to the site during the construction of a sea wall seemed to have destroyed it by about 1900. However records begin afresh in 1914, when A.J. Binet, 'having long ago given up all hope of ever seeing this plant again',[102] discovered a few plants near the old site. But although *Otanthus* subsequently reappeared in another part of the Bay, and seemed for a while to be increasing, the site was a popular beach: children are said to have knocked the plants about with sticks, and – the ultimate humiliation! – the patch was run over by a carter collecting seaweed. The last record is of '3 fine plants Aug 1928'. It is possible that this second phase of records originated from an introduction. Jean Piquet, a local botanist, had reportedly procured seed from Cambridge, cultivated the plant in pans in his garden and secretly planted out seedlings in the old locality. At least one of his plantings persisted and was still there in 1912, when Piquet died. The question of whether these post-1914 plants were of natural or planted origin is therefore an open one.

The record suggests that in Britain *Otanthus* comes and goes in a puzzling manner, but that its range has gradually contracted. During the past century it has been confined to the mild, relatively frost-free coasts of Cornwall, Scilly and Jersey. It seems to be a temporary colonist of strandlines, and only in Scilly and Jersey has it survived in the same place for as long as 60 years. Since *Otanthus* seed must be distributed by wind and tide, and the documented sites were just above the high water mark, the colonies are at the mercy of the sea – or, on accreting dunes, to competition from taller vegetation. It may be, as Lousley suspected, that occasional frosts prevent its permanent colonization of the British Isles. Two other Mediterranean seashore plants, Sea Knotgrass and Purple Spurge, have before now been given up as lost, only to reappear again, usually somewhere else. Admittedly it has been a long time since *Otanthus* has shown any similar capacity to baffle the doomsters, but if our climate really is becoming warmer its reappearance may only be a matter of time.

Hairy or Downy Spurge, *Euphorbia villosa (= E. pilosa)* † 1937

The most surprising thing about the Hairy Spurge is the date of its discovery. A rare and subtle spurge, easily passed over unless you take a good look at it, ought to have been first noticed by a Druce or a Marshall around 1900. In fact, its story begins during the reign of Queen Elizabeth I, when the Flemish physician Matthias de L'Obel (Lobelius) was walking in the woods surrounding Bath and chanced on what was almost certainly this plant. (He probably found it on a visit to Bath in around 1569, though the herbal in which the plant is mentioned was printed in 1576. L'Obel came this way again, a few years later, pausing to enjoy the hot springs at Bath, botanize in the Avon Gorge, where he was presented with a Mandrake, and charter a boat to Steepholm in the Bristol Channel. Perhaps he found time on the way to revisit the Hairy Spurge.) The site seems to have become one of the earliest 'rare plant localities' in history. Thomas Johnson, for example, paid his respects to what he termed the Esula Major Germanica 'by a woodside, some mile south of Bathe', around 1630.[203] And to this day the spurge has never been found anywhere else in Britain.

Hairy Spurge grew in woods, but also persisted in hedge banks along nearby lanes. Its distinguishing hairiness is a soft down found mainly on the leaves. It is also unusual among spurges in being rhizomatous: the flowering shoots, up to a metre tall, arise from a stout rhizome buried deep in the soil, and for that reason the plant was usually found growing in clumps. In exceptionally favourable conditions, the Hairy Spurge could fill a woodland glade. The flowers are petite, not unlike the well-known Coral Spurge of gardens, but in tints of yellow rather than orange. It is quite nice. The classic places to see it were the coppice woods and hedgebanks of Prior Park in the bend of the River Avon, about two miles from the town centre. The best account is by J.W. White, who observed the Hairy Spurge at intervals between 1884 and 1909. It runs as follows:

> In May, 1884, the late Mr T.F. Inman conducted me to the wood, and we found the spurge scattered sparingly over a space of about two acres in rather thick coppice, with plenty of *E. amygdaloides* [Wood Spurge]. A number of plants were also seen under a hedge at some distance (nearer Claverton), perhaps a quarter of a mile from the wood. We were told that its growth is greatly influenced by the state of the underwood, which is cut in the customary way at intervals. Soon after the wood has been cleared the plant appears in great plenty; and then annually diminishes in quantity as the brushwood regains its stature, until in some seasons little or none is to be found. At the date mentioned, and for long afterwards, the wood was unenclosed, at least on the south side, and the adjoining land being ploughed the wood-border remained light and open. On my successive visits, every few years, I always found the plant growing chiefly along the sunny edge. Latterly there has been a change of ownership. A barbed wire fence now surrounds the place, harbouring and protecting a mass of nettles and tall herbage that seems to have choked the spurge. At the end of June, 1909, no sign was visible of its having flowered that season. Still, as it has survived the trials and changes of three centuries it may not easily be stifled out of existence.
>
> My late friend, Mr A.E. Burr, told me that *E. pilosa* grew under similar conditions upon private property near Prior Park Lane, a mile or more from Collett's Wood; but of this I know nothing personally.
>
> When cultivated, this plant proves rather tender and does not last many years. I have seen it die out in four gardens.[203]

This is a well-observed chronicle of a vanished species, and mentions an aspect of woodland ecology of vital importance to many rare plants that is sometimes overlooked: that even woodland flowers require sunlight. They get it by either blooming early in the year, or by taking advantage of wood cutting and flowering in temporary glades, as the Hairy Spurge clearly did. It seems that the woods at Prior Park were cut over on a regular coppice cycle before about 1900 but that thenceforward the coppice became dense and shady – a pattern common to ancient woods throughout England. Unfortunately White's confidence that the spurge would survive was misplaced. Permanent shade, and competition from nettles and other invasives, proved fatal to it. The Hairy Spurge grew scarcer and scarcer, and then ceased. What was possibly the very last plant was photographed by J.E. Lousley in 1937, and 'could not be found when I looked for it nine years later'.[106] In the past few years a limited amount of coppicing has been carried out at Prior Park for conservation reasons, but it may have come too late for the Hairy Spurge, unless dormant seed survives. And century-old seed will be deeply buried.

Although Stace considers it to be naturalized,[180] Hairy Spurge had the appearance of a native woodland plant. Why it was confined to the Bath area we can only guess, but since the Avon valley has a warm, moist local climate, and is rich in rare plants, especially at the Bristol end, it may be a refugium for warmth-loving species. A related species, the Upright Spurge (*Euphorbia serrulata*) is also confined to limestone woods in a single river valley, in this case the Wye, and is similarly dependent on periodic cutting and felling. Bath's mild climate and limestone soils supported three very local flowers, two of which are now extinct (the other was *Carex davalliana*), though the third, Spiked Star-of-Bethlehem or 'Bath Asparagus' (*Ornithogalum pyrenaicum*), still thrives there. Probably its extreme localization and dependance on wood cutting was the undoing of de L'Obel's Hairy Spurge.

Esthwaite Waterweed, *Hydrilla verticillata* † *c.* 1945

In the years before the First World War, a father and son, both named W.H. Pearsall, and both the keenest of botanists, used to rent a holiday cottage in the Lake District, at first by Windermere, and later by Esthwaite Water. W.H. Pearsall junior loved messing about in boats, and it was during these 'Swallows and Amazons' holidays on the lakes, his memoirist tells us, that the Pearsalls became interested in aquatic plants.[28] They constructed a three-pronged, weighted dredger to pull up weeds from the bottom; and while one of them cast and hauled from the stern, the other pulled at the oars and manoeuvred the boat over the sample site. Their routine plant recording at Esthwaite Water grew into a more serious investigation of lakeland vegetation which, after the younger Pearsall had entered university, resulted in one of the classic studies of plant succession from water to dry land: a paper read by all ecology students to this day. It was in 1914, early on in this work, that one of the Pearsalls hauled in his dripping line for the umpteenth time and found, entwined among the prongs of the dredger, a translucent green water weed with whorls of slender leaves, which under a hand-lens were slightly toothed towards the tips. It was new to Britain. It was *Hydrilla verticillata*.

It has to be said that *Hydrilla* is no beauty. Worse, it resembles the now common waterweed *Elodea nutallii* from which it differs by the presence of tiny fringed glands at the nodes, the larger number of leaves per whorl, and in the structure of the male flowers (though the latter character is not very helpful, for *Hydrilla* has seldom if ever produced flowers in Britain). It took W.H. Pearsall again to find *Hydrilla* in a second lake, Rusheenduff Lough in Co. Galway, where it still occurs. In Esthwaite Water, on the other hand, it was last recorded around 1945 and seems to have disappeared. Its nearest continental station is in north-east Germany.

So dim and elusive a weed, growing in six feet of peat-stained water and accessible only by boat, naturally became a great prize for botanical 'twitchers'. Detailed instructions on how to find it circulated privately among the Wild Flower Society during the 1920s and 1930s, and ran as follows: 'Row due south from the Boat Landing to the projecting piece of land and towards the left-hand end of the shore. As you approach it, you'll see a *big stone* ... Off that, close in and you'll find what you desire. The bottom shelves rapidly into deep water and if you throw out your dredges and pull it *up* the slope (towards the shore) you will succeed. If you drag it down the slope it will fail ...' And where could they obtain a boat and hauling gear? Why, 'You can rent a boat from McGarre, in Hawkshead. He is a very nice man and he has fishing lines and a rake, if you take the "meat hooks".' The latter were presumably makeshift weed-drags. And one wonders whether the 'big stone' was put there on purpose to mark the spot!

The reason why *Hydrilla* died out at Esthwaite Water was probably progressive eutrophication.[78] From the 1950s, the formerly clear lake became polluted by effluent from Hawkshead, and the once rich submerged flora, including *Hydrilla* and the rare Slender Naiad (*Najas flexilis*), was starved of light and a firm base. The mystery lies not in the cause of extinction but why such a plant should be confined to this one lake among so many hundreds of others in western Britain. Granted that Esthwaite Water has been an 'outdoor laboratory' for three generations of ecologists, the lake flora of Britain has been well studied, yet no one has refound *Hydrilla*. One possible explanation is that Tutin in the *Flora of the British Isles*[30] confused *Hydrilla* with *Elodea nutallii*. In effect, this left most British botanists of the 1950s and 1960s with no idea how to recognize it. But, most likely, *Hydrilla* is not a permanent species at all. It is very widespread globally, extending from the tropics to northern Europe, and, like some other aquatic 'weeds', it naturalizes readily. In the United States, where it was first recorded in 1960, it has spread rapidly from lake to lake on boat gear and fishing lines, and can block shallow fish ponds and canals, just as did Canadian Pondweed in Britain. According to K.A. Langeland, it is a perfect curse to lake owners, causing 'substantial economic hardships', interfering with fishing and water sports, and elbowing aside the native flora.[101] In chillier Britain, by contrast, it has conducted itself with such excessive modesty as to enter the Red Data Book. There are clues to its probable introduced status in *Hydrilla*'s lack of flowers (when it did flower, one hot summer in Western Ireland, all the flowers were female, suggesting a common clonal origin). The younger Pearsall suggested the plant was carried in on the feet or wing-feathers of wildfowl. Human agency is considered less likely.[148] In the circumstances, it might be premature to write off *Hydrilla* as another bygone species. A degree or two of increased water temperature combined with a passage of weed-entangled duck might be all that is needed to re-establish it.

Peach-leaved Bellflower, *Campanula persicifolia* † 1949

This garden plant was included in the first two British Red Data Books as a possibly native species, probably on the authority of Druce's *Comital Flora*. It is an attractive 'Canterbury-bell' with spreading pale blue flowers and long ovate leaves, rather like those of a peach tree. Probably the reason why Druce and others thought it might be native is that Peach-leaved Bellflower is widespread throughout Europe in meadows and open woods, except in the far north, and it occurred in wild-looking ground in southern England. As far as climate is concerned, it would be quite at home here. The flower occasionally reminds us of the fact by escaping from gardens with ease and establishing itself on road verges, railway cuttings, old walls and waste ground here and there, usually on sandy soils. Several other Campanulas, including the true Canterbury-bell, *Campanula medium*, do the same.

Once you start looking hard at the evidence, Peach-leaved Bellflower's claim to nativeness simply dissolves. Druce considered that it was 'probably native' in Berkshire and 'possibly native' in South Devon and Gloucestershire, to which other authors have added Surrey. This view has been repeated from book to book, but it rested on the slenderest of evidence, and the more recent authors of those county floras have all concluded that it is no more than a garden escape. Druce was fairly generous in awarding the accolade of native, especially on his own stamping grounds, like Berkshire. Humphrey Bowen, however, adduced from the few Berkshire records, all on dry waste ground, walls and rubbish tips, that it was an introduction.[15] H.J. Riddelsdell considered one Gloucestershire colony to be 'possibly native' since the plant grew on 'untouched' land in the parish of Colesbourne, some half a mile from the nearest house.[167] This was a patch of bushy pasture which the owner had enclosed against deer and rabbits.

Riddelsdell saw it there in about 1910, when the flowers were visible from 50 yards away as 'a patch of blue', which suggests that they grew in a dense mass, one of the hallmarks of introduced plants. By the time he returned there, 15 years later, 'rabbits were all over the place once more, and the plants had disappeared'. The authors of the more recent Gloucestershire flora regarded it as a 'presumed garden escape in all cases'.

The stronghold of Peach-leaved Bellflower is in Surrey, where it is still fairly widespread on old walls, road verges and the borders of fields, and occasionally becomes established on sandy ground in open woods and scrubby commons.[108] But neither J.E. Lousley nor his successors in that exceptionally well-recorded county regarded it as anything other than a successful escapist. One means by which it establishes itself was suggested by a Suffolk record, where Francis Simpson found a seedling plant among garden rubbish that had been tipped into an old sand-pit.[179] Although the plant can probably spread naturally by seed and so establish itself some distance from gardens, it rarely seems to survive very long.

The extinction date for Peach-leaved Bellflower is given in the Red Data Book as 1949, a year after the publication of Riddelsdell's *Flora of Gloucestershire*. As a garden escape, the flower is not in fact extinct at all, and seems to be doing very nicely in parts of Surrey. There is also a recent record from a 'wild' site by a forest ride at Hart Hill in Somerset. It is the view of it as a native plant that is extinct. Its probable status was summed up by Clement and Foster as 'an established garden escape, naturalised in woods and on commons in widely scattered localities'.[32]

Purple Spurge, *Euphorbia peplis* † 1949 († 1976 on Alderney)

Many sunbathers in the Mediterranean will have come across this exotic-looking plant, with its red, tubular stems and pairs of curled deep-green leaves snaking over the sand on the upper part of the beach. At one time you could find it among fine shingle or coarse sand on a few beaches in south-west England and the Channel Islands. During the present century, however, it has disappeared everywhere except possibly in Alderney, and is now considered extinct as a British plant.

Purple Spurge is an annual, whose seeds are dispersed by currents and tides. When the pattern of ocean currents changes, so must the fortunes of the Purple Spurge. Even in the past, the plant fluctuated greatly from one year to the next. According to J.E. Lousley the incidence of winter storms could determine whether or not it would have a good season the following year.[107] If a severe storm had blown tons of shingle over its habitat, then it might be years before the spurge could again germinate and grow. Overall there seems to have been a gradual decline from an apparent high point in the 1660s, when it was first recorded by Christopher Merret and John Ray. The pattern of records – and sometimes the exact sites – is strikingly similar to that of *Otanthus*, suggesting that both species are warm-water plants whose world range has shrunk since then, probably through natural causes. It is unlikely that habitat destruction is the reason for extinction in either case, except locally, since there are many beaches which remain suitable, and the Purple Spurge, at least, seems to be fairly tolerant of human beings on holiday.

Purple Spurge (*Euphorbia peplis*) is common on Mediterranean beaches but had only a toehold in Britain. It was last recorded on Alderney in 1976. This one was photographed across the Channel in northern France. [Bob Gibbons]

Well-established colonies of Purple Spurge occurred mainly in Devon, Cornwall, Scilly and the Channel Islands. (There are very old records, some of them doubtful, from Dorset (Bridport Harbour, last seen 1907), Somerset (Burnham beach, nineteenth century), Isle of Wight (Sandown Bay, 1830s), Cardigan (Aberystwyth, 1805) and Kent (eighteenth century).) The loss of this rare spurge was of great interest to the respective flora authors, who report its story in some detail. In each case, the tale is one of a long-term decrease to extinction during the twentieth century.

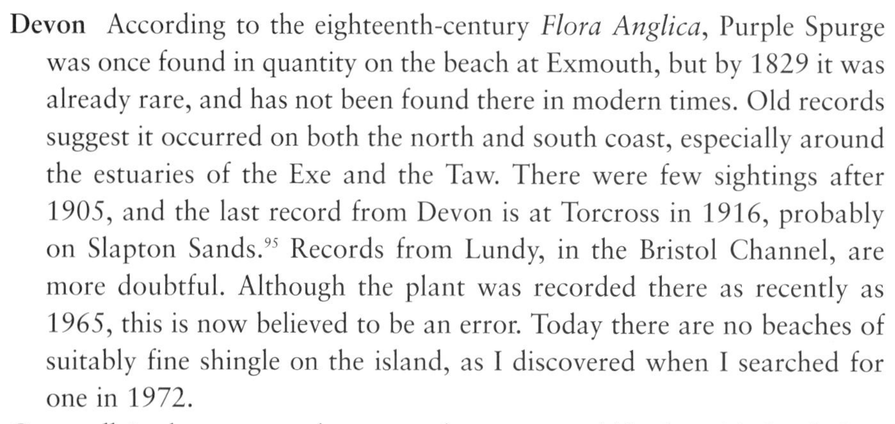

Devon According to the eighteenth-century *Flora Anglica*, Purple Spurge was once found in quantity on the beach at Exmouth, but by 1829 it was already rare, and has not been found there in modern times. Old records suggest it occurred on both the north and south coast, especially around the estuaries of the Exe and the Taw. There were few sightings after 1905, and the last record from Devon is at Torcross in 1916, probably on Slapton Sands.[95] Records from Lundy, in the Bristol Channel, are more doubtful. Although the plant was recorded there as recently as 1965, this is now believed to be an error. Today there are no beaches of suitably fine shingle on the island, as I discovered when I searched for one in 1972.

Cornwall In the seventeenth century, the spurge could be found in local plenty on beaches on the southern coast, especially between Penzance and Marazion, where five plants were found as late as 1934–5. Hamilton Davey indicates that it still occurred in local plenty during the nineteenth century, notably at Lantick Bay.[38] In modern times it has always been a very rare plant in Cornwall, and the last specimen was recorded in 1949 at Downderry, near Looe, by J.E. Lousley.

Isles of Scilly Purple Spurge was first found by Joseph Woods in 1852, and maintained an erratic existence on beaches here and there for about half a century. A specimen was sent to Lousley in 1920, and it was said to occur on Scilly at late as 1936. Interestingly it seems to have disappeared at about the same time as *Otanthus*. Lousley suggested that suitable beaches are now more heavily trampled than in the past, but increased human activity is unlikely to have been the cause of its apparent extinction there in the 1930s.[107]

Channel Islands The spurge has occurred at different times on all the main Channel Isles, suggesting that its seed was freely circulating in the currents. On Guernsey, Babington found it in plenty in August 1837 – but he must have arrived in the nick of time, for there are only four subsequent records, the last (without a voucher specimen) dated 1931.[116] On nearby Sark, which has few suitable beaches, it was last seen in 1838, but the spurge was more at home on the wonderful singing sands of Herm, where it was last recorded as recently as 1968. On Jersey it used to grow on the upper beaches of several bays, but was last reliably recorded in around 1902–3. Possibly, as Frances Le Sueur suggests, the building of sea walls around the island at that time destroyed its habitat.[102] Which leaves Alderney. Although there is less suitable habitat on rocky Alderney than on Jersey, Guernsey, Herm, or Scilly, a colony of Purple Spurge maintained itself there at least until the 1970s – it may possibly still survive, for the plant has always fluctuated widely from one year to the next, but was last reliably recorded in 1976.

It may be premature to write off this ocean wanderer just yet, particularly with warmer weather in prospect. The places to look out for it are unspoiled beaches of fine shingle or coarse sand between the strand line and the pioneer dunes between July and September. If only a tithe of the bird 'twitchers' that congregate on Scilly each autumn could be persuaded to look downwards occasionally, there is a good chance that one day, perhaps soon, the Purple Spurge will reappear.

Narrow-leaved Cudweed (*Filago gallica*), perhaps once native in south-east England. [Tim Rich/ Plantlife]

Narrow-leaved Cudweed, *Filago gallica* † 1955

Here is another diminutive plant that is flattered by a hand lens. Despite its lack of inches, the 'Narrow-leaved Cudweed' is quite a distinctive little 'weed', with its sharp, bristle-like leaves, like fairy daggers. Viewed with sympathy, there is beauty in the silvery pappus-hairs, tufted with brown, and the spears of green arising from cottony stems. Mind you, you need to take your time.

All things considered, it says a great deal about the eyesight of Ray's helper,

Samuel Dale, that he found this particular plant, back in the seventeenth century. He recorded it as 'Gnaphalium parvum. About Castle Heveningham', and the record was first published in the 1696 edition of Ray's *Synopsis*. It then grew 'among corn in sandy ground'. The classic site for it, at Berechurch Common, near Colchester, was discovered later, in 1842.

Stanley Jermyn summarized the rest of the story in his *Flora of Essex*.[91] The Narrow-leaved Cudweed survived at Berechurch Common, and in several of the nearby fields, for just over a century. In at least one place it grew among a fabulous community of annuals, including all the British species of *Filago*, as well as Swine's Succory and Downy Hemp-nettle. It varied in quantity from year to year, but became fairly plentiful just after the Second World War, thanks 'to the war time use of the area as a training ground for tanks and heavy vehicles, which kept the ground disturbed and the brambles and coarser vegetation in check'. Unfortunately the grass soon grew back, and thereafter the plants were progressively confined to a few small gravel pits. In 1954, when Jermyn visited the site with J.E. Lousley and Dr R.W. Butcher, even these pits were grassing over, and only three plants could be found. Next year he found only one, a small one, and it was to be the last, for 'despite many regular searches since, over the whole of Berechurch Common, I have failed to refind any *Filago gallica*'. To make absolutely sure, someone filled in the pits with mushroom compost.

Narrow-leaved Cudweed died out because it ran out of space, but that might not have been decisive if the species was one of those whose seeds can last for decades. Unfortunately it seems that Cudweed seeds germinate all at once. By chance, stock from Berechurch survived, in the garden of the botanist David McClintock, and it was from there that Tim Rich collected seed for a reintroduction attempt. Young plants grown on in pots were planted out on specially prepared ground on Berechurch Common. Some have survived the first two winters, but it is too early to judge whether the attempt has succeeded. The event attracted the media, although what intrigued the reporters most was the apparent eccentricity of introducing a 'weed'. Perhaps, as Tim suggests, we should rename it the Cudwort.

The wild species still survives on Sark, in an old quarry used to store refuse from a hotel prior to shipping it to Guernsey. As insurance, seed from these plants, as well as from Berechurch, is stored in the seedbank at Wakehurst Place. But accidents do happen, as the curators of that establishment discovered after they gave the young Cudweed plants high-release fertilizer by mistake and killed them all!

Was it a native species? Taking its cue from previous floras, the original Red Data Book thought not, but I would be inclined to give Narrow-leaved Cudweed the benefit of the doubt. Although best known from a single site, a scatter of very old records suggest that the Cudweed was once locally widespread in suitably dry, sandy places in south-east England. What is more, the Berechurch locality was a respectably wild site – a sandy common with a rich flora. This may have been another plant of dry open heaths that successfully invaded gravel pits and crop fields. But in the end its hold had became so tenuous that chance events eliminated it.

Downy Hemp-nettle, *Galeopsis segetum* † 1957

As hemp-nettles go, this apparently extinct one was quite attractive, with its large creamy flowers, big floppy leaves and a coat of soft downy hairs. It was clearly much less rare in the past. Ray's Catalogue of 1670 records it 'circa Wakefield, Dafield, Sheffield, *inter segetes*', that is, among corn. In the early nineteenth century, Downy Hemp-nettle seems to have been a fairly widespread casual weed of cornfields and other arable crops in at least 12 counties as far apart as Devon, Essex and Durham. Its appearances were more regular in a small area near Bangor, where local botanists considered it to be a native species. 'The plant is in field edges in corn', ran one unpublished account. 'It has a really large pale cream flower and grows with heaps of the common *Galeopsis*.' Unfortunately the Hemp-nettle depended on regular ploughing, and, when the old sites reverted to grass in the 1950s, it disappeared. The last record in print was in 1957, on arable land near Port Dinorwic, where it had been known for 130 years.[142]

The Downy Hemp-nettle may have thrived best in a 'peasant' economy of smallholdings where land was ploughed and grew crops for a short time, and then was left fallow to recover its fertility. In places it grew in delectable botanical company. At Berechurch in Essex it once rubbed shoulders with Swines Succory and Narrow-leaved Cudweed among a rich assortment of other rare annuals, suggesting something more than a casual 'weed'. It might have been an established crop weed, or, less likely, a native plant that lost its natural habitat of open sandy heaths. Stace considers it to be possibly native, but the view of the Red Data Book is that it was a sporadic alien that established itself only occasionally.

Summer Lady's-tresses, *Spiranthes aestivalis* † 1959

Most accounts of this graceful and well-known orchid mention the disgraceful collecting that went on in the Victorian era. Nothing quite prepares you, though, for the sight of the herbarium specimens, plucked roots, tubers and all from their boggy hiding places in Jersey and the New Forest. Some collectors – reputable men, many of them – physicians, clergymen, army officers – seem to have measured their status by the number of orchids they could cram into their vasculum. But did such profligacy, such perverted love, propel the Summer Lady's-tresses to extinction? Some have thought so. The Red Data Book suggests collecting as at least a contributory cause, and David McClintock, in his *Wild Flowers of Guernsey* noted that 'The number of complete plants from the Grande Mare, pulled up tuber and all, dried and killed, in herbaria is an utter disgrace to the unthinking cupidity of collectors, all of whom had to have at least a sheet of it for themselves'.[116] Frances Le Sueur agreed: 'There is hardly a major museum herbarium in the country which does not contain a specimen of Summer Lady's Tresses from Jersey. Some contain more than a score, most complete with the root tubers'.[102]

Summer Lady's-tresses (*Spiranthes aestivalis*) is our best known extinct plant, perhaps a victim of drainage combined with over-collecting. [Bob Gibbons]

Repulsive as such mindless amassment seems today (but not necessarily then), I do wonder whether this apparent ruthlessness was partly unintentional. The Summer Lady's-tresses grows in wet, spring-fed bogs, or in spongy ground by the sides of lakes, more or less loosely rooted among peat and moss. In such circumstances it is easy to imagine someone reaching down to pluck a stem and inadvertently pulling up the whole plant. The Fen Orchid suffered similarly in the East Anglian fens. However, its glamour as a rare orchid made the Summer Lady's-tresses a particularly desirable plant, especially since it was evidently difficult to grow in gardens. What might have made the situation even worse was the magnetic attraction of the New Forest for collectors of all kinds, especially with the coming of the railway. This was an unusually vulnerable species.

The Summer Lady's-tresses is a central and southern European plant with just a toehold in Britain in the warm, almost frost-free, valley mires of the New Forest, and also on Jersey and Guernsey, where in biogeographical terms, it belongs to the French flora rather than the British. Today it is declining throughout Europe and is on the protected list of several countries, mainly because of the wholesale loss of its boggy habitat. Species near the limits of their range tend to be extinction-prone: *Spiranthes aestivalis* has survived in Brittany, a few hundred miles to the south, whereas in chillier Britain, Belgium and Denmark it has died out. It was first discovered on Jersey, in 1837, and three years later was found in the New Forest. A new orchid was an event, and unfortunately the discovery coincided not only with a new railway station at Lyndhurst but also with the Victorian taste for rarities, bog gardens and pressed plants. The herbarium material suggests that collecting reached a peak in the 1850s but that it continued throughout the nineteenth century and well into the twentieth. And yet the Summer Lady's-tresses seems to have been sufficiently plentiful in the New Forest to withstand at least moderate collecting. The smaller colonies in the Channel Islands may have been more vulnerable. On Guernsey it was confined to a few acres of boggy ground around a lake called the Grande Mare. It had become very scarce by 1906, and by 1914 it had gone. The tiny area of Sphagnum bog there today may not have been able to sustain the plant even in the absence of collecting – though it was undoubtedly collected without mercy. It was much the same story on Jersey, where the orchid was confined to wet sandy ground by St Ouen's Pond. Frances Le Sueur relates a story told by an old man who remembered plucking a spike of it around 1906, and bringing it to Jean Piquet's chemist's shop. 'Ah, my boy, you've got it!' exclaimed Piquet with surprise and delight, and promptly swopped him a bottle of scent for it.[102] Evidently the orchid had become a rare prize by then. The last four plants, complete with tubers, were dug up in 1926. 'Killed by collectors?' asks Le Sueur. It sounds like it, though drainage might have hastened its demise.

In the New Forest there were at least five different sites, all in the Lyndhurst area, but only three were ever named in print. These were all spring-fed valley bogs carpeted by Sphagnum with Bog Asphodel, White Beak-sedge, sundews and other plants of wet acid bogs. The orchid varied in quantity from one year to the next, probably depending on the rainfall. One observer around the turn of the century saw 'half an acre of bog perfectly white with (the) flowers, but the following year only a few spikes of bloom appeared'. Another counted 200 flower spikes in one place. But of course the populations needed to be strong enough to sustain the species in bad times as well as good. The Wild Flower Society seems to have made a brave effort to put visitors off by describing the best site as 'a very deep, dangerous and adder-haunted bog'. Unfortunately, adders or not, they were regularly raided, and there are an irresponsibly large number of herbarium specimens dating from as late as the 1920s and early 1930s. By then there were other problems for the orchid to contend with. One site was drained and planted with coniferous trees. In the remaining two, the orchid grew scarcer and scarcer, until it finally disappeared, in 1940 and 1959 respectively. Although there had been attempts to drain them, both sites remain in surprisingly good condition, and seem capable of supporting the orchid had it survived. Quite possibly its numbers had become so reduced by the 1930s that the Summer Lady's-tresses was eliminated by chance events. Just as a drunkard, meandering down a corridor, will sooner or later collide with the wall, so the handful of weakened plants remaining may have fallen victim to summer drought, hungry ponies or perhaps even an overshading bough.

Opinions have differed on who or what to blame. The Forestry Commission? The late J.E. Lousley was 'certain that its destruction was entirely due to drainage'.[140] Collectors? Lousley thought the amount of collecting 'deplorable', but doubted that it had contributed to the final loss of this species. Francis Rose thought that collecting probably did exacerbate the situation, but he agrees that the primary cause was probably drainage and consequential changes to the habitat.[17] John Cox points out that spring-fed bogs are subject to fluctuating water levels, and that this shallow-rooted orchid may have been vulnerable to prolonged dry spells. Perhaps the hot summers of the late 1940s were its *coup de grace.*

The Summer Lady's-tresses may have been an unusually extinction-prone plant. It was always very local and was evidently unable to colonize new ground. The orchid's habitat was vulnerable to drainage, shading and trampling, and the plant itself was all too easily uprooted. The crunch point seems to have been reached by about 1900 in the Channel Islands, and about 1930 in the New Forest. The only strategy available to protect it – a belated attempt at secrecy – proved useless.

Thorow-wax, *Bupleurum rotundifolium* † 1970

To the herbalists and gardeners of centuries past, Thorow-wax (see picture on p. 204) must have been almost as familiar as poppies and mayweeds. Alone among our extinct flowers it has borne the same familiar name for nearly half a millennium: Turner, who knew it 'In Somersetshire between Summerton and Martock', called it 'Thorowax' 'because the stalke waxeth throwe the leaves' – that is, the leaves clasp the hollow stem, and, like it, are covered by a glaucous waxy bloom. It is a pretty plant, in its weedy way, a medley of greens and yellows – with blue-green leaves and yellow flowers sitting rather sweetly inside their spiky nest of greenish bracts or 'hare's ears'. It was also, at times, a very common plant. On his famous excursion through Kent in 1629, Thomas Johnson passed cereal fields where Thorow-wax grew in such quantities 'that it may well be termed the infirmities of them'. Until the early years of this century one could find similarly afflicted fields over much of England. John Hogg saw one in Durham in 1829: 'Yesterday I found vast quantities of the curious Thorow-wax ... in several stubbles in a clayey soil ... Indeed, parts of several fields were quite thick with it. I send you herewith a small specimen'.[88] In the same clay fields he found another now extinct cornfield weed, *Caucalis platycarpos* (Small Bur-parsley). Then as now rare weeds stick together.

The demise of 'weeds' like Thorow-wax is seldom well documented and has to be pieced together from statements in the local floras. The more general books go on reporting such plants as common long after they have disappeared from the landscape. For example, the 1962 edition of 'CTW' reports the Thorow-wax from no fewer than 53 counties: 'Native or introduced. In cornfields. From Devon to Kent to Wigtown and N Yorkshire, local and rarer in the north'.[30] One would suppose from this that Thorow-wax was still a fairly common species, but in fact it was approaching extinction. By 1962, there were in fact recent records from only Devon, Wiltshire, Norfolk and Rutland. By 1976, it was listed as endangered. Unfortunately the recording of this species was bedevilled by confusion with an alien relative, *Bupleurum subovatum*, which is imported in bird-seed and establishes itself from time to time. The real thing had, in fact, gone already. The last certain cornfield record was near Canterbury in about 1970. I remember making a journey in 1984 to investigate reported Thorow-wax in flowerbeds below some council flats in Oxford, only to find the wretched *Bupleurum subovatum* again, no doubt the product of some dear lady's budgerigar cage. I suggested they tried weedkiller.

What happened to the real Thorow-wax? The picture that emerges from the local floras is of a gradual retreat starting around 1870, disappearing first from the north, then more generally from clay soils, leaving its last redoubts on the sandy or chalky soils of the Midlands and central southern England. In Kent, where Johnson had wandered past acres of Thorow-wax, it had gone altogether by 1900. It was already a rare 'weed' by the 1920s, when the Wild Flower Society regarded it as a good find. The cornfields near Alresford, Hants, were on the botanical itinerary as a good place to see it, especially as they co-starred another beautiful 'weed', Pheasant's Eye (*Adonis annua*). Post-war records seem, despite 'CTW', few and far between, and often in 'casual' places rather than cornfields.

The cause lay in agricultural improvements, most notably more effective methods of cleaning seed. Thorow-wax is among the weeds with relatively large fleshy seeds, like Corncockle, the Bur-parsleys, Larkspur and Field Cow-wheat. Such seed is easily removed by screening and these plants do not seem to have had much of an independent existence outside the farming calendar. Thorow-wax probably lost its 'worst weed' status long ago, for its name was not on an agricultural 'wanted list' published in 1809 which did include Corncockle, Cornflower and Corn Marigold. It seems to have depended on being resown with the crop, which might explain why its disappearance was so sudden and so permanent. Probably herbicide sprays killed off the last few colonies during the 1950s and 60s. There were no semi-natural sites to sustain the plant, as there are for certain other former crop weeds like Corn Parsley or Shepherd's Needle, and this suggests that Thorow-wax is probably an ancient introduction and not a native plant.

Perhaps it is truer to speak not of the loss of Thorow-wax so much as a change in status. In recent years its bright yellow-green flowers have been spotted on seed packets, in bunches at flower stalls outside London stations, and even, rather appropriately in the circumstances, in funeral wreathes (Sorrow-wax?). A seed catalogue refers to 'its fashionable form and colouring', and the plant in question looks like the true wild Thorow-wax and not a cultivar. Like so many rare flowers, Thorow-wax can become exultantly rampant in sunny garden borders, and there perhaps it may have a future, biding the ignominy of confinement, like a farmed ostrich, for the opportune moment to wriggle under the fence and return to the wild.

Thorow-wax (*Bupleurum rotundifolium*) – with mayweeds – once a widespread 'infirmity' of cornfields, is now sold in seed packets and cultivated for funeral wreaths. [Peter Marren]

Swine's or Lamb's Succory, *Arnoseris minima* † 1970

Like Thorow-wax, Swine's Succory was a victim of modern farming. It is a distinctive annual plant, with its tiny buttons of yellow florets and its curious inflated hollow flower stalks – 'clubs' would be a better description. Botanists nicknamed it 'Mumps'. By contrast with its rarity in modern times, the Tudor herbalists were well acquainted with this poor relative of Succory or Chicory, whose bitter root could be used as a diuretic, like dandelion. In living memory the places to find it were cultivated fields on light, sandy soils, notably the Greensand and the Bagshot Beds, and the sandy wastes of Breckland. As with Thorow-wax, botanists were slow to notice that all was far from well with Swine's Succory. A Wild Flower Society outing in 1958 recorded it without noticeable glee, though by this time it was certainly a rare plant. It slipped from view quietly, and only in retrospect did the botanical world wake up to the fact that another rare weed had gone for good.

Sir Edward Salisbury observed that Swine's Succory grew and seeded best in dry seasons.[174] It is one of our 'continental' flowers, and the western half of Britain is probably too wet for it. Even so its distribution was oddly patchy: for example, it used to be locally frequent, occasionally even 'extraordinarily abundant', in Surrey and Berkshire, but was always rare in Sussex. One botanist, who signed the piece 'A.B.', described a long walk through the Kentish countryside in 1861 to procure specimens. In the pleasantly rambling style of the day, he mentioned various finds along the way before at last finding 'no lack of' Swine's Succory in what was now 'a great strawberry-bed' on the site of Bexley Heath. In a nearby field, however, it had been 'quite eradicated, a consequence of careful cultivation'. It had been eradicated from the whole county by 1930.

The decline of Swine's Succory was prolonged. The *Atlas of the British Flora* shows many more pre-1930 hollow circles than solid dots. In recent years, one of its best sites was at Sandy Lodge, now the main office of the RSPB, where it was last recorded in 1953. In another stronghold, the Breck, it had gone by 1955. It hung on in Nottinghamshire, Surrey and Berkshire until the 1960s. A classic place to see it was some sandy cultivated fields by the Basingstoke Canal at Frimley Green in Surrey, where it grew on open ground among an assortment of other good weeds. But the area has been suburbanized since then, and the site is now a recreation area with lawns, flower beds and what are described as 'dog toilets'. At another Surrey site at Gomshall, a few miles east of Guildford, it was still locally abundant in 1945, and 'widespread' in 1950, but by 1958 had become confined to a single field corner. J.E. Lousley searched for it there in vain in 1970 and 1972.[108] Chance events and ill luck may have eliminated the Swine's Succory – the drunkard and the wall again – even when it managed to dodge the weedkiller.

The extinction of Swine's Succory in Britain is an aspect of a much wider decline of such plants throughout western Europe, through intensive (and super-efficient) agricultural practice. In the Netherlands, for example, it was associated with crops of rye, which are no longer widely grown. In Britain it liked crop fields on poor, light soils which, before the advent of chemical

Below left: Swine's Succory (*Arnoseris minima*) disappeared quickly and quietly. There was scarcely time to list it as rare before it had gone altogether. [Bob Gibbons]

Below right: Distribution map showing the decline of Swine's Succory. It was last seen in Berkshire, Surrey and Nottinghamshire in the 1960s. Solid dots indicate post-1930 records; open circles, pre-1930. [Biological Records Centre]

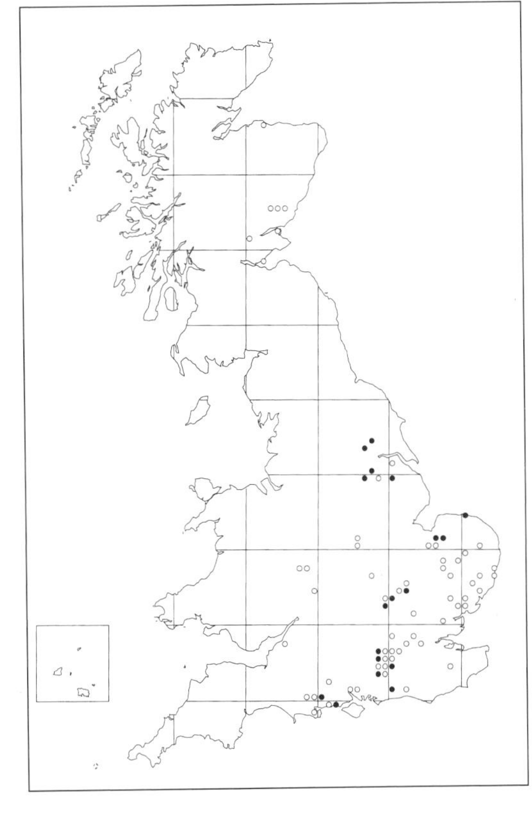

fertilizers, had to be rotated with break crops and periods of fallow. This 'habitat' has virtually disappeared, nor was it ever protected on nature reserves. Today, perhaps, cultivated gardens and allotments would offer the best chance of survival.

We are left to decide on the status of Swine's Succory. The absence of any apparent long-term strategy like seed dormancy suggests that it was dependent on agriculture. According to E.J. Salisbury, it was sometimes imported from the European mainland with seed of Timothy Grass (*Phleum pratense*). On the other hand, some of its sites were former heaths and commons which had probably seen a form of rough cultivation, on and off, for centuries – like our Kentish traveller's 'strawberry patch' at Bexley Heath. It had a regular associate in Smooth Cat's-ear (*Hypochaeris glabra*), a fast declining but undoubtedly native plant. Perhaps we have in Swine's Succory an aboriginal flower of open, well-grazed sandy heaths whose natural range was extended by agricultural practice, including imported seed. The traveller in Kent, quoted above, made a remark whose elegance of phrase might stand as an epitaph to all our vanished 'weeds': 'Clean culture is detrimental to the botanist, although it [is] agreeable to see the evidences of care and the promise of success.'

Interrupted Brome, *Bromus interruptus* † 1972

Britain has very few endemic plants. To allow one of them to die out completely – and that one a grass of economic potential – could be regarded as rather careless. Unfortunately, at the time, *Bromus interruptus* had no influential champions; indeed few people, even agricultural botanists, had ever heard of it. Until very recently, nature conservation in Britain was about preserving natural habitats. And if a species did not happen to live in a recognized natural habitat, then that was too bad. Were it not for the chance that one person took an interest in the 'Interrupted Brome' and maintained a plot of it at a university, this grass would be as irredeemably extinct as an Iguanodon.

The story of its discovery is as follows. In July 1888, George Claridge Druce was wandering through a fallow field on the Berkshire chalk in which a crop of barley had been growing the previous year, when he chanced on quantities of an odd-looking brome-grass. It resembled a head of wheat, with spikelets arranged on the stalk in clusters of three. The gap between the clusters gave the grass an 'interrupted' look, as though some idler had plucked away bits of it. Druce peered at it keenly through his lens, gathered up a quantity for pressing and moved on. Next year he found the same peculiar grass in a clover field, this time on the Oxfordshire side of the Thames near Goring. Druce considered it to be an unusual and unknown form of Soft Brome (*Bromus hordeaceus*), but sent specimens off to Professor Hackel, the then authority on grasses, for his opinion. Hackel agreed that it must be a new variety of Soft Brome, which he named *var. interrupta* Hackel. Later, in 1895, Druce came to the view that this must be more than a mere variety, since the part of the flower known as the upper palea was split to the base, while in all other bromes it is entire. He thereupon renamed it *Bromus interruptus* (Hackel) Druce. As it turned out (and no doubt to his great chagrin), Druce was not after all its discoverer. Herbarium specimens were discovered, gathered by a Miss A.M. Barnard at Odsey, Bedfordshire, in 1849. She had evidently realized they were different, for they were first labelled 'Bromus pseudo-velutinus'. However, when the great H.C. Watson had taken a look at them he decided, as Druce and Hackel would do, that it must be a peculiar form of Soft Brome.

A dry botanical description really does not do justice to *Bromus interruptus*. It looks very odd indeed with its tufts and whiskers and the general sense of a 'grass gone wrong'. Where did it come from? *Bromus interruptus* is known only from grass leys or crop fields in Britain (it did later turn up in Holland, but probably as an import in English fodder). Therefore it must have evolved here, and presumably within a farmed environment. Its usual habitat was a fallow field, the rest period common to many fields on the chalk before 1940, or else in fields sown with Sainfoin or clover for fodder. Though it was sometimes found in quantity, *Bromus interruptus* was not usually associated with a rich arable flora (though its relative *Bromus commutatus* often kept company with it). The most likely explanation is that the species first arose by genetic mutation, probably from *Bromus hordeaceus*, and that it spread over much of lowland England in fodder or seed-grain. Whether this happened fairly recently or long ago is unknown. Possibly it took off in the seventeenth century when Sainfoin was introduced to Britain. What is certain is that the decline of *Bromus interruptus* was dramatic. Between 1900 and 1920, it was recorded from about 65 ten-kilometre squares in 27 vice-counties. After 1935, however, it was confined to a mere six localities, and by the 1950s, there were only two, both in Cambridgeshire. The grass seems to have been lost sight of altogether between 1953 and 1962. Then, in response to a paper about *Bromus interruptus* by Franklyn Perring,[139] searches were made and the plant refound by a farm track at Pampisford, near Cambridge. It persisted there until 1972, then died out, despite the best efforts of conservationists and a rotovator. By coincidence, Pampisford is not far from the site of the first discovery, at Odsey in 1849.

The decline in *Bromus interruptus* must have been linked to agricultural change, most probably the loss of fallow land and fodder crops for horses. Like Corncockle, it seems to have relied on regular resowing, since, like many temperate grasses, it has little or no seed dormancy. It became extinct in the wild because no one knew how to conserve it. Even seed stored at the Cambridge Botanic Garden as an insurance measure was lost – it had been stored at room temperature; we know better now. For several years Interrupted Brome seemed to be at one with the Dodo and the trilobite. Then, quite by chance, it was learned that Dr Philip Smith, one of the party which had refound the grass in 1963, had taken a few seeds and maintained the plant in cultivation at Edinburgh University. At first he was unaware that his was the only living material of *Bromus interruptus*. At a BSBI conference at Manchester in April 1979, Smith dramatically unveiled a pot of the extinct plant as the high point of his talk on the evolution and chemical taxonomy of grasses. To make sure it did not become extinct again, plants and seeds were distributed to leading botanic gardens, the seed bank at Kew and to the Iron Age farm project at Butser Hill. To see it in a 'wild' setting today, the latter is your best bet.

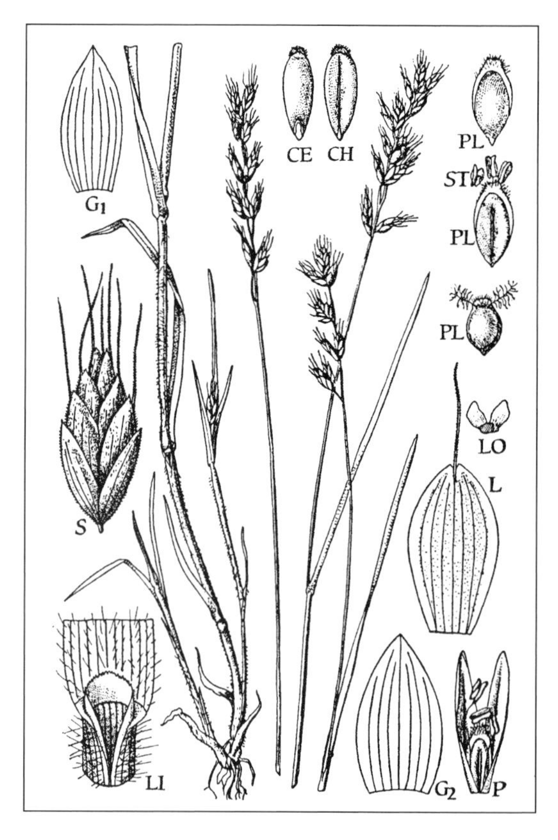

Interrupted Brome (*Bromus interruptus*), an endemic grass, would be extinct worldwide were it not that a scientist had by chance maintained a tuft of it in his garden.

It would be nicer still if the plant was refound somewhere – and 'somewhere' could mean almost anywhere south of the Humber. *Bromus interruptus* could yet tell us how such apparently new species arise, and how they adapt themselves to a farming regime, a matter of interest to crop breeders as well as academics and conservationists. This is happily a Dodo that turned into a phoenix, though not because of any planned conservation scheme but because one person decided, for reasons of his own, to grow it on.

Stinking Hawks-beard, *Crepis foetida* † 1980

If ever a pretty flower was lumbered with an ugly name it is this one. From its portrait in the books you would be forgiven for regarding Stinking Hawks-beard as just another of those interminable yellow composites – 'hawks-beards or hawk-weeds or hawk's-somethings'. I once had the privilege of overlooking a patch of it in a flower bed as I worked, and was able after a while to appreciate its subtle charms. The 'clocks' of Stinking Hawks-beard are as white as lamb's wool; its relatives look dingy by comparison. The flowers droop modestly in bud, then half-open in little spurts of orange and yellow. Even the raggedy grey-green leaves add to the ensemble. The plant is straggly and gypsyish: a connoisseur's plant. Nor does it 'stink'; when pinched, the fresh leaves have a sharp but not unpleasant scent, like bitter almonds or crushed laurel – the sharp, slightly mouthwatering whiff of cyanide.

Stinking Hawks-beard is another southern plant with a tenuous grip on

England. Ray knew it in Cambridgeshire, but outside Sussex and Kent it was usually a transient 'weed' of waysides, chalk pits and cultivated fields, perhaps introduced in straw, manure or impure seed-grain. On the shores of the south-east coast, however, it behaved more like a native plant, as a member of the specialized flora of stabilized shingle. The Revd E.S. Marshall, an immensely experienced and discriminating botanist, considered it to be respectably British, having gathered it in similar places on the opposite shore of the Channel, and seeing 'no reason to distrust its wildness, either there or in Kent'.[81]

It has fared badly in our own century. Gradually, the Hawks-beard gave up its day job as a weed and became more or less restricted to shingle beaches on the south-east coast. Even here, it had gone from its old haunts at Pevensey Bay, Newhaven and Walmer by 1948, and at Dungeness became restricted to a small corner of stabilized shingle. The main cause may have been rabbit grazing, which diminished another rare shingle flower, Least Lettace (*Lactuca saligna*). In recent times, those searching for it at Dungeness knew they were getting warm when they heard a mighty barking of dogs: thanks to the proximity of the Stinking Hawks-beard, a bungalow within sight of the nuclear power station is known to the botanical world as the 'barking dog bungalow'. 'It's always there, near the barking dog bungalow,' they told Rosemary FitzGerald in 1985, but it wasn't, and careful enquiry suggested that no one had in fact seen it since 1980. The colony had always fluctuated from one year to the next: in 1969 over 400 plants were counted, but in the year after only two. What no one realized until it was too late was that Stinking Hawks-beards, like most other annual plants, and certainly other annual hawks-beards, rely on open ground, and hence on occasional disturbance. It apparently did rather well during the construction of the power station, and no doubt in the past it was present in sufficient numbers to take advantage of suitable conditions as they arose. Interestingly, in the bed where I used to gaze down on it, it flowered vigorously the year after sowing, less vigorously the year after that, and then for some time after that it didn't flower at all. No doubt a regular raking over would have sown in the fallen seed and maintained the plant. Something tipped the balance the wrong way for the Stinking Hawks-beard – perhaps there were too few disturbed places for enough seedlings to establish, or perhaps rabbits ate the plants before they could seed.

Fortunately while it yet lived, Dr Brian Ferry of Royal Holloway College had the foresight to collect seed and grow it on in cultivation. Using this stock, he is now undertaking an experimental reintroduction at Dungeness, as part of English Nature's Species Recovery Programme. In the process he, and through him the rest of us, are finding out much more about the lifestyle of Stinking Hawks-beard. It needs surface disturbance to establish, and Ferry is experimenting with different combinations of shingle and fine sand in a series of plots. Through this work, the Hawks-beard has become quite famous locally. The late Derek Jarman lived next door to one of them:

> I have two large stands at the back – they protect the wild flowers like fennel and the stinking hawks-beard that has been reintroduced to the Ness by the ecology department of Sussex University. Most of the hawks-beard has been eaten by the rabbits, though some of the plants rotted in the little squares that the ecologists cut out to plant them in. Next door they seem to have fared better – I have only one plant left.[89]

A plant that attracts the care and attention of an *avant garde* film producer as well as universities and action planners may yet have a future in Britain.

TABLE 5 Extinct British wild flowers

LAST SEEN	SUGGESTED CAUSE	SUGGESTED STATUS
1800–1850		
Cotton Deer-grass	Drainage followed by marl digging	Planted
Davall's Sedge	Drainage followed by building	Possibly native
Northern Bramble	Incapacity to persist	Sporadic bird-borne introduction
1850–1900		
Broad-leaved Centaury	Agricultural enclosure and (?) collecting	Not a species
Marsh Fleawort	Drainage and neglect of fen ditches	Native
Black Pea	Natural causes	Garden escape or planted
1900–1950		
Alpine Butterwort	Cultivation of surrounding land	Planted
Jagged Chickweed	Destruction of old walls (?)	Casual, formerly native
Cottonweed	Natural causes	Intermittent colonizer
Hairy Spurge	Cessation of coppicing and consequent shading	Possibly native
Purple Spurge	Natural causes	Intermittent colonizer
Peach-leaved Bellflower	Rabbit grazing	Garden escape
Esthwaite Waterweed	Eutrophication	Introduction
1950–1980		
Narrow-leaved Cudweed	Loss of sandy crop fields	Possibly native
Downy Hemp-nettle	Loss of sandy crop fields	Casual, possible former native
Summer Lady's-tresses	Collecting and drainage	Native
Thorow-wax	Seed cleaning and herbicides	Long-established introduction
Swine's Succory	Loss of sandy crop fields	Probably native, turned weed
Interrupted Brome	Loss of fodder crops and fallow land	Endemic crop weed
Stinking Hawks-beard	Rabbit grazing and lack of disturbance	Probably native on SE coast, otherwise casual

Conclusion

Of the twenty or so nationally extinct species considered in this chapter, only two were beyond reasonable doubt native wild flowers: Summer Lady's-tresses and Marsh Fleawort. The origin of all the others has been questioned, and some of them are almost certainly introduced crop weeds or garden escapes. Stinking Hawks-beard was probably native on the south-east coast, but a sporadic introduction everywhere else. Hairy Spurge and Davall's Sedge might be considered native in default of a convincing theory to explain how they reached the hillsides and woods near Bath. Jagged Chickweed and Narrow-leaved Cudweed have been dismissed as casuals, but are probably native plants that lost their wild habitats. The same may be true of Swine's Succory. Purple Spurge and Cottonweed are natural colonists that arrive under their own steam. Northern Bramble and Esthwaite Waterweed (and maybe Stalked Sea-purslane) were probably temporary colonists brought in by birds. Thorow-wax is a long-established crop weed with no convincing wild habitat. Interrupted Brome is also a crop weed, which apparently evolved in Britain through genetic mutation. The others owe their persistent place in the floras to tradition as much as anything. Cotton Deer-grass and Alpine Butterwort might have been planted. Downy Hemp-nettle was a casual, which persisted in one or two places. Black Pea and Peach-leaved Bellflower are garden escapes, and, as such, not extinct at all. Broad-leaved Centaury was probably a mutant form or hybrid, not a species.

At the most generous estimate, then, only 13 native or long-established wild flowers have died out in Britain since records began (not counting microspecies). But why those 13? They were not our rarest plants. Marsh Fleawort, Swine's Succory, Stinking Hawks-beard, Interrupted Brome and, especially, Thorow-wax were once widespread and locally common – none would have qualified for a nineteenth-century Red Data Book, except Marsh Fleawort towards the end of it. What seems to have caused their decline was wholesale loss of habitat. In the case of Thorow-wax this is easy to understand. The only flowers that survive in modern cornfields are those with survival mechanisms like seed dormancy (wild poppies) or resistance to selective herbicides (Black Grass) or plants that take advantage of direct drilling (Sterile Brome). Thorow-wax was evidently dependent on impure grain, and was doomed when better seed filters came in, since it had no 'wild' sites to fall back on.

Several other species belonged to a less well-known but equally threatened community of plants, those of open sandy commons and waste land, or of the form of rough cultivation that produced plenty of disturbed ground like cart tracks, sand-pits and field banks. As a class, these plants are in danger. The conservation movement arrived too late to save the key sites, as it did with unimproved meadows and chalk downs. But, in any case, open sandy places have declined even on nature reserves. In today's countryside, only management aimed specifically at the needs of rare annuals would have saved them. Perhaps Swine's Succory was too dependent on poor farms that scratched a living on thirsty, sandy soils. It might seem an unwise strategy, but such farms are in fact ancient landscapes – they have existed in the British landscape for 6000 years.

It is striking that no fewer than five of our extinct plants were members of the family Compositae, most of which have light seeds born on feathery parachutes. There is evidence that some rare Composites have low seed dormancy; that is, most of their seeds germinate at once. Some populations seem also to have very low seed viability: the only known natural population of Fen Ragwort, for example, is now more or less sterile. If this is a characteristic of the family, it would certainly go some way to explain why Swine's Succory, Stinking Hawks-beard and Marsh Fleawort were unable to survive unfavourable periods and why Cottonweed succumbed so swiftly to the closing of vegetation. The lack of an effective dormancy mechanism makes a plant unusually extinction-prone, especially in the uniquely unstable twentieth century.

Most of our extinct wild flowers were short-lived plants dependent on early seral stages, that is open land maintained by regular disturbance. Flowers of 'closed' stable habitats, like downland meadows and woods seem much less likely to become extinct, perhaps because they live longer and are less reliant on seed. They are also better provided for on nature reserves. Even the one woodland species on our list, Hairy Spurge, grew in open glades after coppicing and mowing, and hence relied on practices which eventually became outmoded. The Summer Lady's-tresses is the exception in that its habitat survives, albeit much reduced. This much-mourned orchid was at the edge of a declining European range, and probably in a very delicate balance with its environment. In its case recruitment from seed or root division may have been insufficient to balance loss from collecting and grazing. But for most extinct flowers, some of the mystery disappears if you visit their last-recorded sites. Almost none of them would support the plants today. In the end, extinction is caused by bad luck. These flowers had the ill fortune to grow in the wrong place.

Chapter 10

New natives

Long ago, a venerable and famous tree grew in the middle of the Wyre Forest. It was known as the Whitty Pear. No one knew how it had got there, nor exactly how old it was, though the tree was already considered ancient when a local alderman wrote about it in 1678. Some said it had been planted by a hermit, and there is, or was, a nearby bank which one antiquarian considered to be the remains of a medieval hermitage. A tree still stood there in the nineteenth century, but its life ended abruptly in 1862 when a poacher chopped it down and burned it to avenge himself on the landlord. Fortunately, seed had been taken and cultivated in the nearby grounds of Arley House. One sapling was returned to the original site in 1913; it flourished and has since branched out into a splendid tree, Son of Whitty, so to speak.[154]

The Whitty Pear is a local name for a tree that was seen to combine the leaves of 'whitty', better known as a rowan tree, and the hard green fruit of a pear tree – the tough little pears of the Worcestershire perry orchards, not the succulent ones sold by the greengrocer. But there is no mystery about its identity. The Whitty Pear is a specimen of the Service-tree, *Sorbus domestica*, a little-known tree in Britain. *Our* service-tree is the confusingly-named Wild Service-tree, which is a different species, *Sorbus torminalis*, with plane-like leaves and clusters of round, russet berries. On the continent, where both trees occur, only *Sorbus domestica* is known as Service or 'Sorb' (*Sorbus torminalis* is *Alisier* in France, *Elsbeere* in Germany). The real Service has been cultivated since ancient times for its edible fruit, used to flavour beer – hence the specific name of *domestica*. But in more recent times the tradition has been forgotten in Britain, where the Service was, until recently, only an occasional park tree. The best known one, in Oxford Botanic Garden, is said to be a descendant of the Whitty Pear of the Wyre Forest.

In 1993 the botanical world was amazed to learn that Service had been discovered in a wild location on sea-cliffs in South Wales.[80] It had in fact been found there some ten years earlier by Marc Hampton. The trees were wind-stunted specimens, only 3 to 5 metres high, rooted in an unstable south-facing cliff of lime-rich Lias shales in West Glamorgan. In due course, Marc found a second population two miles away, again growing on inaccessible ledges. Some of these trees were clearly very old; ring counts suggested several hundred years, perhaps more, especially as the tree persists by suckering, so that the roots may be considerably older than the trunk. Despite the cliffs being well known to generations of botanists, it seemed that they had overlooked a native species there – and not a shrinking violet either but a fully formed tree.

It is not as surprising as it may sound. Nothing hides half so well in a wood as a tree, and *Sorbus domestica* is well camouflaged by the cliff scrub of blackthorn, hawthorn and brambles, or, in places, taller woodland of ash, elm and yew. Moreover, for most of the time it looks like a rowan. The characteristic silvery sheen to the young leaves, the larger, pink-flushed flowers, and of course the 'pears' are distinctive characters, but to see them it is very much a matter of being in the right place at the right time. Most of the Glamorgan trees grow out of reach, and they seldom produce ripe pears. Even when you know they are there, the trees can be very difficult to spot. To find them for the first time, as Marc Hampton did, must have required an exceptionally open mind as well as penetrating eyesight.

Since the discovery of *Sorbus domestica* in Glamorgan, Mark and Clare Kitchen have found it in three places around the Severn estuary in Gloucestershire, all on crumbling lime-rich cliffs of Lias or red Keuper Marl. In August 1996, Mark and Clare kindly took me round them, each site with just a single mature tree with a retinue of suckers. Two were out of reach – at least, as far as I was concerned – on steep, south-facing wooded cliffs. The third had been partially uprooted as the soft marl eroded, and the main trunk had jack-knifed downwards into mud at the base of the cliff. It was now projecting into the estuary at ninety degrees to the cliff, its tender shoots nipped by salty breezes. In time the former main trunk will probably be replaced by suckers growing upwards from the displaced rootstock, in a survival process Marc Hampton describes as 'climbing back up its roots' – a crucial adaptation for a long-lived tree growing on unstable cliffs.

Coping with rock falls and mud slides is only one of the problems faced by the Service-tree. It seems that they ripen fruit here only in exceptionally hot summers, like that of 1995. Their normally pear-less condition is one reason why they escaped notice for so long. And when pears are produced, there is another problem. Service-tree pears will soften and 'blet' only after they have fallen from the tree. When fully ripe they have a pungent acetic smell, like pear drops. In an experiment, they were offered to a range of farm animals. Cows, sheep and horses looked the other way; goats trifled unenthusiastically with unripe ones, but the only animal to get seriously stuck in was the pig. In the distant past, windfall Service-tree pears were probably eaten and dispersed by wild boars. In Britain, of course, free-ranging wild boars have been extinct for at least half a millennium. (Roe deer will also eat the pears, but they too died out in southern Britain during the Middle Ages.)

As if all this were not enough, every wild *Sorbus domestica* inspected so far has suffered from fungal disease caused by *Venturia aucuparae*, which leaves a nasty rash of black scabs on the leaves and twigs. The disease is worst on shaded plants, and, significantly, the trees seemed unusually free of it when they produced fertile pears in 1995. Hitherto, this fungus has been confined to rowans in Britain. Possibly fungal damage may be another factor limiting the distribution of Service-tree in the British Isles.

We begin to see why our woods are not full of Whitty Pears. The infertility of *Sorbus domestica* in Britain is shared by much of northern Europe. So is its habitat of warm, unstable cliffs. It is known that other trees, like Small-leaved Lime, ripen fruit less and less often as you travel north, and in places like northern England, hardly at all. Marc Hampton concludes that the Service-tree colonized Britain naturally in the distant past, but has survived only on sunny, lime-rich cliffs, whose instability precludes the growth of taller overshading trees. It must be no coincidence that all the sites found so far are botanically interesting. The Glamorgan sites boast Blue Gromwell and Maidenhair Fern among a rich calcareous flora characteristic of limestone headlands in south-west Britain. Another site has the only patch of Black Bog-rush (*Schoenus nigricans*) in the county, growing below a spring in lime-encrusted moss or tufa. A third site, on the Gloucestershire side of the estuary, contains an astonishing collection of scarce native trees and shrubs, including Small-leaved Lime, the endemic whitebeam *Sorbus eminens*, the sweet-briar, *Rosa micrantha*, and, outdoing its rival in numbers at least, the Wild Service Tree, *Sorbus torminalis*. Very probably, all these places are *refugia* for limestone plants and have never been shaded by dense woodland. In northern Europe *Sorbus domestica* is regarded as an 'indicator species' for such *refugia*. Its survival in Britain against all the odds seems truly miraculous. But, then, that is equally true of many very rare wild plants growing at the extreme limits of their natural ranges.

Which might lead us back to where we started. Isoenzyme analysis of the Glamorgan Service-trees shows that they are more closely related to the Whitty Pear in the Wyre Forest than to continental trees. That indicates that the Whitty Pear is at the very least the progeny of native Service-trees. Such a close genetic relationship implies that the species was once more widespread in the Severn valley, linking the estuary with the Wyre Forest, and it may yet turn up in other parts of southern Britain (there are, for example, old, discounted records from Devon that are worth reassessing). In the mind's eye, one can see that forgotten forest hermit taking a cutting from a

Service-tree he found on the banks of the Severn, and transplanting it outside his front door for good luck, just as rowans were planted out by crofts and farm bothies in Scotland. In so doing he left behind him the only clue that the true Service-tree grew wild in the British Isles.

The hermit, of course, knew this very well. Marc Hampton has found intriguing evidence that our distant ancestors knew that the Service-tree occurred in South Wales, and, like us, thought it remarkable.[79] A ninth-century compilation called *Historia Brittonum*, mentions that 'By the river which is called Guoy (Wye), apples are found on an ash tree on the declivity of the wood which is near the mouth of the river'. The context suggests that this apple-bearing ash tree, clearly the Service-tree, was regarded as a marvel, and was perhaps visited by seekers of such things – in a similar spirit to those who travel far to pay homage to a Monkey Orchid. Quite possibly its progeny survives to this day, for indeed the tree does still grow on a 'declivity' not far from the mouth of the River Wye. It is becoming a romantic species – a 'lost' tree, rediscovered by a different civilization with different beliefs, but perhaps the same ability to wonder and marvel at a plant with such unusual qualities.

Top left and right: The Service Tree, photographed in spring, when it is most easily distinguished, by its discoverer. [Marc Hampton]

Above: Ripe Service 'pears' compared with the berries of the otherwise similar-looking Rowan tree. [Marc Hampton]

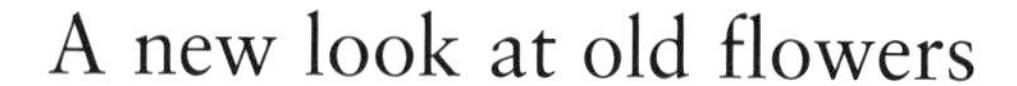

A new look at old flowers

The Service-tree became the only new native species to be added to the 1997 edition of Clive Stace's *New Flora of the British Isles*. At this end of the millennium, new natives are a fairly rare event, unless you count the apparently endless stream of new bramble or dandelion 'micro-species', or the occasional taxonomic revision that carves up a previously known species or sticks a previously split species back together again. I describe the more exciting 'new natives' of recent years later in the chapter and discuss why they were overlooked. Here I want to pause to look at another form of 'new native': plants which have long been known in Britain but which were previously assumed to be of garden origin. What led botanists to change their minds was not so much new information (though that played its part) as new ways of looking at plant distribution. Plant ecology today takes more account of history and past land-use than it did. We know, for example, a great deal about ancient woodland, and how to distinguish it from more recent or planted woodland. We know that woodland and permanent grassland plant communities are normally very stable; may indeed not change much over hundreds of years. We have also become more adept at spotting clues in the landscape, partly because ecology has become a more inclusive subject. The old way of studying plant relationships has broadened out, in one direction towards genetic and molecular analysis in a laboratory, in the other towards geographical and historical research in a records office. Perhaps, as a result, we have also become more imaginative. The discovery that time, as well as space, is fundamental to understanding what is in front of our eyes is one of the great imaginative leaps of British natural history – perhaps the greatest advance of our own time.

May Lily (*Maianthemum bifolium*), known wild for 400 years, but only recently acknowledged as native. [Bob Gibbons]

Take the May Lily (*Maianthemum bifolium*), for example. For a long time this rare flower was among the 'doubtfully native' elements of our flora. The scatter of dots on its *Atlas* map suggests no particular pattern apart from a preference for the eastern half of Britain. What added to the doubt was the fact that some of its sites were plantations of imported conifers. And the Lily has a habit of growing in small dense patches, suggesting that they are clones descended from a single individual, and, if so, are probably introduced. It was hard to be sure since most sites of the May Lily were kept secret. It is a pretty plant, like a modest version of Lily-of-the-Valley, with a pair of heart-shaped leaves and a single spike of virgin-white flowers. At its best-known site, Ken Wood, by Hampstead Heath, there was a suspicion of planted origin. The Lily has been there since 1814, but the original record was made by the steward of the nearby mansion house, who had reported all kinds of extremely unlikely plants.[20] He may have indulged in some wild gardening, not uncommon at that time. Botanists in the north of England had other theories to account for the odd patch of May Lily in wild places. In one place it might have been introduced with fir trees from Scandinavia. In another it might have sprung from the droppings of migrant thrushes or waxwings that had gorged on the scarlet May Lily berries in Norway before setting out. Yet another theory was that it was planted secretly in order to beautify the woods. All these ideas are quite possible.

However, other strands of evidence point to a plant that is at the very least well-naturalized in Britain. The May Lily has been known for a surprisingly long time. Gerard recorded it from 'Dingley Wood and Harwood' in Lancashire, in his Herbal of 1597. And although it has certainly grown under conifers, it is happier without them. At Forge Valley, near Scarborough, where the Lily was first discovered in 1857, it seldom flowered until the overshading conifers were felled during the Second World War, whereupon they burst into bloom. Unfortunately more fir trees were soon planted on top of them, but the Lily repeated the trick some thirty years later after they were cut down again, this time for conservation reasons and, one hopes, permanently.

What really changed people's minds about the May Lily was the rediscovery of ancient woodland (see picture on p. 217). As ancient woods were identified and mapped by teams of NCC surveyors during the 1980s, it was apparent that many less common woodland flowers showed a marked preference for old original woods, so much so that they came to be seen as habitat indicators. Most of the sites for May Lily came into this category, albeit woods often modified by felling and planting. It was also discovered that the Lily had, after all, a 'geography': it was not randomly distributed, but centred on light, well-drained acid soils, similar to its sites in Sweden. In Lincolnshire and Norfolk, it grows under sessile oak, which is almost never planted, often in company with that faithful ancient woodland indicator (in eastern England anyway), Lily-of-the-Valley. Furthermore, its occurrence in Britain makes sense in terms of its world distribution. The May Lily has a virtually worldwide range in northern temperate forests (in America it is replaced by the almost identical *Maianthemum canadense*). In that context, it would be more remarkable if the May Lily did not occur in Britain than if it did – for we would be a gaping hole on the world stage.

George Peterken, the woodland ecologist, has by chance seen most of the British sites for May Lily and has also been able to compare them with their counterparts in Sweden. He has become convinced that it is a native plant, and that its distribution contains a hidden pattern. It is a continental species, confined to the drier eastern half of Britain. The reason it is so rare is that woods on light acid soils in eastern England are a much-diminished habitat – most of them have long been transformed into open heathland by grazing or clear-felled and planted up with conifers. The May Lily is not a plant that can readily colonize new sites. Long observation shows that its populations are stable but extremely circumscribed. The evidence suggests that it is a relict species, confined mostly to extensive areas of ancient woodland. What might have saved it from extinction is its tolerance of shading. George once showed me the Norfolk site of May Lily. From what I had read about it, I was expecting a nondescript bit of woodland, with conifers lurking in the background. Instead, and to my surprise, we found the May Lily in one of the wildest corners of a large ancient wood, under gnarled sessile oak coppice with thickets of holly casting a dappled light over the dainty paired leaves and modest sprigs of bloom. This is not the naturalized plant of old, but a native plant emerging from its historical camouflage.

Until recently, the pretty pink-flowered Wall Germander (*Teucrium chamaedrys*) was also considered to be, at most, a naturalized plant. Its best-known locality is on the walls of Camber Castle near Rye, where it was first found by William Sherard and James Petiver in 1714 during a botanical journey from London to Dover. Very probably the plant had escaped from a former garden. Many of the places where it was found subsequently, like old garden walls at Curry Mallet and East Malling, or near Ruthin Castle in Clwyd, or on a bridge near Wexford in south-east Ireland, also suggest a garden escape. It can however persist a long time: Wall Germander has flourished on rocks near Weston-Super-Mare for at least 30 years and it can still be seen at Camber, nearly 300 years after Sherard and Petiver passed that way. Its origins were probably the gardens of physic maintained by apothecaries and physicians. Wall Germander was a medicinal herb. The dried leaves have a bitter, sage-like taste, and were used in the preparation of tonics, and as a treatment for gout – one of its beneficiaries being Emperor Charles V. Culpepper adds that the plant was 'most effectual against the poison of serpents', and that, when steeped in wine, the flowers helped to get rid of intestinal worms. No doubt that was why the apothecary, Petiver, recognized them immediately.

On the other hand, Wall Germander was described as looking 'perfectly wild' on limestone in two places in Somerset and near South Cornelly in West Glamorgan, where it has been known for a century. In 1945, a new and sizeable population of Wall Germander was discovered in natural chalk grassland between Cuckmere Haven and Beachy Head in Sussex, an area already noted for its rare chalk plants. In this short salt-pruned turf near the cliffs, the Germander is a dwarf, hairy plant, quite unlike the graceful trailing specimens on old walls, and is very difficult to spot unless the plants are in flower. These downs are similar to the places where Wall Germander grows in northern France, as near as the Pas-de-Calais. There may be two reasons why we overlooked it: we were looking in the wrong habitat and for a different plant. I wonder whether the relatively modern name Wall Germander might have created a false impression in Britain. The ancients, more interested in the medicinal leaves than the flowers, called it *chamaedrys* – the ground oak – after its oak-like leaves (the Ground Pine – *chamaepitys* – was named on the same basis). It is not, in fact, a specialized wall plant, but one of hot, dry, calcareous grassland. The bigger, less hairy garden plants are probably the hybrid of Wall Germander and the non-native *Teucrium lucidum*. The real thing is now considered to be an overlooked native plant at Cuckmere Haven. It may prove to be more widespread than we know. For example, Wolley Dod's herbarium contained a specimen collected in 1937 from 'downs at Blackcap', a few miles inland from Beachy Head.

Another wild flower whose native credentials are stronger than it is usually given credit for is Viper's-grass, (*Scorzonera humilis*). The species was not known in Britain until 1914, when it was found near Ridge in Dorset by the schoolboy Noel Sandwith. When I made a trip to Wareham to see it, I was expecting to find a wet field – and perhaps some evidence that this close relative of a garden vegetable, with a long black edible root, had been planted. What I found instead was a piece of wild wet Purbeck heath, a wonderful place of Cross-leaved Heath, Meadow Thistle, moor-grass, numerous kinds of sedge, and with at least one other rare flower, the Whorled Caraway (*Carum verticillatum*) – which grows with Viper's Grass in its native Brittany. It was a field only in the sense that it was surrounded by a fence and broad ditch. The Viper's-grass was visible as little spurts of yellow, flushed reddish on the outside, among the hummocks of moor-grass. It has a short season, and at other times the narrow ribbed leaves might be passed over as ribwort plantain. Moreover, like Yellow Goat's-beard (*Tragopogon pratensis*), the flowers open only in bright sunshine. Most of them are confined to this single field, but a few have established themselves in ditches nearby. But much of its former habitat has been destroyed and it once occurred near Poole on the opposite side of the Harbour.

Given its location at the edge of the fabled Purbeck heaths of Dorset, it is surprising that *Scorzonera* was for so long a 'Cinderella' plant, more or less ignored by the conservation world (its site was on private property, though it is now part of an RSPB nature reserve). In 1996, things suddenly looked very different with the amazing news that Viper's-grass had been found in South Wales, at Cefn Cribwr in East Glamorgan. The population runs to thousands of plants spread over five fields (though mostly in two of them). The habitat is said to be broadly similar to that in Dorset. This second site makes it now virtually certain that Viper's-grass is a native, relict plant. (Stop press: we now have a second site in South Wales!) Given its low

Right: A new site in South Wales discovered in 1997 established the native status of Viper's-grass (*Scorzonera humilis*) beyond reasonable doubt.
[Bob Gibbons]

Far right: Grey Mouse-ear (*Cerastium brachypetalum*) grows by railway lines in Bedfordshire and Kent. But which came first, the railway or the flower?
[Andrew Gagg's Photoflora]

profile in Wales, it could turn up in other places, Devon and Cornwall being the strongest possibilities. But given the scale of destruction of wet heathland and natural meadows in Britain, how many Viper's-grass sites might have been lost, unknown and so unmourned?

The annual Grey Mouse-ear (*Cerastium brachypetalum*) is so-called from its shaggy coating of hair, which gives it a hoary look, not much off-set by the narrow little petals held in the clasp of green hairy bracts. It was first discovered on a railway cutting in Bedfordshire, in 1947, and for a long time was assumed to be just another 'railway alien', perhaps introduced in ballast, or from leaking truckloads of foreign grain. Then, in 1978, a new site was found near Longfield in West Kent, also by a railway line. Although that at first reinforced the impression of a recently introduced species, subsequent searches in the area found six more populations nearby, all in natural chalk grassland. John Palmer concluded that, far from being a railway introduction, the Grey Mouse-ear was a native downland plant that just happens to have spread onto a nearby railway bank and track[133] (or maybe the railway engineers had cut straight through a wild colony?). I have seen a similar situation with the rare Cotswold Pennycress, one of whose Oxfordshire colonies is on the open ground between a road and a railway line. It looked like an 'introduced' waste ground site, until subsequent exploration revealed a much larger population hidden nearby on a natural limestone bank. The question, of course, is which way round was it? Did the Mouse-ear spread from the down on to the railway bank or vice versa? Either seems possible, and some botanists continue to believe that this is an introduction that has 'gone native' in one corner of Kent. But would they, one wonders, be of the same opinion if the Grey Mouse-ear had *first* been found in downland and only later on a railway line?

Native or introduced, the Grey Mouse-ear is an endangered species. The proposed route of the Waterloo link of the Channel Tunnel railway is set to wipe out a significant proportion of the plants. Furthermore, the Bedfordshire colony, described as 'in large quantity over a considerable distance' when Edgar Milne-Redhead discovered it, has been much reduced, and is now confined to two small areas on the west-facing bank. As an annual, the plant probably depended on a regular supply of temporary bare ground from fires started by steam locomotives. With the cessation of fires and other disturbance, the Mouse-ear has been squeezed out by more competitive plants. A recent conservation programme of controlled burning and rotovation may be the last chance for this endangered plant in Britain, unless it proves to be more widespread than we know.

Yellow Whitlow-grass (see picture on p. 224), *Draba aizoides*, the pretty springtime rock plant of the Gower peninsula in South Wales, was long thought to be another naturalized alien. It has been known in Britain since 1803, when it was reported 'growing wild abundantly on walls and rocks around Pennard Castle'. Like Wall Germander, Yellow Whitlow-grass will certainly grow on walls; it is perfectly suited for life on a crumbly stone wall, with its dense leaf rosettes attached to tough cranny-seeking horizontal stems, each one bearing its tight cluster of bright yellow flowers. From a flower's point of view, a wall is just another rock-face, with a generous provision of ledges and crannies, and plenty of available lime in the mortar. Two things long prevented the Yellow Whitlow-grass from immediately entering that select group of native flowers confined to limestone headlands in south-west Britain. First, its continental range is mostly southern or alpine (although there is an isolated colony in Belgium). And, more importantly, ever since its discovery there, people have nearly always gone to Pennard Castle to see the Yellow Whitlow-grass. Although the plant was first found not there but on wild cliffs near Worms Head, its apparent association with historic ruins was enough to label it as naturalized.

The Yellow Whitlow-grass became a better understood flower in 1970, when its ecology and history was described in detail by Kay and Harrison.[94] As they showed, the plant is not confined to Pennard Castle but is characteristic of sunny rock crevices right along the south coast of the Gower, mostly on steep limestone cliffs facing the sea but partly sheltered by an undercliff of grassland or heath. Here, in one of the warmest, mildest parts of the British Isles, it is part of a distinguished flora rich in rare species like Hoary Rockrose, Spring Cinquefoil and Western Spiked Speedwell. Most probably it colonized the castle walls from the wild, not the other way round. Further evidence that it is a native plant was adduced from its genetics. Quentin Kay found that the Gower populations differ from European plants, making a recent introduced origin most unlikely. The relatively low genetic variation suggests that the species was even rarer in the past, and it may have spread along the Gower coast after 'a bottleneck of small population size'. The horticultural raiding of more recent times does not seem to have caused any lasting damage. Here, then, is another example of an undoubtedly native plant tarnished by its ability to colonize ancient stones.

That degree of certitude is not yet possible for Sickle-leaved Hare's-ear (*Bupleurum falcatum*), one of our most endangered plants. Since its discovery in 1831, the Hare's-ear has been known only from a small part of Essex centred on Norton Heath. At one time, it was described as common over several miles of road verge and nearby field borders between Chipping Ongar and Chelmsford. By the 1950s, however, there was only a single patch left, centred on a hedgerow ditch. Even that last patch was not spared. In 1962, the Essex botanist Stanley Jermyn found a scene of ruin: the highway authority had cut down the hedge and cleaned out the ditch while realigning telegraph poles. 'On the site of the main colony,' noted Jermyn,[91] there was a large burnt patch 'which provided ample evidence of the fate of the plant'. That, it seemed, was that. But in 1979 the plant was refound on the same spot, whether from dormant seed from the original colony, or from garden seed sown there subsequently is uncertain. Perhaps the latter, since Sickle-leaved Hare's-ear seed is said to remain viable for only a short time. A year later the highway authority rendered the question academic by spraying the plants with herbicide.

There has always been debate about whether the Sickle-leaved Hare's-ear was a native species or an introduction. George Gibson, author of the nineteenth-century *Flora of Essex*, was in no doubt that it was 'truly indigenous'. Stanley Jermyn, on the other hand, thought it unlikely that earlier botanists would not have noticed it, and suggested that it had been accidentally introduced in fodder by troops returning from the Napoleonic War (there were transit camps in the area). In France, the species is associated with chalk rather than clay, as in Essex. Since then, new evidence has come to light which suggests that the plant may be native after all. While studying fossilized plant remains excavated from gravel pits in Warwickshire and Cambridgeshire, M.H. Field unexpectedly found half-seeds (mericarps) of *Bupleurum falcatum*.[54] Although the finds were difficult to date precisely, they prove that the plant was native in Britain during the warm period preceding the last Ice Age. The archaeological context indicates that it grew by watersides or on marshy ground at a time when the climate was similar to today's. Could the Essex plants be the last survivors of a species that has been with us for thousands of years? Perhaps we shall never know for certain, but it is interesting that Jermyn, who saw only a sadly endangered remnant close to a busy road, concluded it was an introduction, whilst Gibson, who knew a flourishing 'weed' of damp ground, field corners and ditches, thought

Yellow Whitlow-grass (*Draba aizoides*), confined to the Gower peninsula in South Wales. [Bob Gibbons]

it was native. The fact that a rare plant grows near a road does not necessarily mean that it came in with the road. One wonders what Norton Heath was like before arable cultivation and housing had reduced it to a few overgrown acres. Given that many of the 'continental' flowers of East Anglia are extremely restricted in range, it seems at least possible that Sickle-leaved Hare's-ear was a rare native of wet East Anglian heathland, but that it finally ran out of habitat.

Happily Stan Jermyn had had the foresight to collect seed from the original colony, and grow it on in his garden. The Jermyn garden became the 'seed bank' from whence the plant was introduced to botanic gardens and nature reserves in Essex. This stock has multiplied over the years, and a froth of yellowish Hare's-ears in a cottage 'border' is often a good habitat indicator of a keen field botanist. In 1988, the site, by now stranded between the old road and a new bypass, was landscaped and sown with seed descended from the original plants. The site is now looked after by the Essex Wildlife Trust, and though the Sickle-leaved Hare's-ear can no longer be regarded as a naturally occurring wild flower, in the context of its recent history this could be regarded as a happy ending.

Finally, there is *Gaudinia fragilis*, a wispy little grass, somewhat like an oat-grass in appearance, bearing a single row of bristly spikelets on a stem that turns startingly purple in late June. Although it has been known in Britain since 1903, most early records were from allotments or waste ground and seemed to place *Gaudinia* among the hundreds of sporadic aliens in the British flora. In his landmark book on British grasses, C.E. Hubbard[86] gave it just three lines (and no picture) and ever since our Flora writers have unquestionably followed where the great man led, assuming the grass to be just another introduction. Then, in the early 1970s, John Keylock began to find it most unexpectedly in unimproved pastures in Somerset, two of which are now National Nature Reserves! From 1978, it also turned up in damp, grazed meadows near Melksham in Wiltshire, again in grassland of nature conservation interest. In subsequent years, *Gaudinia* was found in more and more fields in North Wilts, and also in fields and verges in neighbouring Dorset and on the Isle of Wight. A comparison of records begins to reveal an unexpected 'geography'. The grass prefers heavy clay soils on the Jurassic Lower Lias series, and the majority of sites are old grassland, rich in local species like Adder's-tongue (*Ophioglossum vulgatum*), Dyer's Greenweed (*Genista tinctoria*), Grass-leaved Vetch (*Lathyrus nissifolia*), Pepper Saxifrage (*Silaum silaus*) and Green-winged Orchid. A regular associate, particularly in Dorset, is the impossibly named Corky-fruited Water-dropwort (the Latin name is just as bad – *Oenanthe pimpinelloides*), which has a strikingly similar distribution to *Gaudinia*, as well as sharing the same grassy habitat. Their principal need is for a rather open turf where the plants have room to establish and set seed. *Gaudinia* cannot compete with tall aggressive grasses, and does best where the soil is naturally poor, thin and unfertilized. Given the right conditions, it thrives in meadows cut for hay in the summer or regularly mown verges or grazed pasture on hillsides. All this suggests an overlooked native species, not a 'casual'.

Could *Gaudinia* really have been missed by our botanical predecessors (to say nothing of the conservation surveyors and managers who must have been over these fields many times)? David Pearman tells me it is hard to spot until you are 'clued in to it' (though the purple stems are a give away), and the season is a short one for the mature inflorescence soon breaks up. On the other hand, the grass is quite common within its restricted habitat and range. It has even acquired a common name, 'French Oat-grass'. Yet there still seems room for doubt. *Gaudinia* is classed as a Mediterranean species, though it occurs on the Atlantic coast as far north as Brittany. Pearman discovered that some of its fields in Somerset had been under the plough for a couple of years during the Second World War, and that they were subsequently resown with seed of French origin. Although much of the native flora seems to have survived the temporary ploughing, so much so that today you would never have suspected an arable episode, it does seem possible that *Gaudinia* could be a wartime introduction. Perhaps further clues will emerge from investigation. As Pearman remarks, 'the detective story goes on'.

Newly discovered wild flowers

The days when a naturalist could go for a walk and discover a new native wild flower before supper are long over. The surprising thing is that probable native species have eluded us for so long, and that they continue to turn up at a rate of one every two or three years. More surprising still, some of these species are attractive, even spectacular wild flowers. I mentioned a few of them in Chapters 3 and 8 – the Diapensia on its shattered summit, the Proliferous Pink on its Norfolk heath, the mysterious Tongue-orchids of the south coast. Here I outline the story of four more remarkable discoveries of very recent years.

The tracking down of what is now known as the Radnor Lily (see picture on p. 228), *Gagea bohemica*, formerly the Early Star-of-Bethlehem, was a prolonged piece of botanical detective work. The story began in 1965 while R.F.O. Kemp was collecting samples of moss at Stanner Rocks, an outcrop of lime-rich volcanic rocks in Radnorshire and a well-known locality for rare plants. It was some time later that he noticed a bulb with a few narrow withered leaves among the moss, which he thought was *Lloydia serotina*, the Snowdon Lily.[211] This was amazing news, for *Lloydia* is known only from high altitude cliffs in Snowdonia. In April 1974, Ray Woods searched the area hoping to find better-preserved specimens, but found only a shrivelled flower with white petals among hundreds of spidery sterile leaves. However, these leaves were covered by short, crisp hairs, unlike *Lloydia* which is hairless. Ray suspected that the mystery species was something even more exciting, a species of *Gagea* new to Britain. Next year, he returned in mid-January when even the Welsh snowdrops had barely begun to burst their buds. He was rewarded by a single beautiful flower, not white but a glorious celandine-yellow. Clearly its removal was undesirable, but an inspection of European floras suggested that the plant was *Gagea bohemica*, a very local species of base-rich rocks in central and south-east Europe. This tentative identification was confirmed in March 1978, when Ray Woods, David McClintock and the *Gagea* expert E.M. Rix met on site and were at last rewarded by 25 specimens in full flower.[168] *Gagea bohemica* is an exquisite plant, virtually all flower, which unfolds from a crocus shape to a star, held in a wreath of narrow curling leaves with a pair of broader, stiff stem leaves spread out below the head like little hands. It is a flower to stop you in your tracks, but it is not surprising that it was overlooked. Stanner Rocks in winter is no place for a casual visit, with its keen winds and flurries of sleet. There is little likelihood of an introduction here. The Welsh plants share the hairy leaves of specimens from western France; those found further east are largely non-hairy. And it grows on the same sunny, summer-dry rocks that support Sticky Catchfly, Rock Stonecrop and other rare flowers. One possible reason for its extreme rarity in Britain is that the plants are given almost no chance to fruit. When mild weather returns, the ubiquitous Welsh sheep soon hoover up any remaining flowers and flower-stalks, though the plant might in any case be a reluctant fruiter so far from its normal, more continental, range.

Another astounding recent discovery is the Fringed Gentian, *Gentianella ciliata*. Not only is it an attractive and (one might have thought) unmistakable species – with big blue flowers, fringed along the edges like pale eyelashes – but it grows in a well-trodden beauty spot in the Chilterns, close to two public footpaths. The plant was discovered in 1982, by Peter Phillipson, and was confirmed as *Gentianella ciliata* by R. Pankhurst at the Natural History Museum. A site visit that autumn revealed a small but flourishing colony with some 50 flowers. The area is natural, sheep-grazed chalk grassland, already scheduled a Site of Special Scientific Interest (indeed it had been recommended as a nature reserve back in 1916). The associated plants, like Horseshoe Vetch and Dwarf Thistle, are typical of the places where the Gentian occurs naturally in Belgium, Holland and northern France.[100]

It transpired that Peter Phillipson was not, after all, the first botanist to find Fringed Gentian in Britain. As long ago as 1875, Miss M. Williams found an unknown gentian on the downs near her home in Wendover and sent a specimen to the British Museum, where it still survives. It was initially misidentified as Marsh Gentian or 'Calathian Violet', but was correctly named by James Britten, editor of the *Botanical Journal*, a few years later. Britten, however, assumed there must have been some mistake. Further confusion ensued in 1926, when Druce dismissed the record out of hand with this comment in his *Flora of Buckinghamshire*:

> *Gentiana ciliata* L. Calathian Violet. Error. On a hill not far from Wendover, Miss Williams in *Journ. Bot.* 1785 [*sic*] but the specimen is *Campanula glomerata* ... There must be some gross carelessness in such a record, as *ciliata* is not likely to occur in England.[46]

In fact the occurrence of Fringed Gentian on the French chalk not far from the Straits of Dover makes its occurrence on the chalk of southern England not at all unlikely. Miss Williams had quite possibly found it in the same spot as Phillipson, and if so the plant must have bloomed over scores of seasons without being noticed. Two more old records have surfaced since 1982. One, from a meadow at Limpsfield in Surrey in 1910, seems unlikely to be the native plant, as this is an area of acid soils – and the meadow concerned, alas, has been 'improved'. A second record, from the Wiltshire chalk, looks much more promising. An unsuspected specimen was discovered in the herbarium at the Natural History Museum by a Chinese botanist, T.N. Ho, while working on gentians. It had been collected by Edward John Tatum in September 1892 at Pitton, near Salisbury, from a down 'at the junction of chalk and Tertiary beds'. Tatum, known to have been a diligent and active field botanist, had sent it to an expert for confirmation, and it had lain there forgotten in a folder ever since.[42]

How could such a beautiful plant have been overlooked? One probable

Radnor Lily or Early Star-of-Bethlehem (*Gagea bohemica*) flowers in late winter, long before the botanical tourists are about. [Ray Woods]

reason is that Fringed Gentian flowers in mid-September, much later than most chalk flowers, and it might have been confused with the Autumn and Chiltern Gentians that flower at the same time. Another may be that the species flowers irregularly, and in some years hardly flowers at all. Like Early Gentian it relies on barish patches of ground, and probably survives unfavourable periods as dormant seed. Furthermore, its flowers open only in direct sunshine. The late John Fisher (1992) gave a nice description of them:

> They are deep purple, and up to 1½ inches across. When open, the long, narrow petals are disposed windmill-fashion, like the four blades of a purple propeller, and each has along its edges a white silken fringe, the 'eye-lash' which gives the plant both its English and its botanical name.[56]

On dull days, the flowers close up tight with the fringes still visible, like a furled banner. When closed, the flowers are almost impossible to detect in the turf, as I discovered during a hands-and-knees search for it in August 1984 with Richard Fitter. The site, though a good example of well-managed chalk grassland, is not exceptionally rich in rarities. It is hard to avoid the conclusion that sooner or later, it will turn up on other steep, sunny chalk escarpments in South-east England.

The Leafless Hawks-beard, *Crepis praemorsa*, was discovered in the Lake District in July 1988, by Geoffrey Halliday, during a search for a related species, the Soft Hawks-beard (*Crepis mollis*). It was flowering on a low bank by the side of a hayfield, in botanically-rich, natural grassland. Though quite a handsome plant, with clusters of big bright-yellow flowers that open fully only in bright sunshine, *Crepis praemorsa* is not sufficiently different from other more common hawks-beards to attract much attention to itself.

Above: Fringed Gentian (*Gentianella ciliolata*) - overlooked for a century. [Peter Roworth]

Above right: Unspotted Lungwort (*Pulmonaria obscura*), probably a rare native plant of ancient woodland. [Bob Gibbons]

However, since Halliday was looking specifically for a *Crepis*, he took a good look at this one and saw it was different to any known British species, most notably in its leafless flowering stems and unusual blunt-ended leaves with a small point at the tip. After studying *Flora Europaea* Halliday realized he had found *Crepis praemorsa*, a widespread Eurasian plant not yet recorded from Britain. The nearest wild plants lie hundreds of miles away, in eastern France. A pressed specimen of it went on display at the BSBI's annual exhibition in 1988.

Initial reaction was sceptical. The plant was assumed by some to be a recent introduction. But it does not behave like one. The two fields in which it has so far been found are at the heart of an area of limestone drift rich in rare plants, like Alpine Bartsia, Bird's-foot Sedge and Bitter Milkwort. Moreover Leafless Hawk's-beard grows in natural grassland, similar to some of its continental sites, which are not the sort of place one expects to find exotic plants. As Halliday expressed it, 'One's natural instinct when confronted with such an isolated occurrence is to suspect introduction but this seems highly unlikely. The site is a piece of grassland in a remote corner of the county ...'[77] The editors of the Red Data Book agree, and the latest edition will include *Crepis praemorsa* as a probable native species.

East Anglian botanists have long known a lungwort with sorrel-shaped leaves that grows in a wood near Burgate in Suffolk. They assumed it was a form of the Garden Lungwort, *Pulmonaria officinalis*, which had escaped into the wild. However, as they realized, this one did not behave like a garden escape but as a settled part of the woodland vegetation – and in any case there were no gardens nearby for it to have escaped from. C.J. Ashfield, who first found it in 1842, considered that 'there can be no doubt about it being a genuine wild locality; for the plant is plentiful, it grows far into the interior of an extensive wood, and has much the appearance of being truly wild as any of the plants near it'. Visitors like W.M. Hind and E.S. Marshall noted that these woodland lungworts differed from the garden forms in several respects, notably in the absence of spotting on the leaves. Like Ashfield, they considered it to be fully wild, and 'the county's unique production'.

Although the 'Burgate lungwort' was visited regularly, and most of those who saw it tended to come to the same conclusion, the plant seldom received much publicity outside its native county. Its identity was subsumed inside descriptions of the Garden Lungwort. Clapham, Tutin and Warburg even worked out a description which combined both the garden and the Suffolk plants, a composite species whose leaves could be either ovate or cordate, and either spotted or unspotted.[30] Moreover, this flexible plant could occur in hedgebanks, as the Garden Lungwort sometimes does, but also in woods, which, on the whole, it doesn't. The true identity of the cordate, unspotted woodland variety began to become apparent in the 1970s, when Oliver Rackham showed that its locality was ancient woodland with a long history of coppicing. Within the wood, the lungwort occupies a precise and stable ecological niche, and is associated with native woodland flowers like Wood Anemone and Primrose. In 1985, Donald Pigott, then in charge of the University Botanic Garden at Cambridge, identified it as *Pulmonaria obscura*, a species he had come across previously in Poland. He considered it to be probably native, partly because this species is rarely cultivated, partly because its occurrence in Suffolk is consistent with its continental distribution in Europe. If so, one would expect the plant to occur in other woods in the area. Sure enough, when Martin Sanford began to search nearby ancient woods, he found a new colony in the neighbouring parish. It was time for *Pulmonaria obscura* to be given an English name. It is a pity that the one it was given, 'Unspotted Lungwort', celebrates a negative (rather like calling *Turdus merula* the Unwhite Bird), but it does at least underscore its most distinctive character, the plain dark grey-green leaves, as opposed to the fancy white-sploshed leaves of the garden plant.

The ecology of the new species, as worked out by Sandford and Chris Birkinshaw,[14] suggests that *Pulmonaria obscura* is not only a native plant but a very good indicator of ancient woodland. It grows on ill-drained calcareous clay in ash-maple woodland, and flowers best along rides and in recently coppiced glades. It seems to have declined during the past hundred years, probably because these woods are no longer coppiced regularly and so have become more shady. The discovery in 1993 of apparently new plants growing on recently created woodland rides suggests however that seed remains viable for many years. On the criteria for native species provided by David Webb (see Chapter 8), it wins easily. Unspotted Lungwort is the latest of East Anglia's continental plants, and a further indication that, climatically speaking, the interior of Norfolk and Suffolk has more in common with Germany or Poland than Wales or Cornwall.

Chapter 11

Plants and people

Tucked away in a fold of the downs not far from the Wiltshire village of Ham is a wonderful place with eight or nine kinds of wild orchids, a colony of Chalkhill Blues with wings the colour of a summer sky, and a kaleidoscope of colour in the autumn when the chalk shrubs are turning red and gold. It is a pleasant place to sit or snooze on the thyme-scented turf (especially after lunching at the pub below the hill) and idly watch the small lives going on in the grass. This little refuge – it is only ten hectares or so – is not exactly a secret, for it has been a nature reserve for some years past. But it is out of sight by a narrow lane, and so little known except to a few walkers, naturalists and, I notice, artists. Or so I had assumed. It was with mixed feelings that I read in the published diaries of Frances Partridge, *Everything to Lose* (1985), that half the literary establishment of England seem to have been there. On 6 July 1954, for example, Frances had taken Gerald Brenan 'to see the orchids now blooming on Ham Hill – the Fragrant, Pyramid, Burnt Stick, and even the Musk, which he had never seen before'. Later, she used to count the tiny green spikes of the Musk Orchid each year for the Wiltshire Botanical Society. On another occasion, she had the botanist Noel Sandwith over to stay, and together they visited several more 'secret' spots in the area.

> Of young middle age and medium build, he walked with a long springing tread, his eyes flashing with expectation behind his gold-rimmed spectacles. He led us to the sites of several rare species – the last known patch of the Monkey Orchid (*Orchis simia*), for instance, saying as we approached it: 'I really ought to blindfold you here.' When I took him to see our local Bath Asparagus (*Ornithogalum pyrenaicum*) he was more interested in an unpleasant reddish dust on its stems and leaves than in the plant itself, exclaiming excitedly: 'A rust, a *marvellous* thing!', pounced on it and sent it off to Kew. It was somehow tempting to tease

> him and even shock him a little. Richard once asked him why he had never married. 'Because of my great love for the flowers,' he replied without hesitation. He fitted in well with all our friends. As Julia wrote to me with her usual acumen: 'Sandwith is a very soothing element, is he not?'[137]

Given our long love affair with nature, English literature and verse is surprisingly thin on botany, or at least, on scientifically accurate botany. David McClintock once summed up poetry about wild flowers in a phrase: 'botanical piffle'.[115] A few poets, most famously John Clare and George Crabbe, did describe real flowers and trees in realistic settings, but that was probably because they were amateur naturalists as well as poets: they wrote from observation, not mere imagination. In general, though, poetry and science seem to have had little to say to one another. The first holds the world to be a theatre, the second a laboratory. To a poet, flowers matter only to the extent that they affect the writer's feelings. A scientist, by contrast, would see them mainly as a source of data. Amateur naturalists ought to bridge the gap, and some of our forebears certainly did. But we have long been in thrall to the scientist. A paper I once wrote for the botanical journal *Watsonia* was returned to me with all the literary flourishes firmly pencilled through. Quite right too: in that context they would have stuck out like an inflamed pistil.

When it comes to rare, and therefore unfamiliar, flowers, English literature is, with a few significant exceptions, even thinner. There is, of course, a tradition of rural writing which often found its way into works of natural history, like those of Richard Jefferies or Edward Thomas, and continues today with writers like Richard Mabey. Popular floras used to include all sorts of interesting detail on local folk-lore and uses that is now routinely excluded. Much of this material was gloriously exhumed in *The Englishman's Flora* by Geoffrey Grigson (1905–85). It is the irrepressible touch of the poet that makes Grigson's work such a joy to dip into. The strange tubular petals of Round-headed Rampion he compares to 'a violet sea anemone – air anemone – closing upon an incautious bee or fly'. For the habitat of Loddon Lily or Summer Snowflake 'imagine a black swamp on the edge of the Thames, alders or willows overhead, a swamp that quivers and soggs and stinks', in which, by contrast, the white flowers of the Snowflake hang trembling from their stalks 'in a severe purity'. The observation is sharp, vivid and *just right*: you can picture them in your mind's eye. But it takes the confidence of an important poet to dare to breach the unwritten rules so flagrantly.

Two books by Andrew Young (1885–1971), *A Prospect of Flowers* (1945) and *A Retrospect of Flowers* (1950), are full of the spirit of amateur botany and 'wild flower chases' as he called them. Though he held a lifelong passion for wild flowers, Young usually manages to create the air of an innocent onlooker when he writes about botanists and their odd ways. But no one has ever written quite so understandingly about flower hunting. This extract, which comes from a chapter called Collecting Plants, must have echoes for most of us.

> The greatest advantage is that, known not to pick rare plants, you can be trusted with the secret of their whereabouts. Not everyone can be so trusted. Sitting one day with a friend in the back of a car, I asked, 'But where did you find the Purple Spurge?' She had been on the point of telling me before we entered the car, but had delayed to explain she had had to swim from one Cornish cove to another, the latter being inaccessible from the land side. So my question seemed natural enough, but its only answer was a sharp dig in the ribs. Turning to see an angry frown, I realised my mistake. It was dangerous even to mention Purple Spurge in the presence of the two men in front. 'Sorry,' I whispered, humbly. Yet I was only half-sorry; she looked so pretty, putting on an ugly face.[214]

Note the throwaway line about this plant-mad girl who had *swum* from one cove to another, all for the sake of the tiny weed in the shingle, as though that was nothing out of the way. That is what wild flower hunters do. My precious copy of one person's plant-finding instructions made between the wars requires almost equally perilous ventures along railway lines ('if you keep left of the line they won't see you from the station'), crossing windy lakes by row-boat and a scramble along a precipice to a rock shaped like a castle.

Non-botanical books about rare flowers are very rare indeed, but two novels of rural life, written within a few years of one another, do use a rare flower to emblemify something important, yet inexpressible. In *Midsummer*

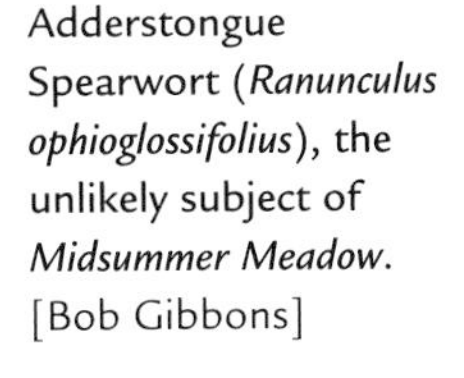

Adderstongue Spearwort (*Ranunculus ophioglossifolius*), the unlikely subject of *Midsummer Meadow*. [Bob Gibbons]

Meadow (1953) by John Moore, the significance of the rare 'buttercup' *Ranunculus ophioglossifolius* lies in its very insignificance. In *The Military Orchid* (1948), the apparently extinct orchid stands for something lost and now unobtainable – a symbol for the author's Housmanesque 'Land of Lost Content'. In neither book is the subject treated scientifically, but each tells us something about the personal reasons why we look for rare flowers and why we try to protect the places where they grow.

Midsummer Meadow is 'the story of one English field', which encapsulates the struggle between the preservation of rural England and the Juggernaut of progress. The novel is set somewhere in the middle of England 'by Shakespeare's Avon' in the fictional parish of Drake's Norton. Doctor Tidmarsh, a 'retired physician, amateur naturalist and incorrigible snapper-up of unconsidered trifles', has bought the meadow to protect a rare buttercup. He also allows a travelling fair to camp there, thus annoying the local council. The story is about his fight to uphold their right – and the buttercup's right – to be there, and it unfolds as a kind of morality play, with the meadow as the stage, and Tidmarsh representing individualism set against the grey, monolithic forces of the state. He had discovered the 'buttercup with snake tongue leaves' in 1939 while 'pottering and poking about the margin of a shallow pool'. An expert identified it for him as the very rare Adderstongue Spearwort (see picture on p. 233). The very lowliness of the flower appealed to Tidmarsh's paternal instinct:

> Quaint things, humble things, curious things, as we have seen, were his special care and delight; a tiny lizard, sunning itself on a rock, would enchant him for hours, distracting his attention from the most magnificent landscape, and a minute beetle crawling beneath his magnifying glass would cause him more wonder than a whole sunset-sky. Smallness in itself made a powerful plea to him; puniness excited his compassion; he was always for David against Goliath. So he took the buttercup under his protection, as later he took Mr Oliver's exiguous Fair. He bought Midsummer Meadow for two hundred pounds and set aside a sum of money, to be administered by a legally-constituted Trust, for its preservation after he was dead. This was his Buttercup Fund, so designated at the Bank where he was thought to be harmlessly and amusingly mad.

John Moore wrote *Midsummer Meadow* at a time when rural life was rapidly changing, and when many old wet meadows were being underdrained and converted into arable prairies using public money. There were no powerful lobby groups to speak up for rare flowers then, and Tidmarsh's own motives were entirely personal. The flower matters to him because, paradoxically, it does not matter.

> 'It means an awful lot to you, doesn't it, this little flower?'
>
> 'Yes; and I really don't know why! It gives pleasure to hardly anybody; it serves no useful purpose at all. But I'm sick to death of the argument that only useful things matter; we're bedevilled by that attitude to-day. And my buttercup's quaint, and it's rare, and this small patch of ground is its last refuge but one in all Britain. I can't find anything else to say in its favour' – He grinned, almost ashamed of his absurdity in loving the small unlovely flower so much – 'except that it's *mine* and if the River Board wants to drain this field I shall fight for it like a she-bear for her young.'

Preserving the last but one colony of *Ranunculus ophioglossifolius* was a small vindication of his own existence, modest and ordinary though it was, like most lives.

> 'I never did anything much except deliver a lot of babies, diagnose a lot of measles, scarlet fevers and mumps, and dabble in every useless hobby from beetle-collecting to brass-rubbings. When I die somebody will have the awful job of sorting the contents of my back room, and if they've got any sense they'll ask the dustman to make a special journey and cart all my rubbish away. But thanks to my buttercup I yet have my little niche of fame. Somebody glancing through the County Flora a hundred years hence will read '*R. ophioglossifolius*, Vill. Discovered 1939 in Midsummer Meadow, Drake's Norton, by E. Tidmarsh who created a Trust for its preservation there'. And he'll think to himself: R.I.P. That's all the immortality I shall get, all the immortality I want, and certainly all the immortality I deserve!'

Adderstongue Spearwort is, of course, a real flower, and its history bore out John Moore's account of a peculiarly luckless plant. It was first discovered by C.C. Babington at St Peter's Marsh in Jersey in 1838. Fifty years later, the marsh was drained and it died out. Then the plant turned up in mainland Britain, when Henry Groves saw it in a wet ditch near Hythe, Hampshire in 1878. A few years later, the ditch was drained, and the Spearwort was extinct for a second time. In 1914, it appeared in a wet meadow in Dorset, and this time it was road-widening that hastened its demise. More fortunate was a strong colony discovered by a shallow pond in the corner of a meadow at Badgeworth, Gloucestershire, in 1890, which survives to this day. Badgeworth lies barely a dozen miles from Moore's home town of Tewkesbury. Plainly his idea for the novel had come from there. At Badgeworth all ended well when local botanists bought the place for a nature reserve (but see Chapter 13). The conclusion of *Midsummer Meadow* is more tentative with the drainage engineers closing in, but a hint that unseen natural forces will protect their own. *Midsummer Meadow* is not a botanical novel, and although the 'buttercup' provides the motive for much of the action, it rarely makes an appearance. It is simply there, standing for the apparently trivial particularities which make one place different from another.

A rare flower lies closer to the centre of the stage in *The Military Orchid*, the first book of a semi-autobiographical trilogy by Jocelyn Brooke (1908–66). From early childhood, Brooke worshipped wild orchids, 'those floral aristocrats, with their equivocal air of belonging partly to the vegetable, partly to the animal kingdom'. The idea of the Military Orchid combined both his botanical pursuits and his longing to become a soldier. The book is not a conventional autobiography, and is concerned mainly with Brooke's childhood on the Kentish downs and his experiences in wartime Italy, where he served in the Royal Army Medical Corps. For him, the orchid provides the quest motif that runs through the book:

> The Military Orchid ... For some reason the name had captured my imagination. At this period – about 1916 – most little boys wanted to be soldiers, and I suppose I was no exception. The Military Orchid had taken on a kind of legendary quality, its image seemed fringed with the mysterious and exciting appurtenances of soldiering, its name was like a distant bugle-call, thrilling and rather sad, a *cor au fond du bois*. The idea of a soldier, I think, had come to represent for me a whole complex of virtues which I knew that I lacked, yet wanted to possess: I was timid, a coward at games, terrified of the aggressively masculine, totemistic life of the boys at school; yet I secretly desired, above all things, to be like other people. These ideas had somehow become incarnated in *Orchis militaris*.

For Brooke, the 'legendary quality' of the Military Orchid took on the proportions of a rural myth, a lost golden age. It symbolized bygone glories, vanished beyond recall like the old red-uniformed soldiers it mimicked: 'gone with scarlet and pipe-clay, with Ouida's guardsmen and Housman's lancers; gone with the concept of soldiering as a chivalric and honourable calling'. They now lived only in the 'land of lost content', of Housman's famous lines from *A Shropshire Lad*:

> What are those blue remembered hills,
> What spires, what farms are those?
>
> That is the land of lost content,
> I see it shining plain,
> The happy highways where I went
> And cannot come again.

Military Orchid (*Orchis militaris*) by Christina Hart-Davies, 1998.

Brooke is a poet, writing perceptively, and with humour and ironic detachment, about favourite things – orchids and fireworks, people and places. The botanical content is skillfully interwoven into the thread of his life, and he shares with Grigson the gift of making his botany intelligible to the unbotanical. He is particularly good on the tricks of temperament and circumstance that turn people into 'botanophiles' – of that odd mixture of unexplained passion, desire and erudition that goes into British field botany. On the heels of the 'Orchid Trilogy', Brooke exorcized his passion for orchids further with a de luxe volume, *Wild Orchids of Britain* (1950), with 40 watercolour plates by Gavin Bone and a hefty price tag. It is now a collector's piece, and it was rather bad luck that the book was eclipsed soon afterwards by Victor Summerhayes' better-known volume in the New Naturalist series.

When Jocelyn Brooke was writing *The Military Orchid*, the real plant was believed to be extinct – it had vanished suddenly and inexplicably. The last reliable record, dated 1902, was from Oxfordshire. A friend of Brooke's made a special journey to photograph it but the flowers had begun to wither. He decided to postpone his photographing until the following year, by which time, of course, the plant had gone, seemingly for good. The nearest that Brooke came to The Military was to find its closest relative, The Monkey, then regarded, improbably, as a subspecies. 'It was nice to have the Monkey; but there seemed to me something slightly inferior about a "sub-species". The very phrases of Edward Step sounded faintly derogatory – "A sub-species *known* as the Monkey Orchid" ... it suggested the subtly insulting phraseology of the police-court: "a woman *described* as an actress".'

Of course, the Military Orchid did not vanish for ever. By coincidence, it was refound in

Buckinghamshire even as Brooke was writing its obituary. The discovery, by J.E. Lousley during a family picnic in May 1947, was acclaimed by John Gilmour[69] as the most successful example of 'unplanned botany' in modern times. Lousley himself described the moment as follows:

> In May, 1947, I rediscovered the Military Orchid! In a way it was just luck. The excursion was intended as a picnic, so I had left my usual apparatus at home and took only my note-book. But I selected our stopping places on the chalk with some care, and naturally wandered off to see what I could find. To my delight I stumbled on the orchid just coming into flower ... Careful plotting showed that there were 39 plants in the colony and that 18 of them had thrown up flowering spikes. Of these, the 5 most exposed had the flower-stems bitten right off – almost certainly by rabbits. The plants most in shade either failed to flower or put up only pale, small spikes ... The smallest was only about four inches tall with two flowers – a miserable depauperate little plant. The largest – the one which is illustrated – was 14 inches (35 cm) tall with no less than 26 flowers. This must be about the finest Military Orchid seen in England; for most of the herbarium specimens are only some 7 to 9 inches in height.[106]

Fearing collectors, Lousley kept the site a close secret, but in the 1960s a nearby colony, if not the same one, was found at Homefield Wood near Marlow. The happy finders sent off a cryptic telegram announcing '*The soldiers are safe in their home field*', which has been a standard item of botanists' table talk ever since. There is a companion jest about another occasion when a party failed to find the Monkey Orchid. As David McClintock recounts, two of the party had returned home, leaving two women, Jocelyn Russell and Nancy Saunders, still searching. On their way back, the former had stopped at a pub near Reading, and from there sent a triumphant telegram: '*Monkey in a vegetative state at 750738.*' The women bought a map specially, found the spot, and excitedly got down on all fours to search for the orchid. Finding nothing but grass, they eventually looked up and only then did they notice the sign creaking in the wind: 'The Green Monkey'. They were on the front lawn of the pub.[115]

Today we need no longer to dream about The Military or The Monkey. They are advertised on Wildlife Trust Open Days, some plants inside little wire hats, others specially earmarked for photography. A warden is there to tell you their story, and Trust merchandise is on sale in case you need a souvenir. All very right and democratic. People should be able to see rare orchids, if they want to, and it is good for the cause. But it takes all the romance out of flower hunting.

Rare flower icons

Has your local rare flower ever appeared on a postcard – or on a shop sign, or souvenir tea towel, or coffee cup? Whenever I pop that question, the answer nearly always seems to be 'no', sometimes with 'thank goodness' added as an afterthought. Perhaps so, but in an age when everybody seems to enjoy shopping so much, someone would surely use rare flowers in that way if they spotted a market. And if there is no market for rare flowers, this seems very surprising in a nation so keen on preserving them. My own research in this field is confined to an expedition to Cricklade to see whether the local rare flower – a famous one, the Fritillary lily – has made any impact on the consciousness of that attractive and historic town. I checked street names, pub signs, and postcard racks. I scrutinized patterns on porcelain and tea towels. I peered at local brand-names and home-brews. I put my head round the door of the magnificent parish church and at last found the Fritillary, on the embroidery of a kneeler. But that was the only hint that, within sight of the church tower, Cricklade has the finest Fritillary meadow in Britain; that it is the only parish left where Fritillaries still outnumber people. And the only information about the flowers is in the nature reserve leaflet, but you have to send away for that.

Admittedly, Cricklade is not a tourist resort – no lake, beach or funfair – and nowadays you can flash past it on the new M4–M5 link road without even noticing its existence. But even in the busiest rural honeypots, such interest as there is in wild flowers seems to be generalized rather than particular. As far as I or my sources know, you cannot purchase a postcard of Tintern Spurge at Tintern Abbey, or a Lizard Clover on a mug at the Lizard, nor even a Cheddar Pink on a tea towel at Cheddar (though at one time you could buy the real thing!). Possibly nature today is seen in terms of generalized causes and campaigns, not actual flowers and places. Or maybe the sort of people who travel to see Fritillaries and Cheddar Pinks just don't buy souvenir tea towels.

Rare flowers *do* occasionally appear on postcards, but mainly ones commissioned and sold by conservation bodies. A very attractive series of cards produced by the National Trust for Scotland in the 1960s displayed rare alpine flowers drawn by Bessie Darling Inglis for the Trust's book about Ben Lawers. By painting in the plant itself, while leaving its background in line, Inglis captured the jewel-like nature of the flowers. The cards were evidently aimed primarily at botanists, for each painting was accompanied by the Latin name only. As far as I know, they were sold mainly at the Ben Lawers visitor centre. To my eye, the charm of these cards lay in their being coloured drawings, not photographs. Purists might argue that *Gentiana nivalis* rarely grows among bare stones, as shown here, nor does *Saxifraga cernua* nor-

Postcards of alpine flowers of Ben Lawers sold by the National Trust for Scotland.

mally enjoy a lush bedding of moss and lady's-mantles. But the artist's spareness and delicacy matched the qualities of the flowers themselves.

Ben Lawers is bound to attract the botanically inclined in the sense that there is little else on offer there, except some rather dull mountain walks. Few other places advertise their special flowers quite so prominently. Most postcards of wild flowers show colourful common species – the ones the average visitor will see, not necessarily the ones which are characteristic of a particular place. In Rosemary FitzGerald's comprehensive collection of them, Red Data Book flowers form a very small minority. J. Arthur Dixon's 'photogravure post card' of the Spring Gentian is one of the few produced by a commercial company. The Kenneth Allsop Trust produce souvenir postcards of the famous Wild Paeony of Steep Holm (see p. 172), designed to be posted from the island to match a flower which has itself been turned into a souvenir in the form of packets of seed. There is also, perhaps, a faint association of flower and place in the Northamptonshire Wildlife Trust's card of Pasqueflowers, or in the Bird's-eye Primrose among a mélange of flowers produced by Yorkshire Dales National Park, or the Round-leaved Wintergreen on a card produced by City of Liverpool Museums (for, though it does not grow in the streets of Liverpool, it does occur by the seaside a short bus ride away). In the 1970s, the Wildlife Trusts published a series of wildlife cards, including a few special flowers like Fritillary, Marsh Gentian or Oxlip, but they were not sold at Fritillary meads or Oxlip woods. Quite often one finds hints that the designers of postcards do not, in fact, know much about their subject. The 'sweet pea' on the card sold in aid of the Friends of Rye Harbour is in fact the rare Sea Pea, and the Skunk Cabbage, though accurately identified, is not usually 'very common in hedgerows', at least not in Britain. A hedgerow lined with Skunk Cabbage would be well worthy of a special postcard.

Perhaps one reason why rare flowers are so seldom seized upon to inspire a sense of place is that the usual photographic close-ups on postcards are so dull and banal. In a close-up, half the story is missing, for you are given no sense of the locality, as though the flower's background is no more relevant than in a herbaceous border or plant pot. Before the camera took over, the artistic quality of cards could be much higher, for example on those produced for the Medici Wild Flower Series, which include a beautiful watercolour of Pasqueflowers in their downland setting. Some years ago, the Hampshire Wildlife Trust commissioned card designs in the same spirit, with compositions of rare flora and fauna, including the Wild Gladiolus of the New Forest, and the Isle of Wight's special Wood Calamint and Field Cow-wheat. But I am told you cannot find an image of the Wild Gladiolus in any of the souvenir shops of Lyndhurst or Beaulieu.

In 1997, the BSBI took the plunge and produced a series of cards which returned to the idea of a flower in its place, based on some magnificent wide-angle photographs by Bob Gibbons. The species chosen are shown in their characteristic setting, and in some cases the marriage of the two is striking and unforgettable – the Downy Greenweed, spilling over the rocks above Mullion Cove (you realize England is a small place when you recognize that very rock!), the Hoary Stock with its backcloth of blue sea, blue sky and white cliffs, or the splash of Purple Saxifrage, like a glowing coal on furnace-black volcanic rocks. Only with the perhaps obligatory Spring Gentian does the set revert to the seed-packet approach – but that is the trouble with Spring Gentians, they grow like lawn daisies on smooth grassy hillsides; and except in Western Ireland you can never find a landscape worthy of them!

Those great collectables of pre-war Britain, cigarette cards, include various sets of wild flowers, but, as with our postcards, nearly all of them show common, everyday wild flowers (or at least, they were then). The subjects of cigarette cards were often themselves collectables – you spotted cars and trains, stuck pins in butterflies and picked and pressed wild flowers. A set of 50 cards issued by Wills in 1936 has an album drawing of a fashionably dressed mother and her two pretty daughters gathering wild flowers on a hillside overlooking the village church – a scene of quiet English bliss. The flowers on the cards were of the sort likely to be picked by this family – common, colourful and easily recognized. There is only one Red Data Book species in the set, and that one is the Corncockle, described as 'unwelcome and troublesome'. The focus on common wayside flowers reflects the popular books of the

Stamps from Guernsey and Jersey featuring rare ferns and flowers. Top: *Asplenium x sarniense*, *Isoetes histrix*, *Asplenophyllitis microdon*, *Ophioglossum lusitanicum*. Bottom: Jersey Fern (*Anogramma leptophylla*), Jersey Thrift (*Armeria arenaria*), Jersey Orchid (*Orchis laxiflora*), Jersey Viper's Bugloss (*Echium plantagineum*).

period, led by Edward Step's *Wayside and Woodland Blossoms* and C.A. Johns' *Flowers of the Field.* They convey a curiously static impression – as lifeless herbarium specimens rather than competing, breathing, reproducing plants. The first two sets of wild flowers, issued by Brooke Bond tea in 1955 and 1959 and painted by C.F. Tunnicliffe, continue the same preoccupation with common plants, although they include just a few less common ones, like Pasqueflower, Marsh Gentian and Henbane. When the third Brooke Bond series appeared in 1964, however, the proportion of uncommon scarce species had grown to nearly one-fifth, including seven plants in the present-day Scarce Plant Atlas (1994), and two more in the Red Data Book – including the Lady's Slipper Orchid, shown blooming in deep twilight, perhaps to indicate doubt whether it still survived. Perhaps, after two series, Tunnicliffe was running out of suitable common flowers. But, by the mid-1960s, more amateur naturalists were becoming interested in rare flowers through works like Keble Martin's *Concise British Flora,* and the growing interest in nature conservation. Perhaps, too, the family car had changed our botanical behaviour. If the Wills cigarette company had issued their series in 1996 instead of 1936, they would have showed that well-heeled lady and her lovely daughters whizzing off to see rare flowers in a country park, and given us a set full of special orchids and alpines.

Picture postage stamps (see picture on p. 241) are another barometer of subjects of popular appeal. Flowers have had a reasonable share in the plethora of picture stamps released by the Post Office since the 1960s, but, as with cigarette cards, the majority are of garden plants. Most of the few wild ones are common species, indicating what is typical of Britain rather than special to Britain – like Keble Martin's stamps of 1967, each one like a snippet from the *Concise British Flora.* In 1997, we were offered a slightly blown Lady's Slipper Orchid based on a vintage painting by Franz Bauer and superimposed on a doom-laden graph, as one of a set of six endangered species. But you have to go back another 30 years to find another rare flower on a British stamp. This is the Spring Gentian, one of a set of four wild flowers painted (and very well painted too) by Michael and Sylvia Goaman in 1964 to commemorate the International Botanical Congress in Edinburgh. The Goamans probably chose the Gentian for its decorative qualities and suitability for miniature format, not because it was rare – though a year or two later we were to hear more about the Spring Gentians of Upper Teesdale when their habitat was threatened by a reservoir.

Interestingly the independent postal authorities of the British Isles have taken the opposite approach, and chosen specifically rare flowers as part of their local identity. In the Channel Islands, flowers have been chosen that have Jersey or Guernsey in their names: Jersey Thrift, Jersey Orchid, Jersey Viper's Bugloss and Jersey Fern. Though people living outside Jersey call some of these plants by different names, the meaning is clear: Jersey Thrift is as much a part of Jersey's distinctiveness as offshore bank accounts and Jersey cows. Finding itself short of equivalent flowers, Guernsey plumped for rare ferns, displaying two hybrid Spleenworts, Land Quillwort and Least Adderstongue, though only the most eagle-eyed botanist would find them. In 1988 it decided to celebrate the bicentenary of a local botanist, Joshua Gosselin. He is remembered by three more Guernsey flowers, local varieties of Small Catchfly and Rock Sea-lavender, and the Hare's-tail grass. The Isle of Man had fewer special species to choose from; for its 1986 Endangered Species set, it scraped the barrel with Dense-flowered Orchid and the lichen *Usnea articulata*, but, oddly enough, neglected its own Isle of Man Cabbage. (On the same occasion, Guernsey poached the Jersey Orchid, surreptitiously changing its name to Loose-flowered Orchid.) Even Lundy has followed suit recently, putting the Lundy Cabbage on one of its local 'Puffin' stamps. I doubt whether many tourists could identify these plants. But that is beside the point – each celebrates local character, each one is a tiny botanical flag.

This approach works best in places where there is a strong sense of national pride and cultural identity. Hence, Ireland has also placed rare native wildlife on posters, cards and stamps, sometimes with considerably more style than Britain has managed so far. In 1978, a set of Irish wild flowers produced not the brambles and snowdrops of the British equivalents, but such things as Strawberry Tree, Large-flowered Butterwort and St Dabeoc's Heath (see picture on p. 244). These have two points in their favour – first they are special to Ireland and second they are not found wild in Britain. The set of Irish Trees in 1984 followed a similar line, stretching the point to include a mere variety (Irish Yew), a very doubtful species (Irish Willow) and a microspecies (Irish Whitebeam) – plus good old Downy Birch as a makeweight. Ireland has also produced postcards of considerable artistic merit, exploring the graphic possibilities of Sea-kale and Irish Lady's-tresses.

If, or perhaps when, Scotland and Wales start issuing their own picture stamps, it is a safe bet that *Primula scotica* and Snowdon Lily will soon be on our envelopes. You can easily imagine either being taken up as a symbol to rank with thistles and leeks. It will be more difficult to decide which wild flower best represents cosmopolitan, multicultural England, especially as we have so few endemic plants – and *Gentianella anglica* seems a poor little plant for a country which once ran an empire. Maybe we'll stick with the dog-rose – but perhaps a rare one like *Rosa agrestis* (it grows near Glyndebourne, and what could be more English than that?).

Kitchen flowers and cure-alls

The plant called Spiked Star-of-Bethlehem or *Ornithogalum pyrenaicum* by botanists and Bath Asparagus by everyone else is perhaps our most delicious

rare flower. I do not mean delicious to look at – its loose spikes give a pallid, strawy impression – but literally delicious. The flower buds not only look like asparagus but they actually taste like it. There is a large patch of them in a hedgerow not far from my house, and, even raw, they are sufficiently crunchy and refreshing to test my botanical restraint. J.W. White, the much-quoted author of the *Flora of Bristol*, found them 'very little inferior to the cultivated esculent'.[203] In his day they were gathered each May from woods around Bath, and sold in the market either as Bath Asparagus, or 'wild asparagus' or simply 'wild grass'. Edward Step says it was also called French Asparagus. The good citizens of Bath have eaten it as a buttered *hors d'oeuvre* as long as anyone can remember. Thomas Johnson found it thereabouts in 1634 on one of his travels as the Botanical Mercury 'in the way betweene Bath and Bradford not farre from Little Ashley'. Johnson enjoyed his food and though he does not mention whether this first record of a British plant ended up on his dinner plate, I am sure that it did. Its culinary fame was well enough known by the end of the following century for Collinson to describe the gathering and sale of the plant in his *History of Somerset* in 1791. Most modern accounts of the species say or imply that Bath Asparagus is no longer gathered for food. Not so, according to local botanist Dave Green, who tells me that it is still sold by two or three greengrocers in Bath and Bradford-on-Avon – at £1.20 per bundle of 30, which strikes me as very reasonable. The plant is still locally abundant in the Avon valley, and picking it will do no harm so long as the bulbs are not disturbed. In fact selling it at the gate could be a novel way of gaining a little income from a nature reserve.

Stamps from Ireland and the Channel Islands featuring local rarities. Top: Spring Gentian (*Gentiana verna*), Greater Butterwort (*Pinguicula grandiflora*), Strawberry Tree (*Arbutus unedo*), St Daboec's Heath (*Daboecia cantabrica*). Bottom: Dwarf Pansy (*Viola kitaibeliana*), Sea Stock (*Matthiola sinuata*), Sand Crocus (*Romulea columnae*).

Above right: Spiked Star-of-Bethlehem (*Ornithogalum pyrenaicum*) is still gathered for food in Avon.
[Peter Marren]

Bath Asparagus is generally considered to be our only native species of *Ornithogalum* – a genus of bulb plants which establish themselves very readily. But its distribution – common in the Avon valley and parts of North Wilts, rare or absent everywhere else – is unusual, to say the least. It is a Mediterranean plant, rare in northern Europe, but much at home on stony hillsides in Italy, Greece and the Iberian peninsula, including, of course, the

Pyrenees, which supply its species name. Dave Green's investigations show a definite pattern of distribution in Britain on south-facing slopes, often near the watershed. Some of its sites are believed to have been Roman vineyards. Do we owe Bath Asparagus to the Romans? The evidence is circumstantial, but it is an interesting theory, conjuring up a delightful picture of growers cultivating grapes and this mock-asparagus side by side.

Only a few of our uncommon wild flowers ever had a similar commercial value, other than to the collector, and those that did were less uncommon in the past. There was once a roaring trade in Sea-kale (*Crambe maritima*). This big cabbagy shingle plant was introduced from the coast of Devon to Covent Garden by William Curtis in 1795, and the demand was such that it eventually outstripped supply and endangered the plant. The trick was to cover the young plants in March so that the crown leaves grow long and pallid, like celery. This is 'blanched' Sea-kale, which you simply boil and serve on a napkin with melted, seasoned butter, like asparagus. The other edible bit is the young flower head, which should be simmered and served as greens while still crunchy, like broccoli. Jane Grigson deplored the failure of supermarket-bound Britons to seek out this 'English contribution to the basic treasury of best vegetables'. It is too labour intensive to grow commercially nowadays, though it is available through seed catalogues. And wild Sea-kale is still gathered by connoisseurs: the trick, apparently, is to pile shingle over the plants in autumn, and uncover the blanched shoots the following spring.

Another 'sea kale' was the Wild Cabbage, which was once sold in Dover market. The leaves are said to taste bitter unless twice boiled, and the result, according to Roger Phillips, is rather disappointing – 'the flavour is that of the shop-bought cabbage: no better, no worse!'.[145] No doubt when it was in cultivation it showed better form. This is another plant believed to have been brought to Britain by the Romans, though it seems very wild on the chalk cliffs of the south coast. Surprisingly there is still a lingering culinary

Below right: Dittander (*Lepidium latifolium*) makes a very acceptable peppery sauce. [Peter Wakely/English Nature]

Below: A striking portrait of Sea Kale (*Crambe maritima*) by Ann Soudain (1991).

interest in a much rarer seaside plant, the Sea Pea. The little, sharp-tasting 'peas' were once gathered as food in times of famine. Phillips was told by a local restaurateur that villagers on the Suffolk coast still occasionally pick them, and either eat them fresh, or dry them within their pods and use the peas like lentils in soups and broths.

The herbalists of old sought several other wild plants that are now rare. An important one was Pennyroyal, whose crushed leaves are just the thing for colds that have gone chesty. Pennyroyal is an ancient name, a corruption of the medieval *pulyol ryal*, that is 'royal (special) thyme'. Another herb with a powerful magic was Wild Chamomile (*Chamaemelum nobile*), whose appley scent was once familiar on village greens and lawns throughout south and central England – proverbially familiar in fact, for the more it was trodden, the better it grew, like virtue and honesty. In William Turner's day, it was 'so plenteous that it groweth not only in gardynes, but also 8 mile above London it groweth in the wylde fielde, in Rychmonde grene, in Brantfurde (Brentford) grene ...'. Chamomile was 'a remedy for all agues'. It could even revive wilting plants placed near it, and so was known as a 'herb doctor'. Several other plants were sought after for their curative properties despite being rather scarce and hard to find. Spignel, Honewort and Hog's Fennel were noted for their hot-tasting roots. The Scots collected Wild Lovage from sea-cliffs (it makes lovely potato soup). Water and Wall Germanders were grown in herb-gardens, as were Deadly Nightshade and Henbane in gardens of physic. It is said that Highlanders gathered the sharp-tasting berries of Dwarf Cornel as an appetizer – for its Gaelic name *lus-a-chraois* means herb of gluttony. These, however, were the exceptions. Few rare plants were special enough to justify the time and effort needed to collect them.

At least one ancient herb, and present-day scarce species, may owe its distribution to past usage. This is Dittander (*Lepidium latifolium*), a cruciferous plant (see picture on p. 245) with a froth of tiny white flowers and broad, green, peppery-tasting leaves. Turner knew it, and Gerard described the healthful properties of its 'extreame hot' roots, which could be chopped up and made into an Elizabethan version of horseradish sauce. It was much better known then than now – the old name was Dittany, while the Germans called it *Pfefferkraut*. A couple of years ago I sampled some Dittander growing in a ditch in Somerset. It is not bad – 'extreme', perhaps, but quite pleasantly so; I would buy it.

Some recent correspondence in *BSBI News*, the always interesting news 'n' snippets magazine of the Botanical Society of the British Isles, casts new light on the status of Dittander. It is commonest on the Essex marshes, but occurs much more widely as a weed of waste ground, colliery tips, old railway lines and the like. In 1990, Nick Sturt found a small patch of Dittander by the River Lavant in Chichester on the site of a medieval hospital.[186] Delving into the records, he discovered it had been recorded on the same spot back in the 1830s, in what was then the garden of St James Hospital. Going back further still, he learned that this hospital had been founded in the twelfth century outside the city walls in order to care for lepers. And both Pliny and Dioscorides, the principal authorities at the time, specifically recommended 'Lepidium' or Dittany for the treatment of leprous sores. Just coincidence? In the next issue, John Palmer weighed in with corroborative information from three sites in the Dartford area, all in the now semi-wild grounds of Victorian hospitals.[134] This could not be coincidence because the plant occurs nowhere else in the area. More associations of the same kind probably await anyone who likes to blend a little history with their botany. It looks as though 'Dittany' is a naturally rare native plant whose range has been extended by its widespread use in houses of healing. But, if so, what a persistent plant it is – the physic garden, even the very stones, of St James's Hospital in Chichester are long gone, and yet here still is the *Lepidium* of Pliny, perhaps on the very spot where it had been planted at the time of the Crusades.

Plants and places

Whenever a new local flora is published, it is always interesting to see what the authors have chosen for the jacket: the plant to represent the county, so to speak. It is only during the past forty years or so that floras have been given colour jackets more or less as a matter of course. Before that, the representative flower generally appeared on the frontispiece as a coloured drawing or a fuzzy black and white photograph. Looking through the collection of local floras in the Druce-Fielding herbarium in Oxford, I am struck by their diversity. Some jackets may show a *typical* flower of the area, often in its wild setting; sometimes, especially more recently, it will show a *rare* species special to the county; and occasionally it does not show any particular flower at all. In the first category is the recent *Flora of Cumbria*[78] which has chosen a dramatic photograph of foxgloves with a background of mountains and lakes. The picture is appropriate to the focus on biogeography in this flora – the plant in its place. The *Flora of Hampshire*[17] has plumped for the alternative approach, and its billboard is a straightforward portrait of the Wild Gladiolus, one of the county's most celebrated wild flowers. With the *Flora of the London Area*,[20] the setting must have been regarded as more significant than any particular species: it is remarkable that any wild vegetation should still grow within the shadow of the Tower of London, let alone the dramatic swirl of purple, yellow and white displayed here. It is not obvious which species they are, nor does it matter: it is the incongruity that is important. The ultimate in modern minimalism was the shoe-string *Flora of East Ross-shire*,[48] which, apart from the title, is 'a perfect and absolute blank'. At the opposite extreme is *Flowering Plants of Wales*[51] which goes the whole hog and illustrates pretty well *all* the special flowers of Wales over a double-spread jacket. The jackets of most British floras are fairly conservative in design terms, based on colour paintings or photographs. There has not, so far, been any attempt to emulate some of the recent Irish floras which depict the local countryside in lithographic, almost abstract designs.

The choice of cover species is sometimes the obvious one. What flower could be more emblematic of the Essex countryside than the Oxlip? – both typical of and special to the county. The Spring Gentian and Durham go together like peaches and cream, and the Early Star-of-Bethlehem was renamed the Radnor Lily just in time to germinate on to the cover of Ray Woods' *Flora of Radnorshire*.[211] In other places the choice must have been more difficult. Leicestershire is probably a case for a setting rather than a species (though, knowing Charnwood Forest, I would have been tempted to portray Bracken). Perhaps the authors of the most recent *Flora of Somerset*[74] had tired of their predecessors' Cheddar Pinks, and so plumped instead for a local all-yellow variety of Mountain Pansy. The most amusing jacket is the *Flora of the Isles of Scilly*,[107] which shows the Hottontot Fig (*Carpobrotus*), a serious threat to some of the native flowers! I suspect the choice was the publisher's, not the author's.

Relatively few wild flowers are named after a place – the botanists who gave rare or obscure flowers their names were generally happier with botanical terms (narrow-leaved, eight-stamened, dense-flowered, etc.) – which are often merely translations of their scientific names. A flower seems more likely to be named after its locality if it grows on an island or at any rate is confined to a well-defined small area. Jersey leads the list, with at least seven eponymous flowers (and the islanders claim several more). Guernsey has to make do with a fern and a centaury, and Alderney with a sea-lavender. Lundy and the Isle of Man have their own special cabbages, Arran its whitebeams, and Shetland a mouse-ear and a pondweed. Of the counties, Cornwall does best, with its own heath, moneywort, eyebright and gentian. Two cities score two plants each: Plymouth's pear and thistle, and Bristol's rock-cress and whitebeam. A few rare flowers take their names after their main localities – Lizard, Loddon, Cheddar, Snowdon, Teesdale, Rannoch; and a few with a slightly larger range have been named (or occasionally misnamed) Breckland Speedwell or Chiltern Gentian or Cotswold Pennycress.

Many of these names are modern fabrications. In my own lifetime Lloydia or 'Mountain Spiderwort' has turned into Snowdon Lily (quite a nice name, I admit) and Ludwigia into 'Hampshire Purslane'. Other species have more venerable names which seem wildly inappropriate today. Portland Spurge and Tunbridge Filmy-fern do at least still grow at or near those places, where they were discovered in the seventeenth century, although they are, of course, much more widespread than these names imply. But there are no Deptford Pinks at Deptford, and possibly never were, nor, unfortunately, are there any more Nottingham Catchflies at Nottingham. It was Thomas Willisell who found the first British Nottingham Catchfly (*Silene nutans*) on Nottingham Castle Rock back in the 1660s. The name has stuck, although the Catchfly is found as far south as Dungeness and as far north as St Cyrus. The plant is said to have spread at Nottingham after the Castle was burned down in the Reform Bill riots of 1830, but had become rare by 1880, and was 'more or less exterminated' when the Castle was restored ten years later. 'In 1934,

TABLE 6 Some rare or local plants displayed on County Floras

Flowering Plants of Anglesey (1982)	Spotted Rockrose (drawing)
Ashdown Forest (1996)	Marsh Gentian (watercolour); Ivy-leaved Bellflower (reverse)
Angus (1981)	Drawings: Alpine Catchfly; Cotton Deergrass (reverse)
Bedfordshire (1953)	Great Earth-nut (frontispiece)
Breckland (1979)	Composite drawing of rare Breck plants incl. *Bromus tectorum*, Spiked Speedwell, Perennial Knawel, Sand Catchfly and Grape Hyacinth
Derbyshire (1969)	Jacob's Ladder
Dorset (1948)	Lizard Orchid (frontispiece)
Durham (1988)	Spring Gentian (painting); Dark-red Helleborine (reverse)
Essex (1974)	Oxlip
Gloucestershire (1948 and 1986)	Pasqueflower (antique coloured drawing)
Hampshire (1883)	Tufted Centaury (coloured drawing: frontispiece)
Hampshire (1996)	Wild Gladiolus
Hertfordshire	Chiltern Gentian
Isle of Man (1986)	Henbane
Jersey (1984)	Hybrid Wild Arum (painting); Jersey Fern, *Anogramma* (reverse)
Atlas of Kent Flora (1982)	Late Spider Orchid
Lincolnshire (1974)	Sea Buckthorn
Lincolnshire supp. (1985)	Burnt Orchid
Liverpool District (1933)	Round-leaved Wintergreen (embossed on board)
Montgomeryshire (1995)	Spiked Speedwell (coloured drawing); Sticky Catchfly (reverse)
Oxfordshire (1998)	Meadow Clary
Radnor (1993)	Radnor Lily (*Gagea bohemica*)
Ross, East (1980)	Norwegian Cudweed (frontispiece)
Rutland (1971)	Yellow Bird's-nest
Scottish Wild Plants (1997)	One-flowered Wintergreen
Shetland (1990)	Autumn Gentian (distinctive northern form); *Cerastium nigrescens* (reverse)
Shropshire (1985)	Rock Stonecrop
Somerset (1896)	Cheddar Pink (drawing; frontispiece)
Somerset (1997)	Mountain Pansy (local yellow-flowered form)
Surrey (1976)	Early Marsh Orchid var. *pulchella*
Sussex Plant Atlas (1980)	Pheasant's Eye
Flowering Plants of Wales (1983)	Composite painting of Welsh plants. Front: Spiked Speedwell, Snowdon Lily, Sea Stock, Welsh Poppy, Welsh Groundsel and Welsh Marsh-orchid. Back: Tufted Saxifrage, Fen Orchid, Marsh Gentian, Radnor Lily, Perennial Centaury, Rock Cinquefoil and Welsh Evening-primrose
Wiltshire (1957)	Tuberous Thistle (frontispiece)
Wiltshire (1993)	Burnt Orchid

one plant was seen by Miss V. Leather flowering in a newly made rockery in the Castle grounds, but was destroyed before it could seed.'[89] Rotten luck for the Catchfly, especially as a much rarer plant, the hawkweed *Hieracium pulmonarioides*, has managed to survive there. Though I am not an advocate of random reintroductions, it would surely be a good idea to sow *Silene nutans* on the rocks of Nottingham Castle.

As for the Deptford Pink (*Dianthus armeria*), that name might have been a case of mistaken identity. We owe it to Gerard: 'There is a little wilde creeping Pinke, which groweth in our pastures neere about London, and in other places, but especially in the great field next to Detford, by the path side as you go from Redriffe to Greenewich.'[67] He goes on to mention the 'many small tender leaves, shorter than any of the other wilde Pinkes; set upon little tender stalks, which lie flat upon the ground, taking hold of the same in sundrie places, whereby it greatly encreaseth; whereupon doth growe little reddish flowers'. As many have pointed out since, this delightful description sounds more like Maiden Pink (*Dianthus deltoides*), and may be, therefore, the Deptford Pink never did occur at Deptford. Thomas Johnson, however, reproduces a convincing woodcut of it, from a plant gathered a little further from London, between Gillingham and Sheppey.

Celebrations

The Fritillary (*Fritillaria meleagris*) is one of the most celebrated wild flowers. It forms a glorious colourful display at the beginning of summer in that most English of landscapes: damp meadows by the river with tall hedges, sleepy willows and a pond in the corner. Its accumulation of local names is exceptional for a rare flower. Some are named after a nearby village (Oaksey Lily, Minety Bells), others are associated with damp (Toad's Heads, Frawcup) or mourning (Dead Man's Bells, Widow's Wail). A particularly colourful name is The Mourning Bells of Sodom, probably a reference to the testicle-like bulb supporting the hanging flower, but you have to dig one up to see the point.[75] Country people recognized a sinister side to its beauty, suggested in the sinewey leaves and flower stalk, and in the way the flower buds sit like a snake poised to strike. Even more potent is the fact that the Fritillary season peaks on or near May Day. At Ducklington in Oxfordshire, a field of them is preserved by the Peel family and is the centrepiece of a festival of the spring. In the church the altar cloth is embroidered with Fritillaries, and there are more of them in a stained-glass window pane. Those wishing to see 'the mottled-purple bells dipping into the grass' are handed a charmingly written leaflet about the field and its flowers. With tea, sunshine and Morris-dancing, it provides, as my correspondent Reg Crossley expressed it, 'a wonderful day with one of our rare wild flowers at its heart'. The Fritillary was once part of the events of May Day in many villages in the Thames valley. In most, the garlands and posies have gone along with the May Pole and the Queen of May, but memories linger. A correspondent of Roy Vickery's recalled 'with fervent nostalgia memories of this lovely event to herald the advent of warm summer sunshine and the promise of abundance of nature's gifts'.[197]

The Fritillary is exceptional among rare flowers in that it is widely seen as a harbinger and symbol, with all the subtle magic and undertones one could wish for (see picture on p. 252). Few others engage us at the same mythic level. Folk memories still linger of the Pasqueflower's association with 'battles long ago' – it grows where blood was spilt, usually that of the Danes in this country. The flower's purplish-red hue is the obvious source of the legend, but the tradition may also be based on its liking for mysterious ancient banks and earthworks. It was said to have bloomed in profusion on the battlefield of Crécy.

Flora Britannica[113] and Roy Vickery's researches reveal a still fairly lively culture of wild flowers and trees just beneath the surface of our urbanized society. We reinvent myths that echo the fashions of the day or give old ones a new twist. There may be local traditions and events involving wild flowers that are quite unknown to the outside world – like the tea party I accidentally stumbled on, held to celebrate the annual appearance of the Spring Snowflake in its secret valley in deepest Somerset. Some of the old familiarity has gone, but nature lies in our bones and constantly reasserts itself. Who, 20 years ago, would have predicted that alternative societies would spring up, living in trees and garlanding their trunks with totems that recall Stone Age religions. I, for one, would never have imagined that hundreds of people, not all of them botanists, would want to travel from far and wide to dance attendance on a rare orchid they could only touch. For that matter, many of us were amazed (but naturally delighted) when a book about the folklore of flowers costing £30 became a best-seller. Rare flowers play only a small part in the ever-changing culture of flowers – but, then, most of them never were 'familiars' or totems. They have become prominent today as never in the past as a new kind of symbol; as the most vulnerable parts of a natural biodiversity widely perceived to be under threat. Exotic-sounding names like Starfruit, Lady's Slipper and Plymouth Pear have been reported in the media, and have caught the public imagination. Images of rare flowers turn up in the most unexpected places, like the Loddon Lilies on Hampshire Health Authority's notepaper or the English Gentians in an episode of *The Archers* (there was also a colony of 'thunder-and-lightning' – I don't know what it was either, but I expect Jo Grundy did). Conservation charities are justifiably proud of rare flowers in their care, and in their literature one can read about the regular progress of a rare orchid or buttercup as though they were favourite animals in a zoo, or pen pals from across the seas. Every age and culture has refashioned flowers, in ways that express as much about ourselves than the objects of our desires and cares. This mythic propensity extends even to nature conservation, for, in properly wishing to preserve rarities, we may overlook the most interesting thing about them – and that is not their apparent vulnerability so much as their ability to *survive*. But more about that in its place, in Chapter 13.

Fritillaries at Coleshill, Oxfordshire. [Bob Gibbons]

Chapter 12

A Whiff of Ground Pine:

species surveys

A member of the Ashmolean Natural History Society monitors a population of Creeping Marshwort on Port Meadow, Oxford.
[Peter Marren]

I once spent a couple of weeks – hot, thirsty weeks as I recall them – searching for a lost plant, well known to the apothecaries and herbwomen of old but seldom seen today. This was the Ground Pine, *Ajuga chamaepitys* so named from its remarkable resemblance to a first-year seedling of Scots Pine, with a sharp whiff of resin to match. The Ground Pine is, of course, no more a relation of pine than is a dolphin to a mackerel. Ground Pine is a flowering plant, a member of the Labiatae, that odoriferous family of mints, dead-nettles and woundworts. Its characteristic narrow leaflets with their thick cuticle are designed to minimize water-loss, and the resinous oils within are a defence against rabbits and other predators – an effective one, I decided, after nibbling a bit of Ground Pine in the interests of science, for it tastes quite unbelievably bitter. This combination of defences is characteristic of many plants of Mediterranean regions, but is unusual among the British flora. Hence I see this attractive little pine-mimic as one of our few 'desert' plants, for it grows on hot dry soils on sun-drenched hillsides in the warmest, driest parts of Britain. Perhaps that is why my memories of Ground Pine hunting are bound up with feeling thirsty all the time.

I first saw it in 1992, when Plantlife director Jane Smart and I decided to follow up an old record from near Wouldham in the Medway valley of Kent. 'Follow up': it sounds so easy, but physically it can be a nightmare. There are two kinds of roads in the Medway valley: fast modern dual carriageways superimposed on a spider's web of winding narrow lanes. Crossing from one to the other is not easy, especially as the car behind you is seldom inclined to let you pull in and consult your map. Eventually, after several wrong turns and circumnavigating half a dozen tractors and a stray cow, you find yourself in roughly the right area. But Ground Pine (see picture on p. 256) is not a plant you can drive up to. The next stage of the quest is a walk, which in this case was another choice between dense woodland and neck-high nettles, or a couple of ploughed fields. Either way there were barbed wire fences, cunningly sited on steep banks, to negotiate. A further problem is that you are chasing a grid reference, not a recognizable spot, and plant recorders are often very shaky on this matter of grid references. It therefore helps to have a good idea in your head of the sort of place where your plant is likely to grow. In this case we guessed right, and headed for a sheltered combe at the base of the wooded down. The last stage of the quest is often a search with your back bent to the ground or a crawl on hands and knees, depending on the plant. Cotswold Pennycress is a crawler, but you can spot Early Gentian at a low crouch. For Sticky Catchfly you need a rope, and for Starfruit you need a child-like love of mud. For Ground Pine you have the luxury of walking upright, so long as you keep your eyeballs glued to the ground. You are looking for a grey-green and ginger furry object about the size of a golf ball. You cannot count on seeing the tiny but intense yellow flowers, which Ground Pine produces rather grudgingly, two or three at a time. Nor can you count on finding a fresh, pristine plant like the picture in the book. As a species of open ground and track sides, Ground Pine often looks rather sorry for itself. When it is trodden on or run over, the pine-scented resin leaks out and the dust sticks to it. This helps to blend Ground Pine and its environment even more effectively.

Despite all this, we soon found it. From the records, we expected to find a small colony of a dozen or so plants, about average for this species. Instead, and to our delight, we found a miniature forest of Ground Pines – an estimated thousand or so plants, and one of the best stands of it seen in recent years. What must have happened was this: recent gales had damaged some of the west-facing woods of the Medway valley, and timber lorries had churned up the Ground Pine patch at the foot of the downs. Since then the soil and chalky rubble had baked hard, and it felt as hot as concrete on an August afternoon. For a few years the area was colonized by a fascinating mixture of chalk and arable plants. As it happened, these conditions, by their nature ephemeral and unpredictable, are perfect for Ground Pine. Probably the lorries and haulage gear had churned the surface sufficiently to bring dormant seed to the surface. But privet, hawthorn and scrambling dog-roses were already closing in. Two years later, I returned to the spot as part of a survey of the species for Plantlife. By then there was little bare ground left, and, out of that memorable 'forest', only three little Ground

Pines were left. It had had its day, and was about to go underground again. But the seed-bank had been replenished during those three or four years, and that was what mattered. Tough, gritty Ground Pine seed has been known to survive at least 50 years in the soil (and so, in theory at least, the plant can be made to appear to order by rotovating or shallow ploughing the surface). What it does *not* do is appear regularly each year and in broadly the same numbers. Ground Pine is not that kind of plant, and a single survey can give a very misleading impression of its status. Though its numbers had diminished several hundredfold in a couple of years, Ground Pine is not heading for extinction at Wouldham. It is just biding its time. If we had visited that combe a few years earlier or later, we might have missed it altogether and concluded it was extinct. Hence it is not enough to track down and count your plant. To make sense of a series of snapshot observations you need to know a little about how plants behave, and – dare I say it? – use your imagination.

Ground Pine is, despite its opportunistic leanings, a rare and declining plant. Since 1990 it has been found in only 24 sites, mostly on the chalk in Kent and Surrey. But many of these populations are small, and in an average year only a few thousand plants appear in all Britain (on the other hand there must be millions of Ground Pine *seeds*). Ground Pine has clearly not prospered since the days of William Turner when it grew 'in good plenty in Kent' (1568), nor even since the turn of the century when W.S. Marshall and C.E. Salmon listed it as still reasonably plentiful on the North Downs. There is no great mystery about it. Ground Pine has suffered in every conceivable way from the twentieth century: from chemical fertilizers, weedkillers, winter cereals, direct drilling, myxomatosis, housing estates and the end of shepherding on downland, to name a few of them. It survives where it can. The advantage of a systematic survey is that it can fill out broad generalizations with detail. The overall picture does look rather gloomy: many former sites are now dense scrub or even woodland, another is sprayed with herbicide every year, another is under polythene for intensive carrot production, yet another has vanished under roadworks. One owner asked for warning if Ground Pine showed signs of spreading, so that he could 'deal with it'. On the other hand, three sites are harrowed or shallow-ploughed to maintain the species, another is lovingly looked after by an employee of a cement company, and a further site has been 'adopted' by the warden of a nearby nature reserve. Ground Pine is not about to disappear as a British plant, but nor is it likely to increase much. If all the best action plans and targeted objectives of the conservation industry actually work out on the ground, it may not decrease much either. At least we have achieved for Ground Pine what many naturalists are doing for other rare plants: securing them a place on today's conservation agenda. If they do die out, it is not for want of plans and strategies.

Changing perspectives

Looking as if a seedling pine had sprouted canary-yellow flowers, Ground Pine (*Ajuga chamaepitys*) depends on open ground in dry, sunny positions on the chalk. [Peter Wakely/English Nature]

Most recent nationwide surveys of rare flowers have been funded by the government's conservation agencies: in England by English Nature, in Scotland by Scottish Natural Heritage, and in Wales by the Countryside Council (CCW), often through the offices of the Joint Nature Conservation Committee (JNCC), responsible for matters affecting all of Great Britain. Commissioning surveyors of rare plants are usually given precise aims. One is to discover the status of the species by investigating the known sites. For that purpose, each population is assessed and sketched on a map so that the survey can be repeated as part of a monitoring scheme. The second aim of surveys is to produce data for a programme of action to conserve the plant. This requires an assessment by the surveyor on the ecological 'health' of the species, and whether its particular environment is threatened. To save a species you need to know quite a lot about it. In the past, action plans were often based on inadequate knowledge. While a lot is known about *where* a species grows, there was relatively little about why and how it grows there. Quite often we completely misunderstood the situation and provided the hapless species with the last thing it wanted, like the trees we planted on top of *Carex flava* ('it's a woodland plant so it *must* like trees') or the cattle we helpfully removed from the Strapwort. Fortunately the situation has improved greatly in recent years thanks partly to the Biodiversity Action Plan, which provides funding for researching the special needs of our rarest species. One of the incidental benefits of the BAP has been to bring fundamental science and ecology together again after twenty years of drift. We are witnessing nothing less than the rediscovery of natural history.

One predictable result of a thorough-going survey is that certain plants are not quite as rare as was thought. In 1993, Simon Leach made a search for the Slender Bird's-foot Trefoil (*Lotus angustissimus*), an annual, trailing plant then considered to be very rare and declining. It turned out that it was not so much rare as misunderstood. Like Ground Pine, it has its good and bad seasons, and moreover has a strange habit of growing with a close relative, *Lotus subbiflorus*, like a slightly different kind of hay in a haystack. According to the Guernsey botanist Marquand, the two are so fond of one another's company that their stems are often intertwined, and 'so interwoven that they cannot be separated without tearing them to pieces'. Fortunately Simon has sharp eyes, and as a result Slender Bird's-foot Trefoil is no longer in the Red Data Book. I give some other examples of 'lost and found' plants later in this chapter. Here I want to illuminate another happy result of putting good botanists among their favourite plants for weeks on end, and that is that new ideas will fly up like sparks. This cannot be planned for or predicted: it simply happens when the circumstances are right. Let us look at some recent surveys that contain this wonderful ingredient, serendipity.

The idea of a rare dock sounds a bit contradictory, like a beautiful wart. The life-style of docks, with their super-abundant seed, deep roots and smothering leaves seems like an aggressive desire to take over the planet. Nature reserve managers, farmers and gardeners alike spend much of their time trying to get rid of them. Shore Dock (*Rumex rupestris*) is the exception. To look at, it has all the obvious qualities of a successful dock – 'looks very docky' was how one expert summed up its qualities – and only a connoisseur would discern through its pervading dockiness the distinctive blue-green basal leaves with their powdery bloom, and the relatively large winged seeds, which float on water like little lifeboats. It is an easily overlooked plant, not only because it looks like a common dock but also grows among them, especially Clustered Dock (*Rumex conglomeratus*) and the coastal form of Curled Dock (*Rumex crispus* var. *littoreus*), which is very similar when in a vegetative state. What has propelled the Shore Dock from relative obscurity to stardom during the past ten years is the realization that, for once, Britain is home to a plant of world rarity. *Rumex rupestris* is often called 'the world's rarest dock'. It is a European endemic restricted to the western fringes of Europe between Anglesey and Galicia in northern Spain. Nearly everywhere it is threatened by coastal development in one form or another, and so Shore Dock is Red-listed Europe-wide as a Vulnerable species, and is also listed under the Bern Convention on threatened habitats and species. In terms of international conservation, this dock is in the superstar league – it has everything going for it except good looks.

The first to survey the Shore Dock and begin to get to grips with its ecology was Plantlife's former conservation officer, Miles King, who chose it as a subject for his MSc thesis. Since then, both English Nature and the Countryside Council for Wales have funded research on the species. Liz McDonnell has surveyed the plant in Devon and Cornwall, Rosemary Parslow in the Isles of Scilly and Quentin Kay in Wales. David Pearman has unflinchingly surveyed its last site in Dorset, which happens to be on a gay nudist beach. Roger Daniels at the Institute of Terrestrial Ecology has carried out isoenzyme analysis to probe the species' genetic variability and look for warning signs of genetic isolation. In the best traditions of British natural history, this work has combined the traditional fieldcraft of the naturalist (the Shore Dock surveyor can be spotted as the only person on the beach wearing an anorak) with the instrumentation and precise measurements of the laboratory scientist. Like so many major conservation projects recently, it is the product of teamwork, involving government agencies, charities and scientists.

The portrait that emerges is of an intriguing plant – both an extreme specialist and an opportunist, and a species which has probably always been rare outside a few exceptionally favoured places. Shore Dock usually grows just above the high tide mark, where freshwater meets the sea. The best place to start looking for it is where water oozes and trickles at the base of a cliff, or over slumped banks of 'head deposits'. Less often it is to be found by the margins of a stream, or by pools in dune slacks. You can often spot a likely looking locality from a distance, though reaching it may involve a precarious descent through bracken and brambles, and then picking your way over slippery rocks. But you have to search hard, because most Shore Dock populations are very small – often less than 10 individual plants – and hidden among more vulgar docks and other tall vegetation. Even in the Isles of Scilly, which provide more opportunities for Shore Dock than practically anywhere in the world, there are only two or three hundred plants. Rosemary Parslow once found quantities of dead stems about to be used as tinder for a beach barbecue! In most places on the mainland the plant is terrifyingly vulnerable. Nearly half of its former sites have been lost. The sheltered coves it likes most are precisely the places where people congregate and build bungalows and hotels, and then demand coastal defences to stop them falling into the sea. This is what happened at Gunwalloe on The Lizard, where one of the best colonies was crushed under a wall of concrete blocks to safeguard a house. Another constant threat is coastal erosion: a former colony at Ringstead Bay in Dorset was probably destroyed during a severe storm. In more sheltered places, the onward march of coarser vegetation can smother Shore Dock to death, as in Constantine Bay in Cornwall. The spread of *Carpobrotus*, encouraged by recent mild winters, also threatens Shore Dock sites, especially in Cornwall and Scilly. And on top of all that, one wonders what sort of effect a Channel oil spill would have. The total British population is only a few hundred plants.

Reading the conservation reports on Shore Dock produced by McDonnell, Parslow, King, Kay *et al.*, one does begin to marvel how such a plant survives at all. The answer is probably that it combines a specialized habitat – restricted by its nature to a few hectares at most – with wide seed dispersal. Much of Shore Dock's copious crop of seeds – and a mature stem is a mass of brown seed – probably finds its way into the sea. There, we must imagine them bobbing along in the drift like corks, awaiting the day when they are driven ashore by storms. That such seeds will disperse many miles from the parent colony is suggested by the results of genetic analysis. The genetic diversity within even small populations is surprisingly high, and for this species physical isolation does not mean genetic isolation. As Daniels, McDonnell and Raybould put it, 'each population is a random selection of genes in a wider gene pool'.[37] The appearance of Shore Dock in new localities, where it may or may not persist, is further evidence of a fairly dynamic plant which takes its chances as it finds them. It had shifted its ground even during the five years between King's survey in 1989 and McDonnell's in 1995. More worrying is the loss of suitable habitat, and there may come a point when the Shore Dock's failsafe mechanism of seed dispersal will break down through the sheer statistical improbability of seed finding a suitable seedbed. That point has not yet been reached, and, although there has been talk of seed propagation and re-establishing lost sites, I think it would be better to accept that Shore Docks come and go in a changing environment, and concentrate our efforts on site safeguard. But it is hard to persuade Shore Dock fans to keep their hands off it, and I must admit that the plan I overheard for a Shore Dock garden, perhaps with a saltwater fountain and parterres of crumbly head deposits, does tickle.

Another species which, of all the world, has chosen England (oh, all right, there are a few in Wales, too), is the Early or English Gentian (*Gentianella anglica*). Its status as one of the few endemic plants that are easy to recognize has ensured the English Gentian a good deal of attention as England's *Primula scotica*. As an endemic it must presumably have evolved during the 10 000 years that Britain has been an island. Evolution has selected English Gentian with unusual speed and assurance. Where did it come from? English Gentian has a similar ecology and a similar chromosome number to the closely related Autumn Gentian (*G. amarella*), which accompanies it in most sites, though the two rarely flower at the same time. Almost certainly English Gentian started off as an early-flowering version of Autumn Gentian. What may have selected it is Britain's mild winters, which enabled the plant to germinate in autumn ready for an early start in the spring. This might explain why it is so much more abundant in the west than in eastern England, where winters are so much colder.

Clues as to the real nature of English Gentian began to emerge during a survey conducted by Tim Rich and Phil Wilson for Plantlife in 1994–5. Like many short-lived plants, English Gentian varies in quantity from one year to the next, and it also shifts its ground. It needs open ground on chalk or limestone grassland, like sheep terraces or the sides of tracks. Even so, what came out of their study was its extraordinarily patchy distribution. There are vast numbers of English Gentians on the Isle of Wight – a survey by Sue Telfer estimated that there were about four million of them. There are a million or two in Wiltshire and Dorset, but very few in the Cotswolds,

Above left: You might not look at it twice, but Shore Dock (*Rumex rupestris*) is a plant of worldwide rarity. [Peter Wakely/English Nature]

Above right: English or Early Gentian (*Gentianella anglica*) is an endemic plant, perhaps evolved from an early-flowering form of Autumn Gentian. [Sue Everett]

Lincolnshire or Kent (where a good annual score is 10 plants). And while there are good sites in East Sussex, there are now none left in West Sussex. What is more, the plant can vary enormously from one year to the next. In 1994, a huge colony estimated at 200 000 plants was found at Beachy Head in Sussex, yet the next year only 50 were found after hard searching, and the year after, none at all. In 1998, I was finding it all over the place on Pewsey Downs in Wiltshire, where, two years previously, it was local and elusive. Presumably the English Gentian sets a lot of seed, most of which goes into the bank in the soil, where it is vitalized by the right conditions, namely disturbance and a wet winter followed by a warm, wet spring. Even so, it is a peculiarly capricious plant, quite unlike the relatively predictable perennial herbs of the chalk downs.

Tim's survey also helped to throw new light on the so-called Cornish Gentian. This plant of sand dunes and short, open turf along the north coast of Cornwall and Devon was previously considered to be a robust and distinctive subspecies of English Gentian, named *Gentianella anglica* ssp. *cornubiense*. But, unlike its relative, Cornish Gentian is a highly variable plant, seemingly intermediate between English and Autumn Gentians, sometimes leaning one way, sometimes the other, and often occurs in swarms. Hybridization between the two had previously been ruled out on the grounds that English and Autumn Gentians never flower at the same time. This is now known to be incorrect – the two do sometimes overlap in southern England and South Wales. Tim Rich, Len Margetts and their colleagues have made a convincing case that the Cornish Gentian is indeed this hybrid – a genetic melting pot where gentian genes slosh about in hybrid swarms which vary from nearly English Gentian to nearly Autumn Gentian. Possibly this readiness to hybridize may threaten the survival of the pure form of English Gentian in Cornwall. But it also indicates the kind of active evolution that must have been going on within these closely related plants. Quite possibly English Gentian has evolved separately at different times and in different parts of the country – hence its peculiar distribution. These discoveries were not part of Tim Rich's survey brief. They are another example of the unpredictable spin-offs that result from paying a surveyor to look into a plant mystery. Yes, we now have conservation data and population counts and a very good idea of how many English Gentians live in Britain (in a good year, four to five million, not counting a few thousand hybrids). But more fundamentally we have acquired a better understanding of what kind of plant it is, and where it came from.

It would be hard to find a more spectacular example of new understanding than the discovery of the gametophyte generation of the Killarney Fern, which we last saw in Chapter 7, being looted out of existence. What everyone previously recognized as the Killarney Fern was the beautiful translucent frond that hides among the rocks and near waterfalls in a handful of secret places in western Britain and Ireland. But that graceful object is only half the species, and quantitatively very much the smaller half. We now know that for every Killarney Fern that looks like the picture in the book, there are hundreds more that resemble nothing so much as the sort of green felt you find on a card table. And whereas the species is normally considered to be one of our rarest plants, the green felt is proving to be unexpectedly widespread. In most parts of Britain, and in continental Germany too, that is what Killarney Fern *is*: not a beautiful frond but a layer of frizz more like a moss than a fern.

As every grower knows, ferns have alternate generations. The familiar one is the sporophyte, which produces spores from small structures embedded in the tissue of the frond. Instead of producing new fronds, these spores germinate into the much less familiar gametophyte. In most species, this is called a prothallus, an amorphous green blob rather like a liverwort. It is the gametophyte that ties ferns to damp places (bracken cheats by spreading vegetatively, using underground rhizomes as tough as street cables). This roundabout way of life is inherited from the fern's ancient coal-forming ancestors which lived half in, half out of the water – for ferns are extremely conservative life forms. Most produce their sporophytes and gametophytes on the same spot, and so the latter is rarely noticed. The Killarney Fern is unique among European ferns in that the gametophyte has become a perennial plant with an independent existence, and which can multiply indefinitely using vegetative structures called gemmae. This fact was long known from plants in cultivation, but was never looked for in the wild. As a result the gametophyte of Killarney Fern was unknown in Britain until 1989, when an American botanist, Don Farrar, found it in the Lake District while on a sabbatical visit to the UK. He knew what to look for because he had been finding the gametophyte of a related species in the United States. Once British botanists knew what to look for, and where, they started to find it. Fred Rumsey, the fern expert at the Natural History Museum, has tracked down the mysterious green felt in places where the fern has never been seen – in the Sussex Weald, the North Yorkshire Moors, on Dartmoor – as well as in its traditional strongholds in West Wales, South-west Scotland and Cornwall.[173] The gametophyte turned out to be much commoner and more widespread in Britain than the sporophyte. The only significant exception is in South-west Ireland where, so far at least, the sporophyte fern is still the dominant form.

Though comparatively widespread, the gametophyte has a very restricted habitat. It is usually found under sheltered underhangs in banks and cliffs, on shaded, north-facing sites. It tolerates low light levels and is one of the few green plants for which you need a torch to find it. Unlike the sporophyte fern, it tolerates fairly dry places, and is not restricted to regions of high rainfall: the main restricting factor is more the availability of rocks facing away from the sun. Most sites are in woodland or, at least, what used to be woodland, and it can probably be regarded as a good indicator of ancient, relatively undisturbed, places. The gametophyte must owe its survival to the sheltered nature of its habitat, among the most stable in the British landscape. Interestingly, it has not yet been found on artificial habitats like stone walls, mine shafts or quarries. It is clearly a survivor, not an opportunist; the gemmae may be dispersed locally through water percolation, or carried by spiders or mites, but clearly they do not go very far.

Why is the frond fern so rare, when its other half is comparatively widespread? The answer is probably to do with rainfall and our seasonal climate. The Killarney Fern we all know and sigh for likes warm places where the rain buckets down all year round. It reaches the apogee of its limited world range in Flores, the westernmost island in the Azores, which is deluged with five metres of rain every year. Only in mild, almost equally rainy South-west Ireland does the fern luxuriate in relatively open places. The lower humidity of Britain restricts it to caves and hollows in rocks and banks. It has also suffered from genetic drift: in South-west Scotland, each colony of Killarney Fern is known to consist of a single genotype and this may prove true of all small isolated populations. The genetic variation within the species is mainly wrapped up in this mysterious green felt. It turns out that the Killarney Fern is a much more flexible strategist than we realized. It probably colonized Britain and Ireland when conditions were wetter and warmer than at present, perhaps during the original Wildwood, perhaps earlier, possibly *much* earlier – in a warm interglacial period, or even back in the Tertiary, when hominids were still on all fours. It has been able to hang on in a less favourable climate by switching generations, retaining the capacity to turn back into a frond whenever the climate allows it. Hence the species is not endangered after all – only one half of it is. If our rainfall was a couple of metres higher, many of these dismal layers of felt would probably burst forth into ferns.

Where does this leave the conservationist, who understandably wants to preserve the prettier half? The legal situation is intriguing, since in theory the law applies to all parts of the plant. It cannot be applied to one generation without the other. Strictly speaking, therefore, the fern no longer qualifies for legal protection since the species has already been found in well over 100 grid-squares: indeed, Killarney Fern is no longer even considered a scarce plant, let alone a Red-listed one. It is a puzzler. Personally, I would leave things as they are. One could make a good case for protecting even the gametophyte on the grounds of its vulnerability to collecting – for the small bare areas left when samples were taken for analysis were still bare five years later. Fortunately it is easy to identify in the field without removal.

As for those staples of rare species protection – relocation, translocation, and management – the Killarney Fern presents a refreshing case for leaving well alone. As Mary Gibby argued at a recent conference, 'the species is safe – so safe that no one found the other half until 1989'.

The challenging plight of the Whorled Solomon's-seal

A rare plant which, until recently, was surprisingly little known despite having been grown in gardens for centuries, is the Whorled Solomon's-seal (*Polygonatum verticillatum*). It is a distinctive species, differing from our other two native Solomon's-seals in its narrow leaves and tight whorls of creamy flowers. Its attractiveness to gardeners lies mainly in the autumn colours, when the leaves turn yellow and hang down like some strange jungle plant exposing the bright scarlet berries at their base. Most of what was known about it came from gardeners, and so there was plenty of information about propagation but virtually nothing on how the plant survives in the wild. By 1980, it seemed to be losing the struggle. Only two sites were known, both in Perthshire, and both were precarious. But in those days, conservationists had different priorities.

As a possibly endangered species, the Whorled Solomon's-seal finally received its due attention in 1992, when Jenny Wright of the Royal Botanic Garden, Edinburgh, began to investigate its ecology and needs.[212] The question she sought to answer was why the species is so rare in Scotland but relatively common at similar latitudes in Scandinavia. A fact-finding tour of Norway and Sweden revealed that the narrow wooded glens which are its Scottish home are untypical. In Scandinavia the plant is widespread in ungrazed subalpine woods right up to the natural tree-line. In Sweden it is characteristic of old wooded meadows, formerly managed for hay and leaf fodder, but which are now largely abandoned and are reverting to woodland. The former practice of raking hay litter into bonfires had often produced a troop of Whorled Solomon's-seals the next year. It seems that the plant can survive unseen by virtue of its underground rhizomes, and pushes up its elegant tiered shoots only when there is some freedom from competing vegetation. This, then, was the optimum habitat for *Polygonatum verticillatum*: lush, flowery vegetation in woodland glades on moist brown-earth soils, traditionally managed by mowing or scything in late summer.

Whorled Solomon's-seal (*Polygonatum verticillatum*) is confined to the Tay river catchment in Perth and Angus, though widespread in Scandinavia and commonly grown in gardens. [Bob Gibbons]

When we move to Scotland, we begin to see why Whorled Solomon's-seal has become so rare. This kind of lush subalpine forest has almost completely disappeared – long since grazed out of existence or replaced with dense blocks of conifers. Where they remain, the soils are often too poor to support lush tall-herb meadows like those in Sweden. Fortunately surveys revealed that the species is not quite as terrifyingly rare as was feared. Some nine sites are now known, all in the catchment area of the River Tay, and all but one in steep, wooded valleys on the fringe of the Highlands (the exception is a strip of riverine woodland). All these sites are in relatively undisturbed natural woodland. As in Scandinavia, the plant prefers base-rich soils kept moist and open by flushing or seasonal flooding, and it grows among open meadow-like swards of tall herbs and ferns. It persists under dense shade, but needs light to flower. It seems, therefore, that the ecology of Whorled Solomon's-seal is the same whether in Scotland or in Sweden; but the sites differ in scale: hectares in Sweden, square metres in Scotland. The plant is vulnerable to drought, grazing and soil leaching. It suffers from the same malignant sawfly that attacks Solomon's-seals in gardens. In Scotland, it must have progressively lost its footing as woods were cleared. Its last strongholds were saved by their steepness and inaccessibility. It has fared even worse in England; there was only one known colony, on the North Tyne, where it has

The Alpine Sow-thistle (*Cicerbita alpina*) looks down from lofty crags to the mountain valleys where it used to grow.
[Peter Marren]

not been recorded since 1866.

Another factor is lack of fecundity. Unlike the garden plants, the wild Scottish ones seem to fruit only occasionally. Possibly they are too shaded, or perhaps they lack suitable pollinators. Some of the populations are thought to be clones, and, since this is an out-breeding species, a clone will therefore be partly or wholly infertile unless crossed with an outside plant. Given the distance between the individual stands of Whorled Solomon's-seal, that is not likely to happen naturally. Yet all may not be lost, for this rhizomatous plant seems well capable of surviving unfavourable periods. For example, the site near Dunkeld where George Don first found it in 1790 still survives – indeed, it is the best stand of Whorled Solomon's-seal in Scotland, with up to 3000 shoots in a good year.

It looks like another case where conservationists should try to control their nerves. At present the plant is considered too rare to indulge in *in situ* management experiments, other than some judicious snipping of shading ferns and the removal of nearby conifers. But should a Scottish colony disappear, there is little chance that the plant will ever recolonize the site by natural means. And the survival of *all* the remaining colonies is highly desirable since such genetic variation as there is lies between the colonies rather than within them. In population terms, each site can be considered a dispersed sub-population of a single large genetic population ('metapopulation') in the Tay catchment. But such is the stable nature of its ancient woodland hideouts that the plant is not immediately endangered. Maintaining the *status quo* should be possible. Providing conditions for it to *increase* is quite another matter.

There is an interesting contrast between the Scottish habitats of Whorled Solomon's-seal and one of its regular associates in Sweden, the Alpine Sow-thistle (*Cicerbita alpina*). Both share the same type of ungrazed subalpine woodland on the continent, yet in Scotland the one is confined to wooded valleys on the Highland fringe, while the other clings to a few high mountain ledges. There is the difference of nearly 1000 metres of altitude between them, yet they still share the same problem – over-grazing. It is sheep and deer that have pushed the two flowers to the margins, and in opposite directions. The Alpine Sow-thistle (see picture on p. 266) did occur on lower ground in the nineteenth century, but, being a nutritious plant, to a sheep as crisp and delicious as an iceberg lettuce, it can now grow only where no large herbivore can reach it.[122] Of the two, it is the Sow-thistle that has lost out more. As a big gangling plant needing deep moist soils, it has almost no habitat left. On just a few broad ledges and gullies it finds conditions that approximate to an alpine glade, with ferns and tall herbs like Red Campion or Wood Cranesbill and even the odd sapling. But it seems poorly equipped for mountain life, prone to frost and mildew, and seldom succeeds in ripening seed. Alpine Sow-thistle can put on a brave show in a good summer, but most of the time you find yourself feeling sorry for it, so seemingly out of place among the alpine flora with its ludicrously big leaves and bedraggled flowering stems. But before we descend on it with well-meant action plans, it is worth pausing to reflect on the special nature of the Scottish flora. Scandinavia may offer *optimum* conditions for plants like this, but Scotland produces what might be called *challenging* conditions, which test a species' inherent ecological flexibility and ability to survive. Is it not partly their tenacity in less than ideal circumstances that makes these species so intriguing? They are interesting *because* they are so rare and yet survive.

Lost and found

If we compare the new third edition of the Red Data Book with the first, compiled twenty years earlier, we find broadly the same species, less some introduced ones which have been weeded out. Relatively few new species have been added, but a number have been banished because they now exceed the required 15 or fewer ten-km squares. The Spreading Bellflower (*Campanula patula*), for example, was known from only 14 ten-km squares in 1977, but is now recorded from 44.[184] But this does *not* mean that it has increased, only that the plant has become better known and more colonies have been discovered. Indeed, this particular plant has almost certainly *lost* ground since 1977. The reasons why plants are under-recorded varies: some, like Narrow-leaved Water Dropwort (*Oenanthe silaifolia*) flower inside a couple of weeks, and are inconspicuous at other times. Others, like Eight-stamened Waterwort (*Elatine hydropiper*) are inconspicuous, period. Others still, like Pyramidal Bugle (*Ajuga pyramidalis*) occur mainly in remote areas; or, like Broad-leaved Spurge (*Euphorbia platyphyllos*), grow mainly in cornfields, which many botanists have given up on. Just occasionally you find a species which has really increased, like Dune Fescue (*Vulpia fasciculata*) or Green Figwort (*Scrophularia umbrosa*). The other 'increases' represent advances in knowledge and understanding, and, above all, increased fieldwork, not, unfortunately, sudden surges on the part of the plant. Present circumstances make it impossible for most rare flowers to increase in fact – though, as we have seen, many of them are good at remaining about the same.

An example of how better understanding can lead to more records is the late John Trist's quest for the Nit-grass, *Gastridium ventricosum*.[193] This grass earned its unflattering name from the white inflated base of the florets, which look like a bad case of head-lice in the bristly heads of the panicle. It is a distinctive grass, known to early botanists like Ray and Dillenius, who were no doubt more familiar with nits than we are. It is an annual, germinating in the autumn and needs bare open spaces which are damp in spring but dry in summer. Until the 1940s, it was fairly widespread as a weed of arable fields, and so became associated in botanical minds with farmland. Trist suggested that the grass was dispersed from farm to farm on the then ubiquitous hessian sacks, which offered ideal lodging places for its sliver-like seeds.[192] Cyril Hubbard, the authority on grasses, decided that the Nit-grass was an import largely on the grounds of its apparent dependence on grain sacks and ploughs.[86] Then, in 1976, someone noticed that there were only six records of Nit-grass in the last twenty years. When John Trist began a search for more on behalf of the NCC, only one arable site was known, on an estate near Lymington, Hants, where game-birds were still valued more than commercial crops. Elsewhere the outlook for Nit-grass looked bleak, and it was categorized as a Vulnerable species – vulnerable to extinction, that is.

Trist noticed that the grass seemed more ecologically 'picky' than most arable weeds, preferring small, damp depressions in fields. He reasoned that farm traffic passing over these slight hollows would further compact the ground thus retaining enough moisture and bare earth for the grass, as well as dispersing seed in the mud clinging to tractor tyres. However, the Nit-grass also occurred in a quite different habitat, on thin bare soil on rocks at the Avon Gorge in Bristol, where it grew best in hot dry weather after a period of heavy rain. Here, too, it had declined – formerly it was found in nine different places, but now in only two. Trist started looking in other warm open places near the sea, on well-drained, rocky slopes and well-rabbited banks, and places where the farmer had been burning gorse. And he began to find it. Some 24 sites for Nit-grass were found over the next two years, and many more have turned up since. In Somerset there has been a virtual explosion of new Nit-grass records on dry south and west-facing slopes, especially on the Polden Hills, though the numbers vary from one year to the next from thousands of plants to just a few, depending on the weather and the amount of exposed soil. The build up of rabbits may have helped the grass in recent years, but it is unlikely that the species has really increased. It was overlooked because its habitat was misunderstood. It is a plant of wild natural grasslands which happened to take advantage of arable fields, not a weed that went native. Moreover we were looking for a relatively large upright grass, like the one in the books. But the Nit-grass of short-cropped grasslands can be only an inch or two long, can grow horizontally and hide in a foot-print. I stepped right over several until David Pearman pointed it out, in the middle of a well-used green track at Durlston Head in Dorset.

A species overlooked for a different reason is Bulbous Foxtail (*Alopecurus bulbosus*), a perennial grass of wet grazing marsh near the sea. At the time of the first Red Data Book there were few recent records from its former stronghold in south-east England.[142] We jumped to the conclusion it had declined drastically, and all kinds of reasons were adduced – habitat destruction, seawater flooding, hybridization with the commoner Marsh Foxtail (*A. geniculatus*), or that good old standby, climate change. On the basis of an apparently catastrophic fall from 60+ sites to only 10, it too was categorized as Vulnerable.

The characteristic swollen flower bases of Nit-grass (*Gastridium ventricosum*) resemble the eggs of head-lice, hence this annual grass's common name.
[Peter Wakely/English Nature]

The first sign that the real situation was not that bad was its rediscovery by John Trist in marshland by the Suffolk coast. These marshes had been flooded by seawater during the war to deter the Germans, and flooded again in 1953. Afterwards the sea walls were rebuilt and the inland drains have since become brackish. Any grass that could withstand periods of immersion in salt water, and all the subsequent earth moving and ecological change, was clearly not one to give up lightly. In fact the little bulbs at the base of the stalk which provide this grass's name are perennating organs which help it to survive through unfavourable periods. When Lady Rosemary FitzGerald made a special search for the plant in the mid-1980s, she found it in a great many places where it had been thought lost, often in abundance.[57] The reason for the apparent decline was not the failings of Bulbous Foxtail but the behaviour of botanists. The grass has a short, early flowering season, from mid-May into June. During this time it is easy to spot, for its slim, pointed spikelets show attractively black against the prevailing green of the marsh. After mid-June, the distinctive spikelets break up, and only someone searching diligently for the tiny bulbs at the base of the stem would be able to spot it. Now, most botanists visit coastal marshland in late July to September, when the greatest variety of species are in flower, and so miss the Bulbous Foxtail's brief moment of glory. Hence it is overlooked because it flowers at

Grass-poly (*Lythrum hyssopifolium*) bears a strong resemblance to the hyssop herb, though it is not related.
[Bob Gibbons]

the wrong time of the year. An additional reason may be that many Bulbous Foxtail sites look like bleak, botanically dull expanses of grass and cows. Although you can spot a likely Foxtail habitat from a distance – look for cows grazing by a grassy sea wall or a flat, wet reach of natural grassland by an estuary – such places attract more birdwatchers than flower-lovers. Fortunately, grasses and sedges are more popular than they were, and that is the main reason why Bulbous Foxtail is no longer in the Red Data Book. In the process, it has changed in our perceptions from a hapless victim of progress to a tenacious grass, capable of surviving full strength seawater, sea-walls, grazing animals and even, in one place, regular cricket matches!

At the time of the first Red Data Book (1977), the Grass-poly (*Lythrum hyssopifolia*) seemed to be one of the rarest and most endangered British plants. Once a fairly widespread 'weed' of bare, muddy ground, it had declined from some 38 ten-kilometre squares to only three. It is an odd little plant with a strong resemblance to the hyssop herb of Mediterranean coasts, though they are not related. No one would suppose, on first acquaintance, that its nearest British relation is in fact the Purple Loosestrife, the relationship being based on the shared botanical structure of the flowers – those of Grass-poly are willowherb-pink, but they open only in full sunshine. For a long time there was nowhere in Britain that you could be sure of finding Grass-poly. Then, in 1958, Gigi Crompton and David Coombe discovered it in good quantity on arable land near Thriplow, Cambs. Or perhaps they refound it, since Ray had recorded the plant from around Cambridge almost exactly 300 years earlier. Here it depends on natural circular hollows known as pingoes, which flood in winter and dry out in summer. Grass-poly likes wet winters and long summers. As an annual, it also likes bare ground, and at Thriplow this is supplied by ploughing. This circumstance has led to a novel form of management agreement with local farmers, whereby the fields are maintained as arable in the interests of nature conservation. One of the farmers is the journalist Oliver Walston, who once wrote a piece of amused irony on the pleasure of being encouraged from *both* sides to plant wheat. Not that standard CAP cereal production would suit the Grass-poly, which also relies on poor drainage and organic methods. It is not really an arable weed at all, but a plant of wet open ground, which takes its chances where it can.

The Grass-poly may have survived, largely un-noticed, at Thriplow for upwards of 300 years. In the past twenty years, it has turned up unexpectedly in other places. In Dorset it appeared in quantity on an arable headland kept damp by a spring. At Cholsey, Oxon, it was found on winter-flooded land near a small pond, where the ground had been disturbed by agricultural machinery. On Thorney Island, West Sussex, its pink eyes were found winking beneath a mass of mayweed in a field corner liable to flooding. Finally, on Jersey it was rediscovered after a long interval at the bottom of a quarry where willow scrub had been cleared. Since then it has even turned up in a garden.

These new sightings suggest either an overlooked plant or a more opportunistic one than had been realized.[24] But if it is an opportunist, how does it reach new sites? The clue was provided by the most spectacular find of all, when hundreds of thousands of Grass-poly plants were discovered, not in a glacial pingo or chalk spring but around Swan Lake at Slimbridge, the headquarters of the Wildfowl and Wetlands Trust. Here, the Grass-poly seems to have found its perfect habitat on muddy, seasonally flooded banks under the feet of countless trampling ducks and swans. It is spreading, and satellite populations have established on the 'Long Ground Scrape' and elsewhere. Swan Lake is visited by wild Bewick's Swans, many of them 'adopted' by swan fans around the world. Perhaps now wild flower lovers, too, will be turning up with field glasses, hoping to glimpse the elusive pink flowers through the wire. It is natural to suppose that the seeds of Grass-poly were brought to Slimbridge in the guts of a wild swan (or goose, or duck). The day before I wrote this, my study was trashed by a pair of inquisitive mallards. While mopping up their gruesome droppings, I was struck by the quantity of undigested seed, which a better botanist than I would have grown on in a pot. That may be how Grass-poly gets about in Britain, though it is also just possible that it was imported to Slimbridge in the grain put out for wintering birds. As a result of these recent finds, the conservation outlook for Grass-poly is now quite favourable. A total of 700 000 plus plants, admittedly most of them by a single pond, places it among the numerically less rare species in the Red Data Book.

Chapter 13

Resurrecting Starfruit:

conservation and rare flowers

Most people would associate bulldozers with development, not nature conservation, least of all with the preservation of rare, delicate flowers. You do not need field craft to detect where a bulldozer has been bulldozing. The banks of earth and mounds of rubble tell you all you need to know. Outside the conservation industry, the idea of using a bulldozer as a conservation tool would have been laughable – like restoring the *Mona Lisa* with a floor mop. A sign of how far conservation has travelled recently was when, at English Nature's 1998 conference on 'Species Recovery', nearly every speaker showed a slide of a bulldozer hired by conservationists. One huge machine had been digging holes in the fen to create pools for Fen Raft Spiders, another was uprooting scrub to create a breathing space for rare orchids, a third was dredging reedswamp from the edge of a lake where a rare water-plantain was in danger. The inevitability of the bulldozer's next appearance became the conference's standing joke, and a ripple of amusement passed through the hall each time another gigantic machine sprang into view, grinding and chugging and changing the landscape on behalf of some threatened bug or violet. The paradox was appreciated – that the mechanical strength which threatened wildlife is also being used to save it.

One species that has benefited from earth-moving machinery is the Starfruit (*Damasonium alisma*), so called from its strangely-shaped seedheads, bearing five or six points like a toy star. At one time it was considered to be the most endangered wild flower in Britain. Though never common, it had once spangled the edges of ponds in 15 counties, mostly in south-east England. By the 1980s, however, it survived in only two, Surrey and Buckinghamshire, and at only one pond in each. The reason for its decline was thought to be the wholescale loss of pond habitat since 1945, and particularly the sort of shallow open ponds you used to find on commons and the wilder sort of village green. These used to be kept free of tall

This classic photograph by Peter Wakely of flowering Starfruit (*Damasonium alisma*) emerging like Excalibur from a Buckinghamshire pond has become a modern icon for the plight of rare species. [Peter Wakely/English Nature]

vegetation by ponies and cattle, or even geese, and the Starfruit used to appear on the warm mud in summer after the water had drawn down. As a result of some NCC-commissioned survey work on rare flowers in the mid-1980s, some management work was carried out at one of its remaining sites, at Gerrard's Cross in Buckinghamshire. A hydraulic digger cleared some of the dense Flote-grass (*Glyceria fluitans*) that had grown up around the now ungrazed margins of the pond to expose the former bare, gravelly bank. The Starfruit responded promptly with a showing of 18 plants – its best effort for some years. But the following year produced only one, and the year after that none at all. The Flote-grass had grown back again (see Table on p. 274).

As often happens, the discovery of a third Starfruit population by Alan Showler in 1989 sparked a great deal of interest in the plant and its mysterious ways. Alan's plants had appeared on an old site after a gap of at least 30 years. The pond had been cleaned out the previous year, and they had almost certainly germinated from dormant seed, buried unseen in the mud. It seemed possible therefore that the Starfruit might not be extinct in some of its old sites after all but merely sleeping through an unfavourable period. Former Starfruit ponds were revisited by Alan, Jonathan Spencer and others, and what they found gave more clues. In most cases it was not surprising that the Starfruit had ceased to appear. Some ponds had simply vanished, drained or filled in, but even where they survived they were seldom in a favourable state. Some had been swallowed up by woodland to become dark, sepulchral pools. At least one had become a village duck pond, planted with weeping willows and water-lilies, and its water level maintained by a ball-cock.

Meanwhile, a study by Chris Birkinshaw at Cambridge shed new light on the Starfruit's way of life.[12] The strange stellar 'fruits' are really bunches of seed pods joined at one end. Close inspection reveals that the pod is shaped like a little boat and contains two seeds of unequal sizes. It seems that while the bigger one usually sinks straight to the bottom of the pond, the other may take a ride in its little boat and be washed up or sunk some distance away. The seeds will germinate only in water, and as the seedling develops, the leaves change shape from narrow and linear to spoon-shaped and finally to oval. These developmental stages once gave the botanist Arthur Bennett an opportunity for some splendid sarcasm at the expense of the German aquatic specialist, Gluck, who had described them as separate forms:

> I watched through the summer of 1887 on a common near (Mitcham). In one pond, *Damasonium* grew pretty abundantly ... In April it was the form *graminifolium* Gluck; in May it began to make itself into the form *spathulatum* Gluck; at the end of June it had become the form *natans* Gluck; flowered through July and part of August; at the end of August the water became very low and the plant became stranded: it was now the form *terrestre* Gluck. The only one I could not say I saw was the form *pumila* Gluck.

It seems then that Starfruit is neither a wholly aquatic species nor a bank plant. It is amphibious, requiring a land stage to produce seeds and a water phase to disperse and germinate them. Since the plant itself cannot move it requires the pond level to do so, partially drying out in late summer and filling up again after heavy rain in the autumn. Hence a deep pond will not do: it must be a shallow, saucer-shaped pond, and moreover, one in a sunny position, free of shade and competition from reeds and tall grass. With its unequal seeds and built-in dormancy, the Starfruit can survive the unpredictability of British weather. What it cannot survive is wholesale permanent change. Warm water, hot mud, sunshine and livestock are the keys for a successful Starfruit. But its old environment has gone. The challenge is to re-invent it.

A Plantlife party clears invading vegetation from a Starfruit pond. [Plantlife]

On the surviving sites conserving Starfruit is largely a matter of traditional conservation management, though of a rather labour intensive kind. But there are too few of them to be confident of long-term success. In 1992 Plantlife began an English Nature-funded project to restore an old site in which the species had not been seen for many years. The obvious candidate was a second pond at Gerrard's Cross called the New Pond. Before the war this had been a heathland pool used by drovers of

TABLE 7 Historic Starfruit localities (compiled by Belinda Wheeler for Plantlife, 1998)

COUNTY	NUMBER OF RECORDS	DATE OF LAST RECORD
South Hants	2	1876
North Hants	1	1910
West Sussex	10	1889
East Sussex	7	1955
East Kent	5	1899
Surrey	**36**	**Extant**
Essex	9	1890
Hertfordshire	3	1928
Middlesex	4	1864
Berkshire	2	1843
Buckinghamshire	**15**	**Extant**
East Suffolk	2	1958
Worcestershire	1	1867
Shropshire	1	1841
South Lincolnshire	1	1836
Leicestershire	3	1890
East Yorkshire	1	1870

cattle on the way to market. But, left ungrazed, the heath had turned first to scrub and then to quite dense woodland. The pond shaded over and filled up with leaves. The Starfruit dwindled and was last seen there in 1966. But since we now know that the dormant seed can survive for decades, there were realistic hopes that restoration might produce a small botanical miracle and recall from history a plant last seen in the year England won the World Cup. In 1992, we began to turn the clock back at New Pond. Enter the mechanical digger.

The arm of a mechanical digger has the same number of joints as a human arm, with an elbow, wrist, knuckles and even fingernails. The worker can control it almost as though it were his own arm, feeling for the original clay lining of the pond, and being careful to remove just the loose accumulated muck on top of it, like ladling broth from a tureen. In deft hands it can be a surprisingly sensitive tool. That autumn our man rode his digger – a small one with a long arm, like a spider crab – to the side of New Pond, and, to the sound of whining hydraulics and crunching gears, he set about saving the Starfruit. It was not easy. To begin with, the soup analogy is all too apt. Although the pond had dried out in late summer, rain the week before had filled it up again, and the sediment was now a terrible colloidal goo. For every bucket-full you managed to remove, another would escape in a cloud of mud as you raised it to the surface. A clean stripping down to the original bottom was impossible. Nor was it any easier to get rid of the

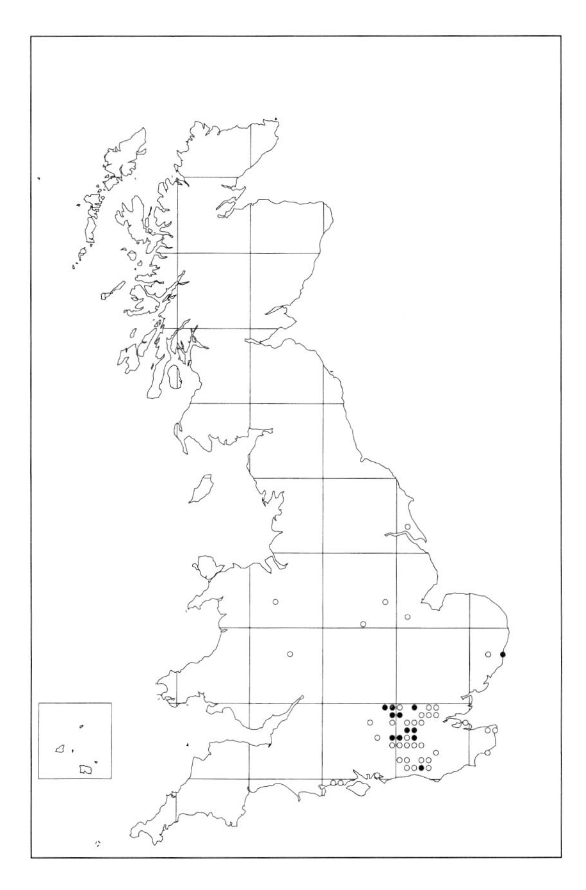

The distribution map shows the extent of decline of Starfruit, mainly due to the loss of muddy ponds on commons and village greens. Solid dots are post-1930 records; open circles, pre-1930. [Biological Records Centre]

reedmace clogging up half of the pond. Reedmace is resistant stuff with roots like steel cables; it is easier to pull them up by hand than dig them out, but wearisome work either way. And pond-scooping is only half the job; the other half is spent dumping the spoil well away from the pond in a custom-made clearing and smoothing it out, using the knuckles of the digger, to form a mulch which will quickly rot down. On the heels of the digger come volunteers with chain-saws, for, unless you feel like repeating the exercise every few years, the over-shading boughs and bushes must come down too, and be cleared away. On top of all that of course, you need the support of local residents if the scheme is to have any long-term benefit. Fortunately, the people of Gerrard's Cross were won over, and not only approved the plan but became active helpers and minders. After the work was finished, Plantlife held a party at the pond. Needless to say, the Starfruit didn't show up. However, it did condescend to appear a year or two later. Though it was half-expected, I, for one, could hardly believe my eyes when the shy trefoliated flower and its spiky fruits once again graced the margins of the pond – in the mind's eye, blinking in the sunlight and wondering what all the fuss was about.

There was nothing original about this approach in physical terms – the man in the digger had cleared out many such ponds in his time. But, for many of us, it was a conceptual novelty, for it turned on its head the notion that rare species grow in refuges from humankind. In fact, on the contrary, plants of bare places, like Starfruit, thrive on regular 'disturbance', and they are dying out precisely because they are no longer receiving it. Pioneer ecologists like E.J. Salisbury had demonstrated as much, but those responsible for protected sites were often slow to absorb the message, being more concerned with birds, otters and other attention-seeking beasts. While one would not want to create reserves that looked like building sites, there is much to be said for looking at the requirements of individual species of plants, and not blandly assuming that overall habitat management will do the job. The digger cannot, of course, 'save' the Starfruit permanently. The best one can do at any given moment is to provide the plant with an opportunity to refresh its seed bank, and try to restore some more ponds where there is a reasonable expectation of it reappearing. It may not be much more than a breathing space, but it is, at least, a metre to two 'back from the brink'.

Rare flowers and nature reserves

Although nature reserves often contain rare flowers, and some rare flowers are more or less confined to nature reserves, few reserves were set up specifically to protect them. The reason lies sunk in the early history of nature conservation in Britain. The first survey of potential nature reserves undertaken in 1912–6 by Charles Rothschild did include the sites of several very rare flowers, including Monkey, Ghost and Lady's Slipper Orchids, Pasqueflower and Spring Crocus. But this plan was scuppered by the First World War, and later progress on nature reserves was dominated by men like A.G. Tansley and W.H. Pearsall who were interested more in natural processes than in individual species. Hence, when the planning for a network of National Nature Reserves did get underway in the 1940s and 50s, the emphasis was on preserving the best wild, near-natural places, and not the small-scale ponds, verges and fields where many rare flowers actually grew.

On the other hand, many of these smaller-scale sites did receive some protection as Sites of Special Scientific Interest (SSSI), and some of these were chosen for their rare flora, with the help of leading naturalists like J.E. 'Ted' Lousley or Professor Vero Wynne-Edwards. When I was professionally involved in such things, my SSSI charges included several small fields, where Fritillary or Wild Tulip grew, and a section of riverbank with a rare hybrid horsetail. The area officers of the conservation agencies devote much of their time to defending such places, generally unsung and out of the public eye. That so many of these places have survived is in large part through their efforts, and it is a pleasure to acknowledge their devoted care here. The resources to protect SSSIs are fortunately greater now than they were forty years ago, when almost your only weapons were stubbornness and persistence.

We are fortunate that many good places to find rare flowers were taken under the wing of the National Trust long before government conservation agencies came into being. However, ownership by a conservation charity did not necessarily help the flowers much. Many species relied on a particular form of husbandry, like coppicing or open-range grazing, which eventually became uneconomic and so ceased. Part of the problem lay in lack of resources, but it was also a matter of ignorance or having other priorities. Much depended on the qualities of the person on the spot. It was local enthusiasm and a nose for opportunity that secured early protection for the sites of the endemic whitebeams on Arran, for example, or the Fritillary meads of the Thames valley. But until the late 1980s, I think it is fair to say that most rare flowers fared rather badly under Britain's nature conservation policies and practices.

When it came to looking after individual species, it was often a matter of 'live and learn'. Slapton Ley in Devon is the only remaining British site for Strapwort, an annual of muddy lake margins. Up until 1968 it grew in the foot-prints of cattle which came down to the waterside to drink and created a lot of barish ground that flooded in winter. Then two unfortunate events combined to bring it to within a whisker of national extinction. A sluice that had created the means to control water levels became inoperative. And the cattle were withdrawn in the belief that their presence was unnatural. The result was that the Strapwort's muddy habitat was replaced by tall sedges and reeds. At one point only six plants could be found. By chance, they found a substitute habitat where fishing boats had been dragged from the water. But they were saved mainly by the timely focus on rare species encouraged by English Nature's Species Recovery Programme. Hundreds of Strapworts cultivated at Kew Gardens have now been planted out; the fence has been removed, and the cattle invited back. In the world of Strapworts, small is indeed beautiful; their niche lies not in great swards of sedge and waterside fen but in hoof prints and boat ruts. To help Strapworts and their like, you need to consider the inches as well as the acres.

Fences round rare species are ultimately self-defeating. They may keep out grazing animals and human intruders but they cannot hold the land in a freeze-frame, and, fatally, they may deny the plant what it needs. Most ecologists recognize this full well, but even so there was, until recently, no shortage of examples of fencing failures on nature reserves – those little squares of coarse grass and thicket where Pasqueflowers or Monkey Orchids used to grow. I remember one fine example in the Cotswolds, where the local water authority had been persuaded to take the Cotswold Pennycress under its wing. Being a responsible authority, they were eager to do their bit to save the plant, and took no half measures when it came to fencing the site. The furlongs of concrete posts and acres of chain link they used would have kept out Attila and all his Huns, but unfortunately the result was thick tall grass – which crowded out all the little Pennycresses. You could still find it outside the fence, especially where stones had evidently been cleared from the fence line, but there were none inside, nor any prospect of any. Personally, I blame television. People are brought up to believe that rare species live in jungles, and that any form of human 'disturbance' must therefore be harmful. Yet, offhand, I cannot think of a single rare British flower that likes dense coarse grass or briar thickets.

Conservation volunteers have done an immense amount of good and necessary work on nature reserves, and it seems a little unfair to mention the occasional disaster which must have upset them more than anybody. But having been unfair already, I will compound the injury and warn that rare shrubs in particular are vulnerable to poorly briefed parties armed with chain-saws

Strapwort (*Corrigiola litoralis*) is confined to the margins of Slapton Ley in Devon, where work to maintain its open habitat is in progress. [Peter Wakely/English Nature]

and matches, intent on clearing scrub to help the environment. What may be our rarest tree, Wilmott's Whitebeam, has been especially luckless in this respect, for both of the best specimens were cut down accidentally by contractors hired by conservationists. Another victim was the fine specimen of wild Cotoneaster which ended up on a conservation task force's bonfire. Ironically their task on this occasion had been to clear over-shading scrub from around the plant. Unfortunately no one seems to have told them what the Wild Cotoneaster looked like. In other respects they seem to have done a splendid job and restored a scrubbed up part of the Great Orme to its pristine glory of bare rock and grass, though, of course, minus the Cotoneaster. There were only some 14 wild plants of this species (if it is a species) in the world. Now there are 13, which sounds potentially unlucky.

It is easiest to think small when your sites are small to begin with. The county wildlife trusts of East Anglia, in particular, have saved innumerable pocket-sized sites for rare and local species from development, from chalk pits and road verges to the sandy field corners of the Breck. Long experience of reserve management, the influence of Cambridge University and a higher than average density of good amateur naturalists enabled them to play a leading part in developing ways of managing land for rare species. Before the biodiversity bandwagon began to roll in the 1990s, rare plant conservation often flourished most in densely populated rural areas. Having mainly small sites means you are more likely to pay attention to small species. It is all too easy to lose sight of little clovers and sandworts when you have a couple of thousand acres of fell or forest to look after.

Though not everyone's idea of a nature reserve, road verges are important refuges for rare plants. In parts of Britain some species have virtually lost their primary habitats except along banks and verges, like the Sulphur Clover (*Trifolium ochroleucon*) once fairly common in East Anglian meadows, but now virtually confined to verges and railway banks. One of the best populations of the very rare Thistle Broomrape is on islands and roadbanks by the busy A1, and new roadworks were routed around it recently, on the advice of English Nature. Some of the largest colonies of Corky-fruited Water-dropwort are on the banks and cuttings of the M5 (though motorway banks are not in general rich in rarities, despite extravagant claims to the contrary by the transport department). The treatment of road verges, then, is a matter of life and death for many plants – as you may recall from the fate of Sickle-leaved Hare's-ear in Chapter 9. After the ruinous days of herbicide sprays, and an almost equally damaging period of total neglect, there is now a network of protected verges which are mown according to the advice of the conservation agencies or the local wildlife trust (though there have been alarming stories of close-mowing and even herbicides recently, perhaps as a result of local authorities contracting out). Much depends on having a resident 'minder' to look after the roadside nature reserve. Only a country like Britain with a large population of dedicated naturalists capable of identifying plants could operate such a system successfully. Hence it is vital that people can identify plants accurately. One sometimes hears that this is not really important any more, and identification has unfortunately taken a back-seat on most environmental courses. Yet we also tend to take it for granted.

The Badgeworth Experience

The classic example of a 'species reserve' set up to protect a rare wild flower is at Badgeworth, Gloucestershire, famous as 'the world's smallest nature reserve' in the *Guinness Book of Records*. For many years the margins of a field pond near the village held the only known colony of Adderstongue Spearwort, *Ranunculus ophioglossifolius*, in Britain. Since its discovery in 1911, there are almost yearly records, showing how the flower could be abundant one year and virtually absent the next, though why that was so no one knew. For years the Spearwort bloomed and seeded, the willow trees were pollarded and cattle drifted down to drink in a hock-high drift of buttercups. Then, in 1932, came sudden consternation when the landowner began to fill in part of the marsh, hoping to sell the site as a building lot. Fortunately G.W. Hedley, one of the authors of the *Flora of Gloucestershire*, reached into his pocket and bought the tiny part containing the flower for £53. He handed it over to the Society for the Promotion of Nature Reserves, and the charmingly named Cotteswold Naturalists' Field Club undertook to look after it. Their main action was naturally to erect a big barbed-wire fence all round the site and put up a sign. What they failed to appreciate was that, as an annual, Adderstongue Spearwort likes open muddy conditions – in other words, it wants cows, not fences. The late Lewis Frost recounted what happened next.

> Devoid of grazing and poaching by cattle, the buttercup rapidly declined. In five of the years between 1934 and 1962, no plants at all occurred in the reserve; in only two of the years were there 100 or more plants.... In two other years there were hundreds of plants outside the reserve, but not one plant inside it. In 1962 ... when Mrs Holland and I visited the reserve, we found it like an overgrown wilderness with the soil surface buried beneath a deep layer of decaying vegetation.[63]

The poor plant had done its best to hint at what was needed. In 1951, Mary Richards found the species growing outside the enclosure in fair quantity, but none inside it where there was 'such a thick tangle of matted grass that nothing else will grow'.[33] In the reserve records book, a visitor remarked that young plants were growing in the footprints of previous visitors, evidently one of the few places where they could find bare, disturbed soil. Fortunately seed survived in the mud. Soon after Frost and Sonia Holland had cleared away some of the thicket and accumulated plant litter in 1962, thousands of seedlings sprang up from the exposed clay. It was their bad luck that they happened to germinate shortly before one of the coldest winters of the century, for the standing water froze solid and killed them all. But, in succeeding years, over a thousand flowering Spearworts appeared as conservation volunteers cleared and weeded a succession of plots. Frost and his colleagues at Bristol University grew the plant in greenhouses to test its responses to different water levels and temperatures. It transpired that Adderstongue Spearwort likes warm, long summers. One unusually mild autumn, the plants flowered in their pots until November and unexpectedly became a greenhouse pest: 'seedlings came up on the staging, in the pots of other experimental plants and even blocking the greenhouse drain'. Frost also discovered the essential role of standing water. The developing plant grows well underwater, sending up floating leaves in autumn and erect snakes-tongue leaves the following spring. By contrast, the seeds need air and light to germinate, and sprout on moist earth after the water has drawn down in late summer. Hence the species is adapted to shallow pools which flood in late autumn and winter but partly dry out in late summer. It will not grow in deep, steep-sided ponds, nor in dense vegetation. And it needs a sufficient depth of water in winter to protect the growth points from frost. Trampling by cattle brings buried seed to the surface, where they are sent on their way with a beneficial blob of cow-dung as fertilizer.

Badgeworth, Glos, is 'the world's smallest nature reserve', home to the Adderstongue Spearwort. [Peter Marren]

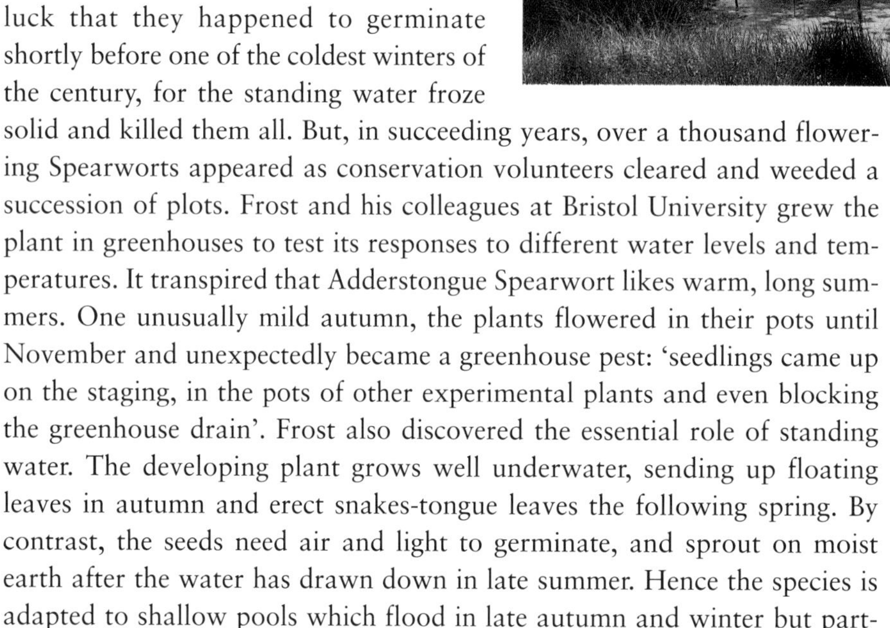

This kind of autecological (single species) work is precisely what is needed if we are to understand how to look after rare flowers on nature reserves. At first sight this particular flower looks almost suicidally pernickety. But shallow cattle ponds provide just the right combination of conditions for winter-wet, summer-dry plants like Adderstongue Spearwort. The trick, as Frost and Holland found at Badgeworth, is to keep down the competing grasses and create plenty of bare mud. Since then, the Gloucestershire Wildlife Trust has brought the management for this species to a fine art that practically guarantees a good display of flowers in late spring. The main threat to it came from outside when a hire-firm bought the patch of ground next door to build a heavy vehicles park with a washing bay. Pollutants from the vehicles would have washed straight into the reserve. The case attracted a lot of media attention and fortunately right prevailed. The village took the 'Buttercup' to their hearts, and for some years a Buttercup Queen was crowned at the village fete with a garland of flowers, including, of course, a few sprigs of Adderstongue Spearwort. A happy ending, then, and an object lesson in conserving rare flowers. As Franklyn Perring has said, 'we need a thousand Badgeworths'.

Managing rare flowers

The Fen Violet, *Viola persicifolia*, is the classic example of a flower that came back from the grave. In the nineteenth century, this pretty May-time violet with long triangular leaves and flowers of the palest shade of lilac often adorned barish patches of peat in the East Anglian Fens where turf-cutters had been at work. It was said to 'abound' at Wicken Fen, for example. Then, mysteriously, it crashed. At Wicken, it disappeared around 1920. For many years, the only place where you could hope to find it was at Wood Walton Fen, and even there its appearances were unpredictable. It was noticed, however, that the violet often turned up in places where the ground had been disturbed by trampling.

In May 1980, Terry Rowell spotted a violet seedling in peaty soil samples he had taken from beneath scrub at Wicken Fen.[172] It was grown on at Cambridge and turned into a beautiful, typical Fen Violet, supposedly extinct at Wicken for the past 60 years. Shortly afterwards, in 1982, a large patch of flowering Fen Violets was found in a different part of the Fen where scrub had been removed the previous year. Furthermore, it had been joined by another rare violet, the fenland subspecies of Heath Dog-violet (*Viola canina* ssp. *montana*). Rowell concluded that they had germinated from dormant seed, long buried in the peat and brought to the surface when the bushes were uprooted. But how do you arrange things so that the Fen Violet flowers every year? Experiment indicated that the violet has a further requirement: it needs seasonally flooded land that dries out in the spring. The drier, more densely vegetated parts of the Fen were now unsuitable for it. Given its past abundance and known seed dormancy, however, there could be a lot of buried Fen Violet seed at Wicken. As part of English Nature's Species Recovery Programme, the reserve's managers decided to cut portions of fen vegetation on a rotation to open up more of the ground to seedling establishment. The results were patchy – a good year in 1990

Fen Violet (*Viola persicifolia*) depends on disturbance, such as turf cutting, to flower and set seed. [Peter Wakely/English Nature]

was followed by no violets at all in 1993, and then a surprise reappearance in 1994 after nearby scrub clearance unrelated to the Violet project! It seemed that mere cutting is not enough for this plant. The vegetation needs to be rooted out and the underlying peat exposed.[198] At Wood Walton Fen, meanwhile, excellent results were obtained by rotovating the ground. Possibly the reason why the violet survived better at Wood Walton is that the latter was grazed by Galloway cattle, so maintaining more suitable open fen. Wicken had no livestock and much of the open fen had been allowed to go bushy and dry out. This strange behaviour of British Fen Violets could not have been predicted by studying the plant abroad. In Ireland it grows in seasonally wet turloughs. In Eastern Europe it is hard to recognize the same plant in the tall riverine grassland mown for hay in late summer. In Britain it hangs on by specializing.

This capacity of plants to survive as buried seed was long overlooked by British naturalists, as was the role of disturbance in releasing seed into growth. The Fen Violet, like the Ground Pine, is rare because it lacks a permanent niche within its already limited habitat. Without a reservoir of dormant seed it would not be able to survive at all. It was much commoner in the past partly because there was more fen then – most of its nineteenth-century sites are now arable land – but also because grazing, turf cutting and other human activities maintained a more plentiful supply of bare hollows.

The Fen Violet is another reminder of the limitations of nature reserves without knowledge, and its plight is typical of a large number of species of temporary conditions. Fortunately this need for *disturbance* is becoming more widely understood. Wherever possible, grazing animals are being introduced to habitats like lowland heaths and fixed sand dunes to maintain the early seral stages where many rare flowers thrive best. The sites of very rare species are much more likely to receive special attention than in the past. This sometimes leads to measures that surprise visitors from countries where there are more wild open spaces than in Britain. A light bash with a rotovator has produced six-fold increases in species like Cotswold Pennycress and Broad-leaved Cudweed (*Filago pyramidata*). The creation of open sandy drifts in one of its Breckland reserves by the Suffolk Wildlife Trust has resulted in fine stands of the much-declined Grey Hair-grass (*Corynephorus canescens*). Judicious removal of reedmace by English Nature using a hydraulic digger rouses hopes for the Ribbon-leaved Water-plantain. Such projects, though now myriad, are usually quite modest – but even one person armed with a rotovator can make a considerable difference locally. In themselves the projects are usually quite simple. But it needs confidence to make a mess on a nature reserve; somehow it runs contrary to instinct. In the mind's eye, nature may seem synonymous with peace, tranquillity and absence of disturbance. Yet, as we have seen so often, this is a fundamental misunderstanding. The variety of plant habitats is maintained by work, not peace. Rare plants are attuned to a world of shepherds, horses, farmers, labourers and low-tech machines. They are gone. We must now find substitutes from our own world of conservation volunteers, contractors, power tools and management agreements.

An important example of the new approach to rare species conservation is Plantlife's 'Back from the Brink' campaign. The project began soon after Plantlife's establishment in 1989, funded partly through its own resources, partly through English Nature, WWF and private enterprise. To identify the plants and places most in need of action, Plantlife conducts detailed field surveys, sometimes in a particular region or county, sometimes nationally. Surveyors try to find out as much about the ecology and behaviour of a plant as they can, and also the factors that may threaten its survival. Often quite a lot of information is available already from survey work carried out by conservation bodies or the BSBI, or published in journals and reports. You may need to visit all the known sites and search for your plant. Depending on the species, you may soon find yourself dripping with mud, or on a rope halfway up a rockface or, as I often find myself, simply lost. This work produces a kind of 'health check' – an indication of the present status of a species and its habitat. Depending on circumstances, Plantlife then considers follow-up action, with the help of volunteers and often in partnership with others, like a county trust or the Scottish Rare Plants Group. During the 1990s they organized a wide variety of shoestring management projects to benefit endangered species like Red-tipped Cudweed, Brown Galingale, Starved Wood-sedge, Hairy Mallow (see picture on p. 284) and Cotswold Pennycress.[121, 161, 40] The work depends on having a network of volunteers, with three often overlapping roles. The 'Finders' are the people who keep a look out for rare plants. The 'Minders' keep an eye on a particular site, and the 'Grinders' put on their wellies and clear bushes or push barrows to create breathing space for some neglected clary or cudweed. Most of these projects are about hauling back natural succession to the point where a rare plant can thrive for at least a few years. The tools – spades, rotovators, occasionally hydraulic diggers – are constant, but the setting varies. Recent projects have included the rotovation of field edges in Kent, clearing acres of invasive scrub in Somerset and gently digging over pondside mud in former Middlesex. In Scotland, Plantlife's Michael Scott has worked with others to restore wet open habitat for Pillwort (see picture on p. 284) and weeded rock faces to benefit Sticky Catchfly.[175] Over in Wales and south-west England, Liz McDonnell and Rosemary FitzGerald have produced detailed conservation plans for Shore Dock, Three-lobed Crowfoot and Toadflax-leaved St John's-wort (*Hypericum linarifolium*). The work included its share of serendipity, like the rediscovery of the second British site of Starved Wood-sedge in 1992.

At the time of writing, the number of Back from the Brink reports has passed the hundred mark. The campaign has also branched out into historical reports of rare species, reviewing their status in time as far back as records permit. The beauty of these projects is that they are cheap and

Top left: Pillwort (*Pilularia globulifera*) owes its English and scientific names to the aspirin-shaped fertile fronds, visible in this colony fruiting on the exposed silt of a dried-up pond. This peculiar fern-relative is in decline throughout Europe. [Peter Wakely/English Nature]

Top right: Hairy Mallow (*Althaea hirsuta*), also known as Rough Marsh-mallow, was one of the first plants to be rescued under English Nature's Species Recovery Programme in concert with Plantlife. [Peter Wakely/English Nature]

'hands-on', and encourage local participation. They depend in turn on the long tradition of botanical recording in Britain. We are blessed with active recorders in every county, organized by the BSBI, a rich legacy of historical data and scientific research, and above all a vast fund of experience and expertise among British naturalists and conservation workers. The catalyst needed to bring all together and in focus was organization and money. And money is at last available through the Biodiversity Action Plan, English Nature's Species Recovery Programme and comparable projects in Scotland and Wales. The resources are not abundant, but are sufficient to do useful work in restoring many of the neglected small sites so vital for the survival of rare plants, and in drawing public attention to the problem.

The special predicament of cornfield weeds

A weed is defined as 'a wild herb growing where it is not wanted'. For weeds of arable land, the preferred term is arable *flowers*, because, after all, we botanists want them. Nevertheless the original pejorative word describes a fact: most weeds do grow where they are not wanted. This is their, and our, dilemma.

The traditional flowers of arable land – the 'cornfields' of the old floras – are the largest group of declining species in the flora. Some are effectively gone already, like Thorow-wax and Corn Cockle. A few more, like Corn Cleavers and Small-flowered Catchfly, seem to be heading the same way. Others, though seemingly fairly widespread on the familiar dot maps, are in fact scarce almost everywhere, like Corn Buttercup and Red Hemp-nettle (*Galeopsis angustifolia*). In many cases, the decline may have been long and gradual, so gradual that hardly anyone noticed. But for at least one, the Corn Marigold (*Chrysanthemum segetum*), it was very sudden, coinciding with the manufacture of a herbicide specially to deal with it. I can remember potato fields in north-east Scotland in the mid-1970s perfectly yellow with Corn Marigold, but a year or two later they were gone. Conserving flowers like these in their traditional cornfield setting is problematical. First each species has an individual ecology and what suits one may not suit another. Second, conservation bodies are ill-equipped to manage cornfield nature reserves on their own. So, in a different way, are most farmers. Growing weeds instead of crops holds little attraction for the farmer: it goes against everything he holds dear. A partial exception could be made for game crops, which must be an important though overlooked reservoir of rare weeds. The Game Conservancy has been promoting the preservation of field headlands with some success. The Countryside Stewardship scheme holds out some hope for weeds, especially in fields near the coast. But the fact remains that no one wants wild weeds. *Sown* ones, perhaps, courtesy of the Ministry – farmers will grow anything as long as they are paid enough. But wild poppies and catchflies, no. It is an insult to farming. (Other people's experience may be different, but that is my own.)

Although the BSBI recorded the decline of arable weeds in its mapping projects, conservationists (with one or two exceptions, like the Suffolk Wildlife Trust) began to take a serious interest in the subject only in the 1980s. Ayla Smith, working for the NCC and the BSBI, was the first to collect detailed information on individual species, and her work has been extended by Phil Wilson and the Northmoor Trust. In some places certain rare weeds are still flourishing. Shepherd's Needle (*Scandix pecten-veneris*), for example, is still sometimes common enough in parts of Suffolk to cause a nuisance, (I really perked up when I learned this – like being plagued with Dodos.) Spreading Hedge-parsley (*Torilis arvensis*), once considered one of the rarest weeds, is unaccountably still fairly common in Somerset. Mousetail (*Myosurus minimus*) has turned up in all sorts of unlikely places, from New Forest pony pens to carriageway verges in Peterborough. However the 'arable flowers' most likely to survive are those which also occur on non-

Below left: Corn Buttercup (*Ranunculus arvensis*) was once a familiar sight in cornfields, known by country names like Devil's Currycombs, Starveacre and even Hellweed. [Peter Wakely/English Nature]

Below right: A century ago, Shepherd's Needle (*Scandix pecten-veneris*) grew abundantly in my Wiltshire parish. In 1996, I had to make a 250 kilometre round journey to find it in a cornfield. [Peter Wakely/English Nature]

crop land, as a surprising number of them do. These are probably the truly indigenous ones – see Chapter 8. Those that rely entirely on corn probably came in with the corn.

At least four conservation mechanisms hold out some hope for rare 'weeds': SSSIs, Environmentally Sensitive Areas (ESAs), experimental plots and nature reserves. Unfortunately only a few cornfields have been designated as SSSIs. Apart from field corners and headlands in East Anglia, I know of only four, all in England. 'Thriplow Hummocky Fields', Cambridgeshire, has the second best population of Grass-poly, growing in depressions on arable land. Cuxton in Kent has a rich arable flora, including the Hairy Mallow, first recorded here in the 1790s. Longmoor in Hampshire has Pheasant's Eye, Rough Poppy (*Papaver hybridum*) and other rarities. And Fivehead Arable Fields, Somerset, was designated because it contains a superb collection of rare weeds and is also, helpfully, owned by the Somerset Wildlife Trust (see below). A quasi-arable SSSI is the allotment on the Isle of Wight where Martin's Ramping Fumitory (*Fumaria reuteri*) grows. With these exceptions, SSSI mechanism is ineffective for protecting communities of weeds in ploughed fields which are, by definition, ephemeral and dependent on farming.

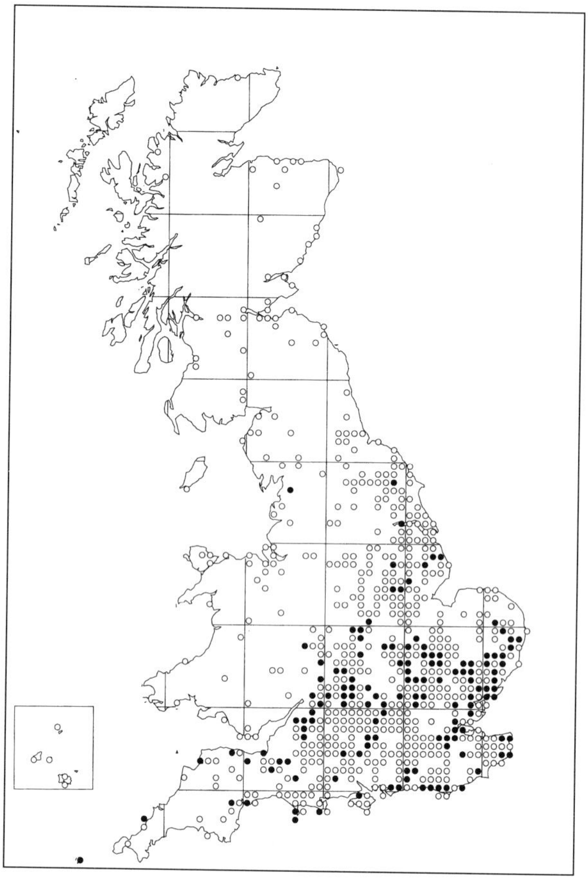

The distribution map of Shepherd's Needle shows the decline of this once familiar flower. Solid dots are post-1970 records; open circles, pre-1970.

Even so there are several, intermeshing bodies that are interested in promoting rare weeds. The Northmoor Trust, an Abingdon-based ecological consultancy, manage a demonstration site and undertake surveys locally. They have investigated organic farms to see if – as many of us had hoped – they were relatively weed-friendly. Preliminary work suggests, unfortunately, that they are not, though several uncommon species have turned up. To farm successfully without chemicals means that you must be exceptionally dexterous with a hoe! In the Thames region, English Nature have a system which links finds of surveys, like the Northmoor Trust's, with local Farming and Wildlife Advisory Groups, who take on themselves the task of selling the charms of weeds to the farmer. In Suffolk, another centre for rare arable flowers, English Nature and the Suffolk Wildlife Trust have taken advantage of the Ministry of Agriculture's ESA scheme to help rare arable flowers survive on field headlands in the Breck. However, as the Trust's Yvonne Leonard told me, 'really successful results will rarely be achieved purely by ESA payments, Stewardship schemes or any other form of financial incentive. Personal contact is essential. Problems arise where there is a lack of communication between farm managers, the estate and the workers on the ground.'

One problem is that some of the plants that benefit most from headlands are not wanted even by conservationists! We can put up with oddities like Common Fiddleneck (*Amsinckia micrantha*), but Springbeauty (*Claytonia perfoliata*) and Sterile Brome have become a serious threat since they monopolize the available space. It is a touch ironic that herbicides are now being used to help arable weeds. One plant that has done well out of headlands both in the Breckland and in Oxfordshire is the wild Grape Hyacinth (*Muscari neglectum*), which always surprises people the first time they encounter it by its deep-indigo flowers, so different from the familiar blue garden plant. At Chadlington, they were so numerous recently that allotment owners were employing children to remove them, at so much per dozen bulbs.

With great courage, the Somerset Wildlife Trust made conservation history in 1992 when they purchased a small farm deep in the countryside that simply teems with rare weeds. This is Fivehead Arable Fields on the calcareous clays at the edge of the Levels. These fields are quite unlike the prairies of eastern England. Each is hidden away behind tall hedges thick with ash trees and approached along narrow winding lanes. They remained unknown to the conservation world until the 1980s. The reason they are so good for weeds is that the land is extremely difficult to farm efficiently. When it rains the fields become quagmires, and in the summer heat the clay sets as hard as concrete. Cultivation is only possible during the 'weather window' between baked ground and mud. Local farmers remember pigs being in one field, potatoes in another, and occasional optimistic sowings of spring or winter sown cereals. It sounds more like the Middle Ages than the Millennium, and crucially, the land was never drenched with herbicides and fertilizers. It is surprising the fields were not left to grass over. One small square field was literally nothing but weeds when I passed by one hot day in June 1996 – and the sort of weeds that botanists dream about. Along one end ran a fringe of Shepherd's Needle, with their strange crowns of spiky seed. With it grew a Red-listed 'weed', Broad-leaved Cornsalad (*Valerianella rimosa*), a relative of the cultivated '*mache*' of French salads, with clusters of tiny pinkish florets on long paired stems. Further into the field were a few Corn Buttercups, with their incredible spiny fruit achenes, once familiar as Devil's Curry-combs. Also there but not yet in flower were Broad-leaved Spurge, Spreading Hedge parsley and Corn Parsley, not to mention masses of commoner weeds like Dwarf Spurge and Scarlet Pimpernel. I half expected to find a Corncockle.

Just at that moment, these flowers were doing very well. But already the field was fast reverting to coarse grassland, dominated by False Oat-grass

(*Arrhenatherum elatius*). To maintain a good crop of weeds means that the land must continue to be ploughed and cultivated, and that is no easier for the Wildlife Trust than it was for the farmer – perhaps harder, for the Trust has to borrow a tractor. A summary report by the Trust's reserves officer, David Northcote-Wright, indicated some of the problems:

November 1992 – All three fields ploughed, became too wet to do any more.
August 1993 – Two fields sprayed with Roundup to try and kill grass weeds; unsuccessful as it was done too late in the year.
October 1993 – One field ploughed and harrowed, became too wet to do any more. Oat-grass clogged the machinery.
August 1994 – All three fields topped to make ploughing easier.
October 1994 – Two fields cultivated and planted with winter wheat. This failed due to excessive competition from grass weeds.
October 1995 – All three fields topped and cultivated.
February 1996 – One field sown with spring wheat.
April 1996 – Sown field sprayed with Topic, a selective herbicide designed to kill black grass and onion couch. This was successful, but has it killed the grass or just retarded it?

It is all proving costly, and rather hit or miss. So far experience suggests that you can have a cereal crop, or a lot of weeds, but not both. Whether the experience of the Somerset Wildlife Trust at Fivehead will encourage other bodies to acquire and manage arable nature reserves remains to be seen. Perhaps, after all, it is better to rely on the farmers, always assuming you can find the right farmer.

Contrary to appearances, the arable landscape is as historic as any other landscape rich in rare flowers. Some downs and moors were arable *before* they were chalk grassland and heather. Here and there, particularly in Cornwall, you can even find ancient arable *fields*, like Forrabury Stiches, near Boscastle, or the small fields at Boscregan near Land's End, the only place in the British Isles outside Jersey where you can find Purple Viper's-bugloss (*Echium plantagineum*). If these fields are turned into grass, the loss to an ancient flora is as great as when you plough up a down or meadow. Rich arable floras may be the rarest plant communities in Britain, with no more than a couple of dozen top-class examples known. But how do you conserve them? Perhaps the best solution available is along the lines of the uncropped strips and conservation headlands promoted by the Game Conservancy Trust, and widely and successfully used in Germany, where farmers are compensated from EU funds. But if enough of us want to preserve wild weeds, we will eventually find a way of doing it.

Rescue missions

'Translocation' is a new word for an old activity – rescuing flowers out of harm's way and transplanting them to a safe refuge. Well-meaning naturalists have been rescuing wild flowers for a long time. A delicious early example was the plant collector who thought he would help the Bristol Rock-cress by scattering its seed every time he dug one up. He apparently helped a great many in this way, which may be one reason why we don't have more Bristol Rock-cresses.[109] William Williams, a mountain guide in Snowdon, dreamed of saving rare ferns from collectors by transferring them to inaccessible rocky islands in a deep lake. For him, too, personal guilt was involved, since Williams was a paid guide who took collectors to the special sites, and he had seen many a rare fern stripped from its rocky hiding place.[93] All too often, however, the chosen safe place proved to be anything but safe. Plant rescue at this primitive level seldom took account of natural succession. Given half a chance, open habitats like to turn into woods, and will do so most expeditiously if you put a fence round them. A classic example is the Monkey Orchid which turned up on a former grass tennis court at Otford in Kent in 1951. It seemed well established and flowered each year until 1956, when it was dug up and transferred to a 'safe place'. Unfortunately, the safe place turned into a dense thicket, while the tennis lawn remained much as it was.[187]

In recent years, 'translocation' has become a much more sophisticated operation with a higher success rate (though claims of success are sometimes made prematurely). Seeds collected from the parent colony have been propagated in botanic gardens, and the young plants sown back to the wild, either to boost the original population or in a new and safer place on the nature reserve. In the process a great deal has been learned about the culturing of rare plants and the identification and management of suitable donor sites. The techniques have been used most widely under English Nature's Species Recovery Programme (SRP), which aims to re-establish

Below left: Fen Ragwort (*Senecio paludosus*), is confined naturally to a single roadside ditch near Ely, but has been introduced to several fenland nature reserves. [Peter Wakely/English Nature]

Below right: Tufted Saxifrage (*Saxifraga cespitosa*), whose dwindling population in North Wales has been boosted by introductions. [Derek Ratcliffe]

populations of endangered plants like Plymouth Pear, Perennial Knawel, Lady's Slipper, Small Alison and Pedunculate Sea-purslane. Some of these projects combine conservation aims with scientific experiment. As Dr Roger Mitchell, in overall charge of the SRP, puts it, 'This form of gardening is really a small-scale disturbance activity which points the way toward larger-scale management. You aim to get it right in a small way before risking all in large-scale management. It is a kind of bioassay.' This is the kind of down-to-earth science which naturalists from the first half of the twentieth century, like E.J. Salisbury, would have enjoyed. However, such translocations and introductions are still somewhat hit-and-miss as conservation tools. The Fen Ragwort, for example, has been introduced to several safe sites from the precarious natural site close to a busy road, but establishing it is proving problematical: a wet site may dry out, rabbits and slugs eat the young plants, Flote-grass moves in and tries to take over. It is currently doing well at Wood Walton Fen, which now has by far the finest stand of the plant in Britain – and in a more natural looking setting than the 'wild' population, despite its planted origin.

One of the more successful translocations is the Tufted Saxifrage (*Saxifraga cespitosa*) on Cwm Idwal in Snowdonia. This attractive cushion plant is one of our true arctic flowers. To see it at its best you have to travel in suitably protective clothing to the gravelly moraines and bleak shores of Lapland, Iceland and the Canadian Arctic. It is one of eight British flowers which grow at Kap Morris Jessup in Greenland, the most northerly point of land in the world.[34] For Tufted Saxifrage, Britain is a southern outpost. Here it is a relict of much colder times, confined to some of the highest mountain peaks in Scotland and North Wales, and nearly always in small quantity. It is one of the most elusive mountain plants. I have only seen it once, and it was while easing towards it, camera in hand, that I slipped and fell down a gully, my worst botanizing experience. This is a flower that likes a view. I doubt I shall ever see another.

The tiny population of Tufted Saxifrage at Cwm Idwal is of importance, both as the southernmost colony in Europe and as the place where it has been recorded on and off for 200 years since its discovery there in 1778. Moreover, these Welsh plants are genetically different from their Scottish counterparts. Unfortunately the whereabouts of its mountain ledge was mentioned in a tourist's guide of 1821, which was a pity because this was one of the alpine flowers every self-respecting collector wanted.[93] By modern times, the population had been reduced by collecting to quite literally a handful of plants. There were only two small clumps left, and one of those frizzled up in the long hot summer of 1976. It looked as though the plant was doomed, since the population had fallen well below the point where natural recruitment could sustain the colony. In 1975, David Parker embarked on a rescue mission for the Welsh Tufted Saxifrage.[135] As part of his PhD research, he collected seed from the main colony and grew them on in 'isolated insect-proof enclosures' at the University of Liverpool Botanic Gardens. This in itself was a challenge, for Tufted Saxifrage is reputedly hard to cultivate, being prone to damp and rotting. But Parker's plants flourished, and provided 30 000 seeds, some of which went into store in the national seed bank at Kew and some reintroduced to Cwm Idwal in a specially prepared transplant site. The aim was to reinforce the wild population until it reached its approximate natural level. The large number of precisely labelled herbarium specimens enabled this to be calculated with a fair degree of confidence – 50 plants would be about right.

It sounds a modest enough aim, but nature is rarely so obliging, least of all on an exposed mountainside. Of about 100 specimens planted out, only half survived the first two years. By 1994, there were only 11 plants in the restocked site, plus a few seedlings (though even this was double the number of native plants). The good news is that the drought of 1976 did not, after all, wipe out one of the colonies: a seedling found a few years later has since matured into a flowering plant. If this project does succeed in saving this very rare saxifrage, it will be due to two things. The first is the detailed reconstruction of the original colony from those faded flowers in the national herbaria – a most original form of historical detective work – which provided the project with a sensible target. And secondly, David Parker, with help from the warden of Cwm Idwal nature reserve, has maintained his commitment to the saxifrage, monitoring its progress annually in minute detail. Unfortunately this kind of persistence is all too rare. Someone may introduce a plant, perhaps as a project for a degree or on contract to a charity, but without a commitment to manage the site in the sole interests of the plant, the chances of long-term success are slim. Staff leave, records are lost, conservation fashions change and eventually the project fades from memory.

Detailed planning and devoted aftercare are paying off in another, more recent attempt to save a classic site for a rare flower. The Sticky Catchfly (*Lychnis viscaria*) is a flower that gains aesthetically from the character of its habitat. In Britain it is associated with ancient volcanoes, and its bright pink blossoms, dissolving into a pink mist at a short distance, form a vivid contrast with the almost black basalts, dolerites and andesites on which it grows. But it is an unlucky plant, which has missed the conservation boat. Few of its scattered sites are nature reserves, and even fewer are in good shape. British Sticky Catchflies are poor competitors, and prefer open grassland or rock ledges in full sunshine (though in eastern Europe it is a common roadside flower). The last thing they want is what they usually get as a result of neglect and natural succession: gorse thickets, thorn scrub and tall grass. Other sites have been quarried away or decimated by fires, treading and even collecting.

All these problems have coalesced at Edinburgh, where the Catchfly once occurred on at least three crags within the city. The best known is Arthur's Seat, in Holyrood Park, on a site known as Samson's Ribs, where the Sticky Catchfly was first discovered 320 years ago by Thomas Willisel, the professional plant hunter. But over the years, botanists, arsonists, gorse bushes and Edinburgh rabbits have given the species a bad time, and by 1990 the colony was teetering on the brink of extinction, with only four plants left.

In such a place, civic pride and local patriotism may come to the aid of an endangered flower (though apparently not a moss – the very rare *Grimmia anodon*, which once shared the crag with the Catchfly, dwindled to extinction unmourned). Phil Lusby of the Royal Botanic Garden was given the task of restoring its fortunes.[110] To do so with any reasonable prospect of success, a great deal of groundwork needed to be done, and it had the effect of bringing this rather neglected species into focus for the first time. A nationwide survey involving several conservation charities revealed only 20 known colonies, ranging between 2 and about 400 individuals – just a few thousand plants in all – with the largest population in the Ochils, in Stirlingshire. Although a fairly long-lived plant, the Sticky Catchfly probably relies on seed to replenish its numbers, and so small populations may fall victim to the laws of population dynamics, where recruitment falls below mortality. At Arthur's Seat it had fallen well below the danger mark, and Lusby decided to try to build up a back-up population by collecting local seed and cultivating the plant at the Botanic Garden. But planting back the young Catchflies proved anything but straightforward. In the wild probably only a tiny proportion of ripened seed produces a flowering Sticky Catchfly, and that process takes several years. To establish the pot-grown plants needed, in Lusby's words, 'more attention than young plants in a garden'. Hand watering was necessary, but on these thin, bone-dry soils, not only did most of the water run off, but threatened to carry away with it what little soil there was. Wire cages were necessary to keep away the rabbits. And although the lower slopes

A surveyor records the progress of Sticky Catchfly (*Lychnis viscaria*) on crags near Edinburgh. [Marie O'Hara]

Inset: Sticky Catchfly

offered better prospects of establishment, any Catchfly there would almost certainly succumb to a combination of fire, scrub invasion and vandalism. As it was, and using the most extreme and labour-intensive care, Lusby succeeded in nursing 12 out of 20 transplants through to flowering. This quadrupled the existing population, and has made it more likely that the Sticky Catchfly will continue to be an Edinburgh plant. But this kind of gardening is scarcely an option for the more remote sites, and would be extremely costly to employ on a large scale. At Edinburgh, translocation is justifiable in terms of historic interest, and as the city's special wild flower (there are not many Red Data Plants in the middle of cities). Most conservation work on Sticky Catchfly, on Breidden Hill and at Glen Farg, has been of the far more conventional kind of habitat management, notably scrub cutting. With healthy populations, that is quite enough. The trick is to wield the pruning shears before the colony sinks to the danger point.

There is one obvious objection to translocations and introductions as a method of conserving a species, and that is that the flower concerned is no longer strictly 'wild'. It will sometimes have been nurtured in artificial conditions, and it will have been taken to a place selected by a scientist and not one selected by nature. In the case of critically endangered species like Strapwort and Fen Ragwort it may be the only realistic alternative to early extinction in the wild. But I suspect that part of the appeal of translocations lies in the involvement of the rescuer in an act of personal benevolence, like rescuing a cat up a tree. At a time when much of nature conservation has become somewhat corporate, abstract and legalistic, it must be satisfying to point to a flourishing patch of orchids or ragwort growing safely on a nature reserve and be able to say 'I did that'. In my opinion it is acceptable at least at an experimental level, so long as good records are kept (and see below for what happens when they are not). But the practice is seductive, and if it becomes a substitute for on-site conservation of wild plants, then we have moved from being conservationists to gardeners. Developers, of course, will happily pay for translocations of species – and sometimes whole habitats – since it enables them to get on with developing.

Brown Bog-rush (*Schoenus ferrugineus*), believed to be wiped out by a reservoir, was rediscovered in Perthshire in 1979. [Michael Scott]

Lessons learned from the Brown Bog-rush experience

The most elaborate translocation mission of any British plant to date was probably the BSBI's attempt to save the Brown Bog-rush (*Schoenus ferrugineus*) from national extinction in 1950. Four years earlier, J.E. Lousley had read in *The Times* that Loch Tummel, Perthshire, was to be dammed and its waters raised to power a hydro-electric scheme. The shores of the loch were the then only known locality for this spare, bristly rush, and, having been there, Lousley realized that raising of the level by even 5 metres would wipe out the plant. That June a team of botanists led by Miss Maybud Campbell gathered at the lochside to perform one of the first recorded rare species surveys. Although the bog-rush was more widespread than they had realized, most of it grew among rocks and wet gravel close to the water's edge. They decided to rescue it, once work had begun on the dam, and transfer as many plants as possible to a specially built enclosure just above what they thought would be the new water level.

In the event they were given little warning. It all had to be done in a great hurry, and in atrocious weather. The work of moving tufts of Bog-rush was undertaken by volunteers. Thirty years later, one of them still had vivid memories of one of the wettest Highland summers on record:

> We took different sides of the Loch, and each day we started where we had left off the previous day ... During the whole fortnight we suffered the worst weather the Highlands had ever had, yet we stuck to it. The tough navvies – out of work miners and dockers – were amazed at our staying powers. Though thinking us daft, they helped us no end to get on with our job and out of their way. They were building the dam all that year, and stopped to cut brushwood to spread under our cars and heave us out daily, the roads being non-existent.
>
> Cynthia Longfield, 1982 (letter in *BSBI News*)

Snow was lying when the last plants were carefully dug up, and frozen fingers replaced wet clothes as the Curse of the Bog-rush. By now, the water was rising with unexpected speed, and the rescuers realized that their transplant site was, after all, too close to the loch. As Brian Brookes recounted later, 'the specially prepared site was completely submerged, and during the following winter the retaining wall was destroyed by wave action'.[19] Insufficient allowance had been made for the effect of wind and waves. It was the cold North Wind, on a shoreline no longer sheltered by boulders and shrubs, that removed *Schoenus ferrugineus* from the British list.

Let us now move on thirty years, in the manner of Shakespeare's *A Winter's Tale*. In 1979, Brookes, then warden of Kindrogan Field Centre, discovered to his great surprise an unknown and healthily large population of Brown Bog-rush on a wet hillside some miles from Loch Tummel. It appeared to be a new site but here lies another problem with translocations. A good many volunteers involved in the abortive rescue mission had come away with their own caches of Brown Bog-rush, and had planted them out in whatever locations seemed safe and suitable to them. Brookes had to try and find out where these locations were.

> Large quantities were planted on the shore of Loch Rannoch and Loch Tay. There is a map made by [A.J.] Wilmott in 1947, now in the botanical library of the British Museum (Natural History), that indicates some of the transplant sites.Wilmott himself transplanted some further individuals to Loch Fincastle in 1949, and on 18 July 1949, D. McClintock and R. Graham took large clumps of *Schoenus* to the north shore of Loch Rannoch. In all some 14 transplants were made, involving hundreds of plants. Also plants were sent to the University Botanic Garden, Cambridge, and to the Royal Horticultural Society's Garden at Wisley.
>
> Only two plants survived ...[19]

Ironically one of the two survivors had got the rescuer into trouble because he had planted it in a different vice-county without permission. But at least he recorded in print exactly what he had done, and where he had done it. The whereabouts of most of the other transplants were jotted down on odd scraps of paper, if at all. It took Brookes a full year of intensive correspondence to track them down. 'Surely,' he wrote, with the potency of frustration, 'it should not be as difficult as this. Unless records are kept in some kind of co-ordinated system, they get lost because people forget, change their names, change their addresses, or die. Are we making it any easier for botanists in thirty years time to know what we are doing?'

A good question. Brooke needed to know whether the new population of the Brown Bog-rush had been there since the Ice Age, or whether someone had planted it in 1950. By the end of his investigations, he was convinced that it was a natural population. The new site resembled the wet moorland flushes in which the plant grows in Scandinavia and Eastern Europe, but not its lake shore habitat in Britain. It was not the sort of site the rescuers would have chosen on the basis of what they knew then. And here lies another lesson: if you wish to rescue a rare British flower, first take a trip to Sweden or Portugal to see how it lives there. It may be that some of our rarest flowers occur where they do not because of some extraordinary quality of habitat but through the chance events of history. In the case of the Brown Bog-rush at Loch Tummel, events seem to have stranded the plant in an untypical habitat.

Bricks, mortar and rarity

By this stage you may be wondering when we will start blaming bypasses, housing and industry for destroying so much of our botanical heritage.

So far as the rarest flowers are concerned, however, there is not a lot to say! Almost every county flora has examples of a cherished species that was destroyed by road schemes or housing development or a rubbish dump. However in most cases the victims are not our rarest flowers but the second tier of *diminishing* species. Roadworks in the Wye valley demolished a colony of Sword-leaved Helleborines (*Cephalanthera longifolia*) for example, while Junction 5 of the M3 near Hook sits squarely on top of what used to be a fine display of Marsh Gentians (perhaps the ministry could call it the Gentian Roundabout?). But reported cases of Red-listed flowers disappearing under bricks and tarmac seem to be relatively few. A redevelopment scheme at Maryport in Cumbria wiped out most of a colony of Purple Broomrape (*Orobanche purpurea*) despite a promise from the developer to preserve it.[78] There are sometimes accidents – like soldiers on exercises near Folkestone who just happened to choose a colony of Late Spider Orchids as the place to dig a trench. But in nearly ten years as a local conservationist with the Nature Conservancy Council I cannot recall a single case of a rare species being deliberately obliterated by developers. One reason is that naturally rare plants tend to live in places that are not liable to development – crags, swamps and mountains. Another is that the sites of the majority of Red-listed flowers are protected as SSSIs. And although SSSIs are liable on occasion to be ploughed, polluted or shamelessly mismanaged, it is rare nowadays for planning permission to be given to cover them in concrete. The earlier motorways smashed through SSSIs and even nature reserves almost as though the planners were playing join-the-dots (count the number of bisected woods on the M2), but more recent road planners tend to steer round them,

Above left: The remaining wild colony of Field Wormwood (*Artemisia campestris*) is now on an industrial estate. Fortunately it seems to like the change. [Peter Wakely/English Nature]

Above centre: The penned-in 'Brandon Artemisia Reserve'. [Peter Marren]

Above right: Field Wormwood escapes. [Peter Marren]

knowing the likely trouble from 'eco-warriors'. Development is not, in fact, the main reason for the decline of wild flowers: farming and commercial forestry have caused far more damage. Today the main problem is more subtle and fundamental – it is change. We have seen many examples in this book – the ponds that turn into swamps, the swamps that turn into thickets and the thickets that turn into woods. We face change at an accelerating rate, and unfortunately most change is for the worse as far as rare flowers are concerned. But though it is a pervasive and universal threat to our wildlife, it happens so gradually that people barely notice. Development, on the other hand, is immediate and visible, though it destroys fewer rare flowers.

One case that did reach national attention in the 1980s was the Affair of Martin's Ramping-fumitory (*Fumaria reuteri*). At the time, the only known population of this attractive but rather critical plant was some allotments at Lake on the Isle of Wight. The allotment holders tolerated and even cherished their unique 'weed' as it threw up its annual froth of purple-and-white flowers and segmented leaves in the furrows between the tomatoes and the dahlias. Unfortunately the local Council had their eye on it and were

The last stand of wild Fingered Speedwell (*Veronica triphyllos*): banks and flower beds on a high-density housing estate. [Peter Marren]

Inset: (Close-up of the plant) [Bob Gibbons]

buying up lots as they became vacant with the intention of selling the land to a builder. The only hurdle in their way was the allotment's status as a Site of Special Scientific Interest. Two public inquiries were fought over the Martin's Ramping-fumitory, which the conservation side eventually won. It is fortunate, though, that such cases are rare, because they are incredibly time-consuming and expensive in legal fees. A few more Ramping-fumitories would swamp the system.

Development does not invariably destroy the site of a rare plant. Indeed an industrial estate can in some circumstances be a safer refuge than a modern farm! The only remaining large colony of the Field Wormwood (*Artemisia campestris*), for example, lies on a patch of waste ground surrounded by roads and warehouses (see picture on p. 296). Until the 1970s, this was open sandy ground bordered by a line of pines – typical Breckland scenery. When the builders moved in, the best that could be done for the poor Wormwood was to set aside a plot of an acre or so and give it to the Suffolk Wildlife Trust. The 'Artemisia Reserve' (see picture on p. 296) became a kind of zoo for rescued wormwoods, planted into the sand behind the wire. My expectations were not high when I went there in 1996 to see this perfunctory nod from business to botany. Yet, unexpectedly, I came away whistling. The plant has fought back with a vigour that must have surprised everybody. Its silvery, wispy foliage is appearing on kerb stones, scattering cigarette butts and dog dirt, and shooting up where it is not wanted with all the virility and confidence of a Common Mugwort. There are Artemisias in the middle of the road; its slight wiry tassels of florets peep from behind sleek offices through cracks in the concrete. It has even turned up as a colonist in nearby gardens.

Even more remarkable is the place amid a high density Thetford housing estate which forms almost the last stand of wild Fingered Speedwell in Britain (see picture on p. 296). The Speedwell lingers on on a bank overlooked by houses and also in community flower beds where its tiny blue 'cat's eyes' peep from the shadow of sprouting daffodils. The Speedwells did not choose these unpromising surroundings: they were there already. Their bank is all that is left of a green lane, once described by a member of the Wild Flower Society as 'a lovely walk in summer with fields on either side'. The plant grew in open places by the sandy track and sometimes along the edge of a nearby field the year after ploughing. From the perspective of the flower, perhaps not all that much has changed, since it still has its open, sandy verges, albeit in a context that has changed utterly. Marvelling at the incongruous sight, it set me wondering what the effect of overhead street lights might be, and whether we would soon find Fingered Speedwells in bloom at Christmas. One also wonders how many Fingered Speedwells might be flourishing out of sight in tiny private gardens behind discreet wooden fences.

However protected a plant may be, an urban environment is one where accidents will happen, often and repeatedly. The example that comes to mind is the sad story of Wurzell's Wormwood. 'Wurzell's Wormwood' is the distinctive hybrid between the native Common Mugwort (*Artemisia vulgaris*) and the naturalized Chinese Mugwort (*A. verlotiorum*). It is known from nowhere else in the world apart from three sites in the London area, and is hence a purely *urban* rare flower. The first one was discovered on the banks of the Thames by the naturalist Brian Wurzell. He sent it off to the authority on Mugworts, who pronounced it new to science. A month or two afterwards, Wurzell took a party of botanists to see it. Need I go on? Some road contractor had decided to dump spoil there and the local authority had of course forgotten to inform them of the unique Wormwood. At the time I was thinking of sending this story to the Guinness Book of Records. Some plants have survived on earth for 100 million years. Wurzell's Wormwood had lasted about six weeks. If you are ever tempted to feel over-confident about the future of rare flowers in this giddy world, think of the instructive case of Wurzell's Wormwood (all right, it did subsequently turn up somewhere else, but let us not spoil the story).

A future for rare flowers

The 1990s have seen more activity to conserve rare British flowers than any previous period of history. 'Species recovery' is indeed essentially a nineties phenomenon. It began on the dot in 1990 with Tony Whitton's 'Recovery' report on protected species, and soon flowered into practical programmes by English Nature and its sister agencies, and those of charities like Plantlife and the BSBI. The decade has also seen the construction of a Millennium Seed Bank to store the seeds from the entire British flora (and ultimately from large parts of the world). It produced the first detailed Red Data Book of British wild flowers, with a database incorporating information on populations, localities and research (this is currently being worked up into a Threatened Plants Database, complete with electronically reproduced sketch maps!). And there have been scores of small-scale projects aimed at managing the sites or rarities, or acquiring them for safe keeping, or at boosting failing colonies with plantings or artificial pollination techniques.

Most significant of all has been the new emphasis on biodiversity, which has set the overall agenda of nature conservation in the nineties. The Biodiversity Action Plan (BAP) is a programme to save the most endangered species of native plants and animals and their habitats through a concerted series of action plans, each of which sets out precisely what it wants to achieve. The genesis of the BAP involved partnership between the voluntary bodies, the academic world and the government, via its statutory agencies. This produced the UK Steering Group Report (1996), which sets out a detailed agenda for the conservation bodies to translate into action. It assigned for each threatened fern or flower a target stating what 'recovery' would amount to in each case. For example, recovery for the Alpine Rock-cress would be reached 'when the present sites have populations that persist for 10 years; when increasing or stable populations can be introduced to nearby ledges as part of a research programme; and when seed stocks have been lodged at Wakehurst Place'. In the case of the Wild Cotoneaster, the target was to be 'a total population of 200 plants over five consecutive years'; for the Lady's Slipper 'five self-sustaining populations of at least 20 plants over five consecutive years'. Some of these aims have been modified as time passes, but they remain very specific definitions of what recovery should mean in each and every case.

The use of plans and targets underlined the businesslike approach of modern conservationists. It was clear that the work outlined would be considerable, and that it would require an enormous amount of research and organization. Fortunately, funding has matched the scale of the task. Recently English Nature has spent more than one million pounds per annum on BAP work, as the list of action plans has grown and grown. Its sister bodies, SNH and CCW, in Scotland and Wales, have developed comparable schemes and funding, with national botanic gardens playing an important part. In a few cases private businesses have pitched in with donations, though maybe not to the extent that was hoped. As a result, a large number of self-employed researchers or research institutes delightedly found themselves being paid to study an interesting species – something we had hitherto done as a hobby. And a network of alliances and partnerships have sprung up like mushrooms, engendering a genuine buzz of enthusiasm and excitement. When so much in conservation has become rather abstract and statistical, it is refreshing to return to one's roots and deal with real species in real places.

The buzzing labyrinth of industry produced by the UK Biodiversity Action Plan ought to delight those of us who once despaired of ever seeing adequate attention and resources devoted to rare and declining species. But there is a new concern, which is evidently widely felt though seldom expressed publicly: that the envisaged level of intervention in the BAP may in some cases come to threaten the concept of wildness itself. We have seen this already in the case of the Lady's Slipper, for which 'recovery' has involved micropropagation, fertilizer, cages and slug pellets; it has in fact received more 'gardening' than any prize-winning rose or hothouse orchid. For many naturalists, wildness means a sense of 'otherness' and free existence outside the human domain, whether the plant's habitat is a mountain top or a hedgerow. It is this intangible 'otherness' which becomes threatened when we wade in too boldly with trowels or technological paraphernalia. You can save an animal by locking it up in a zoo, but at the price of its natural biology. Is it any different if your flowers originated in a garden, or when you are obliged to peer at them through wire mesh? And does it matter, so long as the species is saved?

In *Future Nature* (1996), Bill Adams argued that conservation in Britain has grown up without a coherent philosophy. To him, conservation consists of 'a cultural and scientific rag-bag of passion, insight and good intentions'. Its activities have become strongly institutionalized, and pegged to practical targets, without a corresponding agreed code of ethics. The BAP follows the NCCs *Nature Conservation in Great Britain* (1984) in its statement of why nature is *important*: for 'aesthetic value, recreational activity and for science and education'. But, as Adams noticed, for all that, we no longer feel nature and wildness to be in our bones, as did many writers and poets on the rural scene earlier in this century. Absent from the conservation business world of targets and achievements is any sense of the reverence and awe that was clearly felt, for instance, by John Ray as he explored the meads of Cambridge in the springtime. The novelist John Fowles has criticized the way we can so easily turn a wild flower or a beautiful tree into a 'thing' – something we use – and, in the process, lose 'its presentness, its seeming transience, its creative ferment and hidden potential'. There are times as you look at the statistics displayed on the screen – the monitoring data of some devoted naturalist on a distant summer day – when the pessimistic words of D.H. Lawrence have a special resonance: 'There is nothing to look at any more / Everything has been seen to death.'

Grounds for optimism? The very rare Jersey Cudweed (*Gnaphalium luteoalbum*) has colonized rotting waste from the old Decca gramophone record factory near the edge of Poole Harbour.
[Peter Wakely/English Nature]

The distancing of our selves and our lives from nature has been bad for art and literature. Is it bad for rare flowers? In the most direct sense, probably not. It is impossible to disagree with the conservationist's argument that it is better to rescue a species than to watch it die out. When a conservation body cries that it will soon be TOO LATE for certain plants, then almost any level of intervention – and the more the better – seems justified. But we should nevertheless take time to ask ourselves what we are conserving rare flowers <u>for</u>. For us or for them? We can enjoy flowers in gardens, in windowboxes, in a vase. But wild flowers need to remain wild: it is their patrimony, the reason for their existence. Do we still have the restraint, the humility, to say to ourselves: thus far but no further? Has mankind become so dominant that even wildness itself is our possession too?

I think we can afford not to become too panic-stricken about the survival of rare flowers in Britain. We have been kinder to flowers during the past ten years, and their preservation is much more firmly rooted in the national agenda. The emphasis on action is better than the previous, unwarranted assumption that species take care of themselves on nature reserves. We understand the science better, and perhaps work together better too, helped by advances in information technology. If I raise doubts about the direction of current work, it is nonetheless with the consoling thought that the places where rarities grow are safer than they were, and that the flowers bloom as they always did, unobtrusively on distant hillsides and vales, in hidden quarries and dells, at the spot marked X on the secret map. If, just occasionally, you feel deflated to find your sought-after flower enclosed in a muslin cage, or accompanied by a scrum of fellow admirers, at least it is there. Each generation draws solace and inspiration from our homely native flora, and enmeshes them with some of its own current concerns and fashions. But on finding a rare flower we all still share the same moment of sympathetic contact with nature, which transcends the external world and finds a niche in our minds and our memories.

Appendix 1

First records of rare and Red Data Book flowers and ferns

The discovery date of a wild plant is deemed to be the first record of it in print. This is usually a paper in a scientific journal, or a learned publication. Some of the plants first mentioned in old herbals may have been well known already. Even today there can be a considerable time-lapse between a discovery and its commitment to print – more than 10 years in the recent case of *Sorbus domestica*. The discovery dates and places of rare British wild flowers and ferns below are presented in approximate chronological order (the first time it has been done this way as far as I know). The list is based on *The Comital Flora* (1932) by G.C. Druce, but updated where necessary. Taxonomically difficult plants are omitted. It includes nearly all the Red-listed wild flowers and ferns, plus a few other less rare ones where they took my fancy. This appendix supplements Chapters 2 and 3, where the broader context of plant discovery in Britain is outlined.

Listed in the works of William Turner (1508–68)

Pasqueflower	'About Oxforde, as my frende Falconer tolde me,' *Herbal*, 1551.
Pheasant's Eye	As the 'red mayde weed'. Another Elizabethan name was 'Rose a rubie' (Gerard).
Wild Cabbage	As 'Sea cole' (kail) 'in Dover cliffes, where as I have onely seene it in al my lyfe', *Names of Herbes*, 1548.
Dittander	'By a water called Wanspeke' (R. Wansbeck, Northumberland). *Names*, 1548.
Corncockle	'Coccle aut pople.' *Names*, 1548.
Wild Pear	'Wylde Pere tree ... well knowen.' *Herbal*, 1562.
Thorow-wax	'Thorowax. In Somersetshire between Summerton and Martock.' *Herbal*, 1568.

Spignel	'Once at saynte Oswarldes' (Northumberland). *Names*, 1548. Gerard knew it from 'Roundthwaite betwixt Appleby and Kendal' – it is still there despite being bisected by the M6 and the main London to Glasgow railway!
Honewort	'I found a root of it at Saynt Vincentis rock, a little from Bristowe.' *Herbal*, 1562.
Shepherd's Needle	'Groweth in ye corne.' *Herbal*, 1562.
Blue Pimpernel	He believed this to be the female form of Scarlet Pimpernel.
Greater Broomrape	'In Northumberlande ... newe chappel floure.' *Names*, 1548.
Pennyroyal	'So exceedingly well knowne to all our English nation that it needeth no description' (Gerard). 'Besyde hundsley upon the heth' (i.e. Hounslow Heath). *Herbal*, 1562.
Water Germander	'*Scordium*. I heare say that it groweth besyde Oxforde.' *Names*, 1548.
Ground Pine	'*Chamaepitys*. I heare that it is founde in diverse places in england,' *Herbal*, 1551. 'In good plenty in Kent,' *Herbal*, 1568.
Wild Daffodil	'Narcissus herbaceus ... is after my judgement our yealowe daffodyl.' *Names*, 1548.
Darnel	'Lolium ... Darnell.' *Names*, 1548.

Listed by Matthias de l'Obel (1538–1616)

Sea Pea	Gathered and eaten on the Suffolk coast during the famine of 1555. Caius, in L'Obel's *Stirpium Historium*, 1570.
Small Fleabane	'In Bernard greyn' (Barnard's Green, Worcs, where it survived until 1936), *Stirp. Hist.*, 'At Islington,' Gerard, 1597.
Annual Beard-grass	'Alopecuros altera ... Zout-hamptoniae proxime salinas ... Essexiensis ... juxta Thamesis.' At Southampton near saltings and in Essex by the Thames estuary. L'Obel's *Stirpium Adversaria Nova*, 1605. Thomas Johnson also knew it from 'South-Sea Castle'.

Listed by John Gerard (1545–1612) in *The Herball or generall Historie of Plantes*, 1597

Baneberry	'Herb Christopher groweth in the north parts of England, neere unto the house of the right worshipfull Sir William Bowes' (now Bowes Museum).
Tower Mustard	As Tower Cress, 'at Pyms by a village called Edmonton neere London'.
Medlar	'Often-times in hedges among briars and brambles.'

Greater Water-parsnip	'In moorish and marshie grounds.' 'By Redding,' How's *Phytologia*, 1650.
Hog's-fennel	'Hogs Fennell ... on the south side of a wood (near) Waltham on the Naze, in Essex, and Whitstable in Kent.'
Cottonweed (*Otanthus*)	'At a place called Merezey (Mersea), sixe miles from Colchester.' Near Glynllifon (Caernarvon) – 'we went to the sea shore ... and there we dug up the *Gnaphalium marinum* (Cottonweed)', Thomas Johnson, 1633.
Swine's Succory	Or Swine's Chicory, known to Thomas Penny and Gerard as an arable weed with a useful bitter-tasting root.
Red Star-thistle	'Carduus stellatus. Upon barren places neere unto cities and townes.'
Bird's-eye Primrose	First recorded by the Revd Thomas Penny, 1581. Its yellow centre 'hath mooved the people of the north parts to call it Birds eine'. First localized by How, 1650, on 'a very low and squalid meadow near Knaresborough'.
Green Hound's-tongue	'Dwarfe Houndstoonge ... betweene Esterford and Witham in Essex.'
White Mullein	'Blacke Heath next to London.'
Smooth Rupturewort	'Herniaria. In barren and sandie grounds.'
Stinking Goosefoot	'Atriplex olida ... upon dung hills.'
Box	'Groweth upon sundry waste and barren hils in Englande.' The first botanical reference to Box Hill is in Merret's *Pinax*, 1666.
May Lily	'Monophyllon. Lancashire, in Dingley wood, sixe miles from Preston ... and in Harwood, neer to Blackburne.'

Listed in the works of Thomas Johnson (*c.* 1604–41)

Sea Stock	'At Aberdovye' (Aberdovey, Gwynedd), collected by George Bowles. Revised Gerard's *Herball*, 1633.
Sea Heath	'Byrselden ferrey' (Bursledon nr Southampton), coll. John Goodyer. *Herball*, 1633.
Deptford Pink	'Between Gillingham and Sheppey.' Johnson, *Iter Plantarum Investigationis* 1629. Gerard's 'little wilde creeping Pinke ... in the great field next to Detford' was probably the Maiden Pink.
Touch-me-not Balsam	'On the banks of the river Kemlet at Marington in Shropshire,' George Bowles. *Herbal*, 1633.
Yellow Vetchling	'Near Gravesend.' Johnson, *Descriptio Itineris Plantarum* 1632.
Sickle Medick	'*In montosis*.' *Mercurius Botanicus*, 1634. First localized 'between Linton and Bartlow', Cambs, by Ray, 1660.
Grass-poly	At Dorchester, Oxfordshire, George Bowles. *Herbal*, 1633.
Great Bur-parsley	'In the corne fields about Bathe (Bath), Mr Bowles.' *Herbal*, 1633.
Western Spiked Speedwell	'St Vincents Rocke by Master Goodyer.' *Mercurius Botanicus*, 1641.

Downy Woundwort	'Wilde Stingking Horehound ... wilde in Oxfordshire in the field ioyning to Witney Parke, a mile from the Towne,' Mr Leonard Buckner. *Herbal*, 1633.
Spiked Star-of-Bethlehem	'In the way betweene Bathe and Bradford (on-Avon) not farre from Little Ashley' *Merc. Bot.*, 1634.
Starfruit	'Plantago aquatica minor stellatum. Beyond Ilford (Essex) ... Mr Goodyer had also found it upon Hounslow Heath.' *Herbal*, 1633.
Lady's Slipper	'In the North parts of this kingdom' (i.e. England). *Herbal*, 1633. First localized 'in a wood called the Helkes in Lancashire neere the border of Yorkeshire,' Thomas Parkinson in *Theatrum Botanicum*, 1640.
Lizard Orchid	'Nigh the highway betweene Crayford and Dartford in Kent.' *Merc. Bot*, 1641.
Burnt Orchid	'In montosis pratis' (hill pastures), *Merc. Bot*, 1634. Localized in How's *Phytologia*, 1650: 'On Scosby-lease' (nr Doncaster), Mr Stonehouse.

Listed in the works of Thomas Parkinson (1567–1650)

Coralroot	'At Mayfield in Sussex in a wood called High-reede,' John Goodyer, 1634. *Theatrum Botanicum*, 1640.
Narrow-leaved Lungwort	In Hampshire, John Goodyer. *Paradisus Terrestris*, 1629.
Autumn Squill	'Hyacinthus autumnalis minor. In many places of England ... at the hither end of Chelsea.' *Paradisus*, 1629.
Bog Orchid	'Bifolium palustre. In the low wet grounds between Hatfield and St Albones; in divers places of Romney Marsh.' *Theatre*, 1640.
Scots Pine	In Scotland 'as I am asured.' *Theatre*, 1640.

Listed in the works of William How (*Phytologia*, 1650) or Christopher Merret (*Pinax* 1666–7)

Pillwort	'Gramen piperaceum, near Petersfield, (Hants),' John Goodyer. *Pinax Rerum Naturalium Britannicarum*, 1666.
London Rocket	'Ubique fere in Suburbiis London' (almost everywhere in the suburbs of London). First recorded by Goodyer *c.* 1650. *Pinax*, 1666. Morison later described its spectacular abundance after the Great Fire of London.
Hoary Rockrose	'Cistus mas breviore folio'. *Pinax*, 1666.
Childing Pink	'In the grounds 'twixt Hampton Court and Tuddington.' *Phyt.*, 1650.

Spanish Catchfly	Newmarket Heath by Mr Sare. *Phyt.*, 1650.
Nottingham Catchfly	'3 miles from Dover in the way to Rye on the Beach.' *Pinax*, 1666. Nottingham Castle, T. Willisel, in Ray's *Catalogue*, 1670.
Berry Catchfly	'Alsine Baccifero major ... In sylvis udis.' *Phyt.* 1650. First localized at Isle of Dogs, London 1837, by G. Luxford.
Perernnial Flax	Newmarket Heath by Mr Sare. *Phyt.*, 1650.
Marsh Pea	'In a wet marsh ground on the left hand of Peckham Field from London.' *Pinax*, 1667.
Hampshire Purslane	'In a great Ditch neer the Moor at Petersfield, Hamshire,' John Goodyer. *Pinax*, 1666.
Spreading Hedge-parsley	'Amongst wheat plentifully near Petersfield, Mr Goodyer.' *Pinax*, 1666.
Marsh Fleawort	As 'Hoary Fleabane' 'a stones cast from the East end of Shirley Poole ... in Yorkeshire. Mr Heaton.' *Phyt.*, 1650.
Fen Sow-thistle	'In the medows betwixt Woolwich and Greenwich.' *Pinax*, 1666.
Spreading Bellflower	'At Effaton, a mile from Wigmore,' Herefordshire. *Pinax*, 1666.
Jacob's Ladder	'On the rocks betwixt Mawwater (Malham) Tarn and Mawanco (Malham Cove).' *Pinax*, 1666.
Bastard Balm	'In Mr Champernon's wood neere Totnes, Mr Heaton.' *Phyt.*, 1650.
Coral Necklace	'Sent me from Cornwall.' *Pinax*, 1666. Ray found it near Penzance in 1662 (*Catalogue*, 1670).
Asarabacca	Einsham (Eynsham) Common, Oxfordshire. *Phyt.*, 1650. This is the first localized record, but L'Obel knew it from Somerset, and both Gerard and Parkinson described it.
Birthwort	'Beyond Redding.' *Phyt.*, 1650. Long known as a medicinal plant.
Purple Spurge	'Cornwall.' *Pinax*, 1666. Ray found it at Penzance in 1662 (*Catalogue*, 1670).
Early Spider Orchid	'Upon an old stone pit ground ... hard by Walcot a mile from Barnack' (probably Barnack Hills and Holes). Dr Bowle. *Phyt.*, 1650. Gerard's 'wasp orchid ... the colour of a dry oken leafe' may have been this plant.
Monkey Orchid	'On several Chalkey hills near the highway from Wallingford to Redding on Barkshire side the river', Mr William Browne. *Pinax*, 1666.
Military Orchid	Found apparently at the same time and place, and by the same person, as Monkey Orchid. Merret's 'Orchis militaris polyanthos on Gads-hill in Kent' was probably Lady Orchid. The three species were confused by early botanists.
Triangular Club-rush	'At the Horse-ferry at Westminster.' Dr Dale. *Pinax*, 1666. Ray records it 'by the river Thames side both above and below London' (*Catalogue*, 1670).

Listed in the works of John Ray (1627–1705)

Oblong Woodsia	'*Filicula alpina Pedicularis rubris*' on Clogwyn, Snowdon. Edward Lhwyd. *Synopsis Methodica*, 1690. The two Woodsias were only properly distinguished by James Bolton, *Filices Britannica*, 1785.
Forked Spleenwort	'Filix saxatilis. On the rocks in Edinburgh Park.' Thomas Willisel. *Catalogus Plantarum Angliae*, 1670.
Bristol Rock-cress	St Vincent's Rock, Bristol, John Newton. *Historia Plantarum*, 1686.
Isle of Man Cabbage	'Plentifully going from the Landing-place at Ramsey to the Town' (Isle of Man). *Catalogus*, 1670.
Cotswold Penny-cress	'Among the stone pits between Witney and Burford' (Oxfordshire). Jacob Bobart. *Fasciculus Stirpium Britannicarum*, 1688, *Synopsis*, 1690.
White Rockrose	On 'Brent Downs' (Brean Down, Weston-super-Mare). Leonard Pulkenet. *Fasciculus Stirpium Britannicarum*, 1688.
Small-flowered Catchfly	'Found by Mr Dent among corn near the Devil's-Ditch' in Cambridgeshire. *Catalogue*, 1670.
Sticky Catchfly	Upon the rocks in Edinburgh Park, Thomas Willisel. *Catalogue*, 1670.
Little Robin	'Swannich, Dorset.' W. Sherard. *Synopsis*, 1690.
Shrubby Cinquefoil	'Ad ripam Meridionalem fluvii Tees,' T. Willisel. *Catalogue*, 1670.
Rock Cinquefoil	Craig-Wreidhin (Breidden Hill). *Fasc. Stirp*, 1688.
Field Eryngo	'On a rock which you descend to the Ferrey, from Plymouth over into Cornwall.' Ray, 1662, in *Catalogue*, 1670.
Moon Carrot	Gogmagog Hills, Cambridge. *Synopsis*, 1690.
Lesser Bur-parsley	'In the corn about Kingston wood and elsewhere.' Cambridge Flora (*Catalogus Plantarum circa Cantabrigia nascientum*), 1660.
Corn Cleavers	'Inter segetes passim.' Cambridge Flora, 1663.
Narrow-leaved Cudweed	'Gnaphalium parvum. About Castle-Heveningham (Essex), Mr Dale.' *Synopsis*, 1696.
Fen Ragwort	'In many places about the Fens as by a great ditche side near Stretham ferrey.' Cambridge Flora, 1660.
Field Fleawort	'Park Mountain Ragwort. On Gogmagog Hills and Newmarket Heath.' Cambridge Flora, 1660.
Stinking Hawks-beard	'In Cambridgeshire.' Cambridge Flora, 1660.
Spotted Cat's-ear	On Gogmagog Hills and Newmarket Heath. Cambridge Flora, 1663.
Least Lettuce	Near Cambridge. Cambridge Flora, 1660.
Cornish Heath	'Juniper or Firre-leaved Heath, with many flowers. By the way-side going from Helston to Lizard-point in Cornwal, plentifully.' Ray 1667, in *Catalogue*, 1670.
Oxlip	In Kingston and Madingley Woods. Cambridge Flora, 1660. Long confused with 'False Oxlip (primrose/cowslip hybrid). Darwin decided matters in 1869.

Yellow Centaury	'Cornubiae' (Cornwall). *Catalogue*, 1670.
English Gentian	What is probably this plant surfaces in Ray's *Synopsis* 1696, but it may also be Johnson's Gentiana fugax minor, 1633. It was only distinguished as a separate, endemic species in the twentieth century, by H.W. Pugsley and E.F. Warburg.
Purple or Blue Gromwell	On a 'bushy hill near Denbigh-town.' Ray 1662, in *Catalogue*, 1670.
Hoary Mullein	'Circa moenia Norvici urbis' (Norwich). *Catalogue*, 1670.
Spiked Speedwell	'In a close near the beacon on the left hand of the way from Cambridge to Newmarket in great plenty.' Cambridge Flora, 1660.
Fingered Speedwell	Rowton, Norfolk, Thomas Willisel. *Catalogue*, 1670.
Alpine Bartsia	'Near Orton (Westmorland) by a stream running across the road to Crosby.' *Catalogue*, 1670.
Crested Cow-wheat	Madingley and Kingston Woods, near Cambridge. Cambridge Flora, 1660.
Purple Broomrape	In Hampshire, John Goodyer, 1621. *Synopsis*, 1689. Not found again until 1779.
Downy Hemp-nettle	Among corn near Wakefield, Darfield and Sheffield. *Catalogue*, 1670.
Fringed Rupturewort	'In promentorio Cornubiensi The Lizard Point dicto.' Ray 1667, in *Historia Plantarum*, 1686. Not recognized as distinct species until 1838.
Perennial Knawel	Ssp. *prostratus* 'About Elden (Elveden) in Suffolk.' *Catalogue*, 1696.
Wild Asparagus	The plant Ray found near The Lizard in 1667 was probably the first genuine Wild Asparagus. Gerard knew a saltmarsh variety 'wilde in Essex' (Herbal, 1597).
Wild Leek	'Insula Holms' (Steepholm and Flatholm in the Bristol Channel). *Historia*, 1688.
Snowdon Lily	'Bulbosa juncifolia.' 'Trigyfylchau Rocks' above Cwm Idwal, Edward Lhwyd, 1688. In *Synopsis*, 1696.
Fen Orchid	'In the watery places of Hinton and Teversham Moors', Cambridge. Cambridge Flora, 1660.
Round-headed Club-rush	'In comitatu Somerseti,' D. Stephens (probably the Burnham and Berrow dunes, where it still occurs). *Historia*, 1688.
Nit-grass	'Near Tunbridge Wells, Kent.' Mr S. Doody, 1688. *Fasciculus Stirpium*, 1696.
Creeping Dog's-tooth or Bermuda Grass	'Plentifully ... between Pensans [Penzance] and Market-Jew', Cornwall, Mr Newton. *Fasciculus Stirpium*, 1688.

The Eighteenth Century

Meadow Clary	Cobham, Kent. D. Watson, 1699 (and still there). In Morison's *Historia Oxoniensis*.
Cornish Bladderseed	Cornwall, coll. Revd Lewis Stephens of Menheniot, and described as new species by James Petiver, 1713. *D. Raii Catalogus cum Iconibus*.

Sand Catchfly	'A little to the north of Sandown Castle' (Isle of Wight). William Sherard, 1715. In J.J. Dillenius's new edition of Ray's *Synopsis Methodica*, 1724.
Marsh Saxifrage	On 'Knotsford-moor' (Knutsford, Cheshire). Dr Kingstone. *Dill. Synopsis.*, 1724.
Cheddar Pink	'On Chidderroks' (Cheddar Gorge). Samuel Brewer. *Dill. Syn*, 1724.
Narrow-fruited Cornsalad	Chislehurst, Kent, Du Bois, 1700. *Dill Syn*, 1724.
Balm-leaved Figwort	'About St Ives' (Cornwall). Edward Lhwyd, *c.* 1712, in *Dill. Syn.* 1724. Known from Jersey since 1690.
Field Cow-wheat	'In the corn ... at Lycham in Norfolk,' W. Sherard, in *Dill. Syn.*, 1724.
Killarney Fern	'*Felix humilis repens.* At the head of a remarkable spring' near Bingley, Yorkshire. Dr Richard Richardson. *Dill. Syn.* 1724.
Bulbous Foxtail	J. Sherard in *Dill. Syn.*, 1724, without locality.
Spotted Rockrose	'*Cistus flore pallido punicante.* Nere Grosnez Castle, Jersey.' William Sherard in *Synopsis*, 1690. First found in mainland Britain near Holyhead, Anglesey by the Revd William Green and identified by Dillenius, 1727: 'It is a *Cistus* and seems to be new.' Dillenius named it in honour of Samuel Brewer, whom he mistakenly believed to be its discoverer.
Fritillary	'In Maud-fields near Ruislip Common, observed over forty years by Mr Ashby of Breakspears.' John Blackstone, *Fasciculus Plantarum circa Harefield*, 1737. Known previously as a garden plant.
Early Sand-grass	'Found in Wales by D Stillingfleet.' In Hudson's *Flora Anglica*, 1762.
Purple Oxytropis	'Near Loch Leven.' Dr John Walker, 1761. James Robertson in *Scots Magazine*, 1768.
Creeping Spearwort	West end of Loch Leven, Dr Parsons, 1764. In Lightfoot's *Flora Scotica*, 1777.
Health Lobelia	'Supra Shute Common inter Axminster et Honiton.' D. Newbery, 1768. In William Hudson's *Flora Anglica*, 1778.
Pipewort	In a small lake on Isle of Skye, James Robertson, 1768. In *Philosophical Transactions*, 1770.
Spring Speedwell	Near Bury (St Edmunds), Suffolk, Sir John Cullum. In *Elements of Botany*, 1775.
Hairy Greenweed	Icklingham, Suffolk, Sir John Cullum, 1771. In *Elements of Botany*, 1775.
Jagged Chickweed	'On the city walls of Norwich.' Mr John Pitchford, 1765. In *Elements of Botany*, 1775.
Grape Hyacinth	In the Breckland of Suffolk, 1776, Sir John Cullum. In *Elements of Botany*.
Small Cow-wheat	Gathered by John Lightfoot 'going from Taymouth to the hermitage' in 1775. *Flora Scotica*, 1777.

Wild Chives	'By Fast-Castle (in) Berwickshire,' Dr Parsons. In Lightfoot's *Flora Scotica*, 1777.
Net-leaved Willow	'Upon many of the Highland mountains.' In Lightfoot's *Flora Scotica*, 1777.
Four-leaved Allseed	'Circa Lymston, Devonia (Lympstone nr Exmouth) et in Insula Portlandica.' In Hudson's *Flora Anglica*, 1778. Known from Jersey since 1724.
Pyramidal Bugle	'Supra montem Ben Nevis.' John Hope in Hudson's *Flora Anglica*, 1778.
Tufted Saxifrage	'Alpine rocks above Lake Idwell in Caernarvonshire.' J.W. Griffith, 1778, first published in Smith's *Flora Britannica*, 1804. Probably gathered on Ben Avon in 1771 by James Robertson.
Highland Cudweed	As a variety of Heath Cudweed 'upon the Highland mountains'. In Lightfoot's *Flora Scotica*, 1777.
Purple-stem Cat's-tail	On Newmarket Heath. Relhan's *Flora of Cambridgeshire*, 1785.
Starved Wood-sedge	Charlton Wood, Kent by Mr Curtis Stokes. In *Withering's Arrangement of British Plants*, 1787.
Strapwort	'Found by Mr Hudson on Slapham Sands, beyond Dartmouth and near the Start Point' (Slapton Ley, Devon). Mr Curtis Stokes, 1784. In *Withering's Arrangement*, 1787.
Summer Snowflake	'Undoubtedly wild betwixt Greenwich and Woolwich ... close by the Thames side, just above high-water mark ... also in the Isle of Dogs.' William Curtis, *Flora Londinensis*, 1788.
Alpine Fleabane	Ben Lawers. James Dickson, 1789. *Transactions Linnaean Society*, 1790.
Rock Speedwell	Ben Lawers. J. Dickson, 1789. *Trans. Linn. Soc.*, 1790.
Alpine Speedwell	'In montibus prope Garway Moor et in Ben Nevis.' J. Dickson, *Trans. Linn. Soc.*, 1790.
Highland Saxifrage	'In monte Ben Nevis Scotiae.' Dr Robert Townson, 1790. In *Trans. Linn. Soc.*, 1792.
Drooping Saxifrage	Ben Lawers. Robert Townson, 1790. J. Dickson, *Trans. Linn. Soc.*, 1794.
Alpine Gentian	Ben Lawers. J. Dickson, 1792. *Trans. Linn. Soc.*, 1794.
Starwort Mouse-ear	Ben Nevis. J. Dickson, 1792. *Trans. Linn. Soc.*, 1794.
One-flowered Wintergreen	Near Brodie House, Moray. Mr James Brodie. *English Botany*, 1793.
Whorled Solomon's-seal	'Den Rechip, a deep woody valley, 4 miles north-east of Dunkeld.' Arthur Bruce. *Eng. Bot.*, 1793. Still there.
Hairy Mallow	'Prope Cobham in Cantia (Kent), 1792.' *Symon's Synopsis*, 1798. Still occurs at Cobham.
Mountain Sandwort	Ben Lawers. George Don and John Mackay, 1793. *English Botany*, 1796.
Creeping Marshwort	'Peat beds on Bullingdon Green,' Oxford. Sibthorpe's *Flora Oxoniensis*, 1794.

Narrow-leaved Water-Dropwort	'Banks of the Isis beyond Ifley,' Oxford. Sibthorpe's *Flora Oxoniensis*, 1794.
Irish Saxifrage	'Rocks of Cwm Idwell.' J. Wynne Griffith. *Withering's Arrangement*, 1796.
Alpine Catchfly	'On rocks near the summit of Clova (Culrannoch) in Angusshire.' G. Don, 1795. J.E. Smith in *Trans. Linn. Soc.*, 1811.
Dwarf Mouse-ear	'On dry banks near Croydon.' J. Dickson in Curtis' *Flora Londinensis*, 1795. Then new to science and still sometimes called Curtis' Mouse-ear.
Twinflower	'In an old fir wood at Mearns near Aberdeen,' 1795. *Trans. Linn. Soc.*, 1796.
Red Helleborine	A 'steep stoney bank' on Minchinhampton Common, Glos, Mrs Elizabeth Smith. *Eng. Bot*, 1797. The Revd Lloyd Baker claimed he had found it 'some years' earlier.
Spring Gentian	Upper Teesdale, John Binks, identified by the Revd John Harriman, and published in *Eng. Bot*, 1797. Known since seventeenth century in Ireland.
Cotton Deergrass	'Restennet Moss,' Forfar. George Don and Robert Brown, 1791. *Eng. Bot.*, 1796.

The Nineteenth Century

Rannoch-rush	In Lakeby Car, near Boroughbridge, Yorks. The Revd James Dalton, 1787. *Eng. Bot.*, 1801.
Tuffed Sedge	Upper Teesdale. The Revd J. Harriman, 1797. *Eng. Bot.*, 1805.
Yellow Whitlow-grass	'Near Worms Head,' Glamorgan. John Lucas, 1796. *Eng. Bot.*, 1804.
Downy-fruited Sedge	'In meadows near Merston Measey, Wilts, Mr Teesdale,' 1799. *Trans. Linn. Soc.*, 1800.
Slender Bird's-foot Trefoil	'Among the rocks near Hastings, Mr Dickson.' *Smith's Flora Britannica*, 1800.
Wavy Meadow-grass	Ben Nevis. Mr Mackay. *Smith's Fl. Brit.*, 1800.
South Stack Fleawort	Ssp. *maritima* 'on cliffs near Holyhead' by the Revd H. Davies. *Smith's Fl. Brit.*, 1800.
Alpine Sow-thistle	'Discovered on the Aberdeenshire mountain of Lochnagore [Lochnagar] by Mr G. Don, Sept. 1801.' *Eng. Bot.*, 1810.
Fly Honeysuckle	'In a coppice called the Hacketts' (Hacketts Wood, Sussex), W. Borrer. *Eng. Bot.*, 1801. Still there.
Few-flowered Sedge	'Among the mountains of Clova,' G. Don, 1802. *Eng. Bot.*, 1813.
Small Hare's-ear	'Devonshire.' The Revd Aaron Neck, 1802. *Eng. Bot.*, 1812.
Alpine Forget-me-not	Summit rocks of Ben Lawers, G. Don, 1804. *Herbarium Britannicum*, 1805.

Mountain or Little Tree Willow	'In alpibus Scoticis.' D. Dickson. *Smith's Fl. Brit.*, 1804.
Wild Peony	On Steepholm, Francis Bowcher Wright, 1803. *Eng. Bot.*, 1805. The Peony of How and Gerard is the garden plant, *Peonia officinalis.*
Davall's Sedge	'In a boggy place on the slope of a hill' at Lansdowne, nr Bath. Mr Groult, 1807. E. Forster, *Eng. Bot.*, 1809.
Crested Buckler-fern	Holt, Norfolk, R.B. Francis. *Eng. Bot.*, 1810.
Narrow Small-reed	As Arundo stricta, White Mire, Forfar. G. Don. *Eng. Bot.*, 1810.
Saltmarsh Goosefoot	Near Yarmouth. Mr Lilly Wigg. *Eng. Bot.*, 1811.
Scorched Alpine-sedge	Ben Lawers. G. Don, 1812. *Eng. Bot.*, 1812. Not refound until 1892.
Woolly Willow	Glen Callater, Aberdeenshire, G. Don, 1812. J. E. Smith in *English Flora*, 1828.
Tuberous Thistle	'In a wood ... called Great Ridge, near Boyton House, Wilts.' A.B. Lambert, 1812. *Eng. Bot.*, 1813.
Holy-grass	'In a narrow valley called Kella,' Angus. G. Don. *Hooker's Flora Scotica*, 1821.
Yellow Oxytropis	'On a high rock ... at the head of Clova, Angus.' G. Don, 1812. *Eng. Bot.*, 1813.
Hair-leaved Goldilocks	'On a rocky cliff of Berryhead, Devon.' The Revd C. Holbech, 1812. *Eng. Bot.*, 1813.
Blue Heath	On 'a dry moor ... near Aviemore.' James Brown. *Eng. Bot.*, 1812. James Robertson probably discovered it in the Monadhliaths in 1771.
Brown Galingale	'Little Chelsea.' A.H. Haworth. W.J. Hooker's *Flora Londinensis*, 1821.
Scottish Primrose	Holburn Head, near Thurso, Caithness. Mr Gibb. W.J. Hooker, *Flora Londinensis*, 1819. Previously misidentified as *Primula farinosa.*
Wild Cotoneaster	Great Orme Head, Mr W. Wilson. J.E. Smith, *English Flora*, 1825, but J.W. Griffith said to have discovered it in 1783.
Small Restharrow	'On a steep bank by the sea to the north of West Tarbet', Professor Graham. *Botanical Magazine*, 1835.
Late Spider Orchid	'On the southern declivities of chalky downs near Folkestone.' The Revd Gerard E. Smith, 1828. J.E. Smith in *English Flora*, 1829.
Clove-scented Broomrape	On chalk downland near Folkestone in Kent. The Revd Gerard E. Smith, 1828. J.E. Smith in *English Flora*, 1829.
Dorset Heath	Rather a puzzle. Evidently known by 1768 since Miller includes it among the four known British heaths. Sir Charles Lemon described it in *Eng. Bot.* supplement 1812, but first localized site is 'Near Truro.' The Revd J.S. Tozer, 1828 in *Lindley's Synopsis*, 1829. Earliest Dorset records 1848.
Spiked Rampion	Hadlow Downs, near Mayfield, Sussex. The Revd Ralph Price, 1825. *Eng. Bot. Suppl.*, 1829.

Sharp-leaved Pondweed	Sussex, in dykes at Amberley, Henfield, Lewes. W. Borrer. W. Hooker in *Eng. Bot. Supplement*, 1829. Herbarium specimens collected previously from Essex, Yorks, Lincs, etc.
Hare's-foot Sedge	'Loch na Gar'. Mr Brand, 1830. *Bot. Mag.*, 1839.
Close-headed Alpine Sedge	'Among some precipitous rocks which surround a small loch about two miles above Loch Callader' (Corrie Kander, Aberdeenshire). *Eng. Bot. Suppl.*, 1830.
Eight-stamened Waterwort	SE side of Llyn Coron, Anglesey. J.E. Bowman, 1830. *Eng. Bot. Suppl.*, 1830. Still there.
Alpine Milk-vetch	'In Glen of the Dole, Forfar' (Glen Doll, Angus). Mr Brand, Dr Greville and Dr Graham, 1831. *Eng. Bot. Suppl.*, 1831.
Sickle-leaved Hare's-ear	Norton Heath, between Chelmsford and Ongar, Essex. Thomas Corder, 1831. *Eng. Bot. Suppl.*, 1833.
Alpine Butterwort	'In the bogs of Auchterflow and Shannon, Rosehaugh,' on Black Isle, Ross and Cromarty. The Revd George Gordon, 1831. *Eng. Bot. Suppl.*, 1832.
Italian Catchfly	Greenhithe, Kent. *Eng. Bot.*, 1832.
Fen Violet	'Near Lincoln.' John Nicholson, 1833. *Annals Natural History*, 1839.
Sand Crocus	'On the Warren between Dawlish and Exmouth, Devonshire.' W.C. Trevelyan and John Milford. *Loudon's Magazine of Natural History*, 1834.
Slender Cottongrass	Near Halnaby, Yorks. *Botanical Magazine*, 1835.
Mountain Bladder-fern	Ben Lawers. W. Wilson, 1836. Edward Newman in *Phytologist*, 1844.
Sea Knotgrass	'Christchurch Head on the sandy shore towards Muddiford.' W. Borrer, 1836. C.C. Babington in *Trans. Linn. Soc.*, 1837.
Shetland Mouse-ear	Keen of Hamar on Unst, Thomas Edmonston, 1837 (then aged 11). Edmonston proclaimed its distinctiveness, but H.C. Watson considered it a variety of Arctic Mouse-ear. At present it is considered to be a separate species.
Toadflex-leaved St John's-wort	Banks of the Teign, Devon. The Revd Thomas Kincks, 1838. *Ann. Nat. Hist.*, 1840.
Spring Snowflake	Hethe, near Bicester, Oxon, 'known to grow there for more than a century'. Baxter, in *Gardener's Magazine*, 1836. The possibly native site near Bridport, Dorset, is described in *J. Bot.*, 1866.
Orange Bird's-foot	Tresco, Scilly, Miss Matilda White, C.C. Babington in *Report Royal Society of Cornwall*, 1838. Babington found it on Guernsey, 1837.
Twin-headed Clover	On a wall top at Cadgwith, Cornwall, W. Borrer and C.C. Babington, 1839. *Phytologist*, 1842.
Lizard Clover	Near the Lizard lighthouse, Cornwall. The Revd W.S. Hore, 1839. *Phytologist*, 1842.
Arctic Sandwort	Unst, Shetland. Thomas Edmonston. *Hook. Br. Flora*, 1838.
Dickie's Bladder-fern	Sea-cave near Aberdeen. J. Dickie, 1838. *Gardener's Journal*, 1848.

Summer Lady's-tresses	'Along a stream on a small tract of sphagnous bog' near Lyndhurst in the New Forest. Joseph Janson, 1840. *Phyt.*, 1841. Found on Jersey in 1837.
Great Pignut	Near Cherry Hinton, Cambridge. The Revd W.H. Coleman, 1839. *Eng. Bot. Suppl.*, 1841.
Irish Spurge	Countisbury, Exmoor. Ward. *Phyt.*, 1841.
Upright Spurge	'Between Tintern and the Wind-cliff' (Wye Valley), W. Borrer. *Hook. Br. Flora*, 1842.
Small-flowered	'Between Sheffield and Halifax,' Yorks. W. Borrer, 1842.
Winter-cress	C.C. Babington, *Manual of British Botany*, 1843.
Wood Calamint	'In a wooded valley near Apes Down and Rowledge,' Isle of Wight. Dr William Bromfield. *Phyt.*, 1843.
False Cleavers	Saffron Walden, Essex. G.S. Gibson, 1844. *Phyt.*, 1844.
Cut-grass	'In three places in the Henfield level,' Sussex. W. Borrer, 1844. *Phyt.*, 1844.
Teesdale Sandwort	Widdybank Fell, Durham, by G.S. Gibson and J. Backhouse, 1844. *Eng. Bot. Suppl.*, 1844.
Cut-leaved Germander	'In a wild stony locality ... at the back of Box Hill in Surrey.' Thomas Ingall and William Bennett. *Phyt.*, 1844.
Broad-leaved Cudweed	Saffron Walden. G.S. Gibson, 1843. *Annals and Magazine of Natural History*, 1848. Previous unlocalized records from Sussex and Dorset by Joseph Woods.
Red-tipped Cudweed	'Cantley, Rossington etc near Doncaster.' The Revd G.E. Smith. *Phyt.*, 1846.
Round-headed Leek	'On steep declivities of the cliffs' at St Vincent's Rock, Bristol. H.O. Stevens, 1847.
Upright Clover	'Between the Lizard Head and Kynance Cove,' Cornwall. The Revd C.A. Johns, 1847. *Phyt.*, 1847. Known previously from Jersey (1826).
Oxtongue Broomrape	Near Comberton, Cambs. The Revd W.W. Newbould. C.C. Babington in *Phyt*, 1848.
Bitter Milkwort	Cronkley Fell, Upper Teesdale, J. and J. Backhouse, 1852. *Ann. Nat. Hist.*, 1853.
Newman's Lady-fern	Glen Prosen. J. Backhouse. Described by E. Newman in *Phyt.*, 1853.
Ghost Orchid	Near a brook at Tedstone Delamere, near Bromyard, Herefordshire, Mrs W. Anderton Smith. *Phyt.*, 1855.
Italian Lords-and-Ladies	'In the Undercliff,' Isle of Wight. Dr William Bromfield and Albert Hambrough. *Phyt.*, 1854.
Wild Gladiolus	New Forest, Hampshire, 'at least a mile from any house' among bracken 'which overtops it before it comes into flower'. The Revd W.H. Lucas, 1856. Babington in *Ann. Nat. Hist.*, 1857.
Wavy-leaved St John's-wort	Near Plymouth. T.R. Archer-Briggs, 1861. *Journal of Botany*, 1864. Observed up to 20 years previously by James Cunnack and the Revd W. Moyle Rogers.

Teesdale Violet	Widdybank Fell 'upon the sugar limestone'. J. Backhouse, 1862. Babington in *J. Bot*, 1863.
Snow Pearlwort	Glas Maol, Grampians, J Backhouse, 1848. *Phyt.*, 1849 (but probably *Sagina saginoides*). More certainly identified on Ben Lawers, 1863.
Hairy-fruited Cornsalad	'Between Henley Castle and Barnard Green, Worcs.' Edgar Lees, 1845. *Syme's English Botany*, 1865.
Plymouth Pear	Near Plymouth. T.R. Archer-Briggs, 1865. *Flora of Plymouth*, 1880.
Rootless Duckweed	'Piece of water which probably communicates with the Thames, but looks like a pond, near the railway bridge on Shortwood Common.' Henry Trimen. *J. Bot.*, 1866.
Dwarf Rush	Land's End, Cornwall. W. Beeby. *J. Bot.*, 1872.
Pigmy Rush	Near Kynance Cove, Cornwall. W. Beeby. *J. Bot.*, 1872.
Slender Naiad	Perthshire, A. Sturrock, 1875. *Scottish Naturalist*, 1876. Known from W. Ireland since 1850.
Purple Viper's-bugloss	'Near St Just in corn and potato-fields,' J. Ralfs, 1873. Known from Jersey 'in the sandy grounds near St Hilary' since 1690.
Small Tree-mallow	Scilly. W. Curnow and J. Ralfs, 1873. H. Trimen in *J. Bot.*, 1877.
Dwarf Pansy	On Tresco, Scilly by J. Ralfs, 1873. Recognized as separate species by E.G. Baker, *J. Bot.*, 1901.
Bird's-foot Sedge	Millers Dale, Derbyshire. J. Whitehead and H. Newton, 1874. *J. Bot.*, 1874.
Shore Dock	Wembery, South Devon. T.R. Archer-Briggs, 1875. *J. Bot.*, 1876. W.H. Beeby collected it on Scilly in 1873.
Adderstongue Spearwort	A wet ditch west of Hythe, Hants. Henry Groves, 1878, *J. Bot.*, 1883. Known from Jersey since 1838.
Boyd's Pearlwort	Unknown site, possibly Ben Avon by W.W. Boyd, 1878. *Trans. Bot. Soc. Edinburgh*, 1887.
Slender Centaury	'Isle of Wight.' F. Townsend, 1878. *J. Bot.*, 1879.
Cut-leaved Selfheal	Old pasture field at Birstal Hill, Leics. F.T. Mott, 1878. Long thought of as a variety. Published as new native species by J.W. White in *J. Bot.*, 1906.
Cambridge Parsley	'Near Broughton Woods,' Lincs. The Revd W. Fowler, 1880. *Rep. Bot. Ex. Club*, 1881, A.F. Lees, *J. Bot.*, 1882.
Western Ramping-Fumitory	Lelant, Cornwall. Mrs Gregory, 1881. Named new to science by H.W. Pugsley, *J. Bot.*, 1902, though previously recognized as distinct by C.C. Vigurs.
Holly-leaved Naiad	Hickling Broad, Norfolk. Arthur Bennett. *J. Bot.*, 1883.
Estuarine Sedge	Sand-banks by Wick River, Caithness. J. Grant. H.N. Ridley, *J. Bot.*, 1885.
Scottish Small-reed	As '*Deyeuxia strigosa*', Loch Duran, Caithness. Robert Dick. A. Bennett, *J. Bot.*, 1885. As *D. scotica*, Rep. BEC, 1914 and renamed *Calamagrostis scotica* by G.C. Druce, 1926.

Brown Bog-rush	'Beside Loch Tummel,' James Brebner, 1884. F. Buchanan White in *J. Bot.*, 1885. Possibly found by James Robertson near Rannoch in 1771.
Alpine Rockcress	'On steep slopes of the Cuchullins' (Black Cuillins, Skye). H.C. Hart, *J. Bot.*, 1887.
Interrupted Brome	As *Bromus mollis var. interruptus*, Fallow-field, Berkshire. G.C. Druce, *Rep. BEC*, 1888. Described as new species, Druce, *J. Bot.*, 1904.
English Sandwort	By roadside near Ribblehead Station, Lister Rotheray. *J. Bot.*, 1889.
Mountain Scurvy-grass	Ben Lawers. The Revd E.S. Marshall, 1887. *J. Bot.*, 1894.
Loddon Pondweed	In River Loddon, Berks, G.C. Druce, 1893. Named as *Potamogeton Drucei* by Alfred Fryer, *J. Bot.*, 1898.
Club Sedge	'Arisaig.' *J. Bot.*, 1895. Known from Northern Ireland since 1835.
String Sedge	Altnaharra. Revd E.S. Marshall, *J. Bot.*, 1897.
Limestone Woundwort	In a wood near Wootton, Glos 'where undergrowth had recently been cleared'. Cedric Bucknall. *J. Bot.*, 1897.

The Twentieth Century

Welsh Mudwort	Kenfig Pool, Glamorgan. E. S. Marshall and W. A. Shoolbred, *J. Bot.*, 1901.
Somerset Hair-grass	Found by Jacob Dillenius in 1726, but not published. Refound by G.C. Druce near Uphill, Somerset. *Rep BEC*, 1904. *J. Bot.*, 1905.
Martin's Ramping-fumitory	Gilly Tresamble, Cornwall. F.H. Davey, 1904. Described as new to science by H.W. Pugsley, *J. Bot.*, 1912.
Fen Wood-rush	Wood Walton Fen, Hunts. E.W. Hunnybun, 1907. *Report, Botanical Exchange Club (BEC)*, 1907.
American Pondweed	Salterhebble Bridge, Halifax. 1907. A. Bennett, *Naturalist*, 1908. This was probably an introduction. First native site in S. Uist, W.A. Clark and J.W. Heslop Harrison, 1943.
Thistle Broomrape	Hetchell Crags, nr Witherby, Yorks. H.E. Craven, 1907, though collected previously by J.F. Pickard, 1902. G.C. Druce, *Rep. BEC*, 1909.
Greek Spurrey	Aldeburgh, Suffolk. 1911. G.C. Druce in *Rep. BEC*, 1912. Found in Jersey, 1906, also by Druce.
Yellow-Sedge	Roudsea Wood, Cumbria. D. Lumb, 1913. *Rep. BEC*, 1914, though collected much earlier.
Viper's-grass	Near Ridge, Dorset. Cecil Sandwith, 1914. *Rep. BEC*, 1915.
Esthwaite Waterweed	Esthwaite Water, Cumbria. W.H. Pearsall, 1914. A. Bennet in *J. Bot.*, 1914.
Perennial Centaury	Cliffs near Newport, Pembs. T.B. Rhys, 1918.
Irish Lady's-tresses	Found on Coll 1921, but not identified until 1939. Otherwise, Colonsay, 1930. Known from Ireland since 1810.

Ribbon-leaved Water-plantain	Westwood Great Pool, Droitwich, 1920.
Pigmyweed	Adel Dam, Yorks. R.W. Butcher. *Rep. BEC*, 1921.
Bristle Sedge	Near Ben Lawers. Gertrude Bacon and Lady Davy, 1923. G.C. Druce in *Rep. BEC*, 1923.
Dune Gentian	Damp, sandy pasture near Tenby. H.W. Pugsley, *J. Bot.*, 1924.
Pond Bedstraw	Hatchet Pond, New Forest. J.F. Rayner, 1924.
Pale Forget-me-not	Near Cantley Crag, Sedbergh, Cumbria. A. Wilson, 1892. Recognized as new species by C.E. Salmon, *J. Bot.*, 1926.
Sharp Club-rush	Southport, Lancs (Formby Slacks). R.E. Baker, 1928. G.C. Druce, *Rep. BEC,* 1928. Known from Jersey since 1836.
Breckland Speedwell	Near Tuddenham, Suffolk. J.E. Lousley and A.W. Graveson, 1933. *J. Bot.*, 1933.
Scottish Dock	Balmaha, by Loch Lomond. Robert Mackechnie, 1935. J.E. Lousley, *J. Bot. Lond.*, 1939.
Lundy Cabbage	On cliffs of Lundy. F.R. Elliston White, 1936. Described as a new species by O.E. Schulz in Wright, *J. Bot. Suppl.*, 1936.
Land Quillwort	Caerthilian Cove on The Lizard, Fred Robinson, 1919. Disregarded until more plants found in the same place in 1937. Known on Guernsey since 1860.
Shetland Pondweed	Described by Arthur Bennet, 1900, but species parameters not sorted out until J.E. Dandy and G. Taylor, *J. Bot.*, 1938.
Plot's Elm	Druce set out to find the elm that Robert Plot had described [in his Natural History of Oxfordshire, 1677]. He failed. He found instead another very distinctive elm, Lock's Elm (*U. minor var. lockii*). Then, with mistaken confidence that this was indeed the elm described by Plot, he upgraded his discovery from variety to species and renamed it *Ulmus plotii*.' (Richens, *Elm*.) What is now understood by that name was described by R. Melville in *J. Bot.*, 1940.
Grey Mouse-ear	Railway cutting between Sharnbrook and Irchester, Beds. E. Milne-Redhead, *The Naturalist*, 1947.
Branched Horsetail	In long grass by River Witham, near Boston, Lincs. H. K. Airy Shaw, 1947. A.H.G. Alston, *Watsonia*, 1948.
Least Adders-tongue	St Agnes, Scilly by John Raven, 1950. *Watsonia*, 1950. Known on Guernsey since 1854.
Iceland Purslane	The Storr, Skye, coll. 1934 but misidentified until 1950. B. L. Burtt, *Kew Bulletin*, 1950; J.E. Raven, *Watsonia*, 1952.
Diapensia	Mountain ridge near Glenfinnan, by C.F. Tebbutt, 1951.
Purple Coltsfoot	George Don 'in the high mountains of Clova'. Refound A. Slack, 1951. B.W. Ribbons in *Watsonia*, 1952.
Early Meadow-grass	Scilly and The Lizard, Cornwall, by John Raven, 1950. *Watsonia*, 1950. Known in Channel Islands since 1910.

Norwegian Mugwort	Mountain ridge in Wester Ross by Sir Christopher Cox, 1950. R.A. Blakelock in *Kew Bulletin*, 1953.
Welsh Groundsel	Roadside at Ffrith near Wrexham. H.E. Green, 1948. Described as new species by E. M. Rosser, *Watsonia*, 1955.
Northern Spike-rush	Upper Wharfedale, Noel Sandwith, 1947. Identified 1960, S. M. Walters, *Watsonia*, 1963.
Western Clover	On Lizard cliffs, D.E. Coombe, 1957. New to science, Coombe, *Watsonia*, 1961.
Dense-flowered Orchid	Dunes at Ballaghennie, Isle of Man, 1966. D.E. Allen, *Watsonia*, 1968. Probably a recent colonist. Known in W. Ireland since 1864.
Northern Yellowcress	Differences from Marsh Yellowcress clarified by B. Jonsell of Lund, Sweden, 1968; R.E. Randall, *Watsonia*, 1974.
Alchemilla gracilis	Old lime quarry at Cockplay, Northumberland. M.E. Braithwaite, 1976. G.A. Swan & S.M. Walters, *Watsonia*, 1985. It was gathered (though not recognized) by Druce at Langdon Beck, Teesdale in 1924.
Radnor Lily	Stanner Rocks, Radnor. R.G. Woods, 1975. E.M. Rix and R.G. Woods, *Watsonia*, 1981.
Young's Helleborine	Mine spoil by South Tyne, Northumberland. A.J. Richards, 1976. Richards and A.F. Porter, *Watsonia*, 1982.
Scandinavian Smallreed	Esthwaite Water, Cumbria. Gathered from various sites in Scotland but not recognized until 1980. O.M. Stewart, *BSBI News*, 1988.
Unspotted Lungwort	Birgate Wood, Suffolk. Known since 1842, but identified only in 1985. C.R. Birkinshaw and M.N. Sanford, *Watsonia*, 1996.
Fringed Gentian	Downs near Wendover. P. Philipson, 1982, though probably first found 1875 by Miss M. Williams. P.R. Knipe, *Watsonia*, 1988.
Service-tree	Sea cliffs in West Glamorgan. Marc Hampton, 1983. Hampton and Q.O.N. Kay, *Watsonia*, 1995.
Leafless Hawk's-beard	Edge of a hay meadow in Westmorland by Geoffrey Halliday, 1988. *Watsonia*, 1990.
Proliferous Pink	Cranwich, West Norfolk, J.E. Gaffney, 1992. J.R. Akeroyd, *Watsonia*, 1995. Previously misidentified as Childing Pink, known in Norfolk since 1835.

Appendix 2

Legally protected flowering plants and ferns in Britain

The following 111 Red-listed species are protected under the Wildlife and Countryside Act. This makes it an offence 'to intentionally pick, uproot or destroy' any of these plants without a licence. It is also illegal to sell or advertise them for gain. The plants receive such protection because they are very rare or considered to be in danger of becoming extinct. There is a separate list for Northern Ireland.

Adderstongue Spearwort
Alpine Catchfly
Alpine Fleabane
Alpine Gentian
Alpine Rockcress
Alpine Sow-thistle
Alpine Woodsia
Bedstraw Broomrape
Blue Heath
Branched Horsetail
Bristol Rockcress
Broad-leaved Cudweed
Brown Galingale
Cambridge Milk-parsley
Cheddar Pink
Childing Pink
Creeping Marshwort
Cut-grass
Cut-leaved Germander
Deptford Pink
Diapensia
Dickie's Bladder-fern
Downy Woundwort
Drooping Saxifrage
Dune Gentian
Dwarf Spike-rush
Early Gentian
Early Spider Orchid
Early Star-of-Bethlehem (Radnor Lily)
Fen Orchid
Fen Ragwort
Fen Violet
Field Cow-wheat
Field Eryngo
Field Wormwood
Fingered Speedwell
Floating-leaved Water-plantain
Fringed Gentian
Ghost Orchid
Grass-poly
Greater Yellow-rattle
Green Hound's-tongue
Ground Pine
Holly-leaved Naiad
Jersey Cudweed
Killarney Fern
Lady's Slipper
Lapland Marsh Orchid
Late Spider Orchid
Least Adderstongue
Least Lettuce
Limestone Woundwort
Lizard Orchid
Lundy Cabbage
Martin's Ramping-fumitory
Meadow Clary
Military Orchid
Monkey Orchid
Northroe Hawkweed
Norwegian Sandwort
Oblong Woodsia
Oxtongue Broomrape
Pennyroyal
Perennial Knawel

Perfoliate Penny-cress
Pigmyweed
Plymouth Pear
Purple Coltsfoot
Red Helleborine
Red-tipped Cudweed
Ribbon-leaved Water-plantain
Rock Cinquefoil
Rough Marsh-mallow
Round-headed Leek
Sand Crocus
Sea Knotgrass
Shetland Hawkweed
Shore Dock
Sickle-leaved Hare's-ear
Slender Centaury
Slender Cottongrass
Slender Naiad
Small Alison
Small Fleabane
Small Hare's-ear
Small Restharrow
Snowdon Lily
South Stack Fleawort
Spiked Rampion
Spiked Speedwell
Spring Gentian
Stalked Orache
Starfruit
Starved Wood-sedge
Stinking Goosefoot
Stinking Hawk's-beard
Strapwort
Teesdale Sandwort
Thistle Broomrape
Triangular Club-rush
Tufted Saxifrage
Viper's-grass
Water Germander
Weak-leaved Hawkweed
Welsh Mudwort
Whorled Solomon's-seal
Wild Cotoneaster
Wild Gladiolus
Wood Calamint
Yellow Marsh-saxifrage
Young's Helleborine

Under the EC Conservation (Natural Habitats etc) Regulations 1994, the following plants are protected throughout the European Community.

Creeping Marshwort
Early Gentian
Fen Orchid
Floating-leaved Water-plantain
Killarney Fern
Lady's Slipper
Shore Dock
Slender Naiad
Yellow Marsh-saxifrage

Cheddar Pink (*Dianthus gratianopolitanus)* smiles from its limestone crag near Cheddar Gorge, Somerset [Bob Gibbons]

Appendix 3

Projects on rare vascular plants under English Nature's Species Recovery Programme and the UK Biodiversity Action Plan (1991–8)

SPECIES	'CONTACT POINT'	'LEAD PARTNER'
Starfruit *Damasonium alisma*	English Nature	Plantlife
Hairy Mallow *Althaea hirsuta*	English Nature	Plantlife
Plymouth Pear *Pyrus cordata*	English Nature	Royal Botanic Gardens, Kew ('Kew')
Fen Ragwort *Senecio paludosus*	English Nature	Institute of Terrestrial Ecology (ITE), National Trust
Ribbon-leaved Water-plantain *Alisma gramineum*	English Nature	ITE
Strapwort *Corrigiola litoralis*	English Nature	Field Studies Council
Lady's Slipper *Cypripedium calceolus*	English Nature	Sainsbury Orchid Trust, Kew
Stinking Hawks-beard *Crepis foetida*	English Nature	Royal Holloway College, Kew, RSPB
Fen Violet *Viola persicifolia*	English Nature	ITE, National Trust

SPECIES	'CONTACT POINT'	'LEAD PARTNER'
Fen Orchid *Liparis loeselii*	English Nature	Norfolk Wildlife Trust, Broads Authority, Kew
Ground Pine *Ajuga chamaepitys*	English Nature	Plantlife
Perennial Knawel *Scleranthus perennis*	English Nature	Mrs Yvonne Leonard, Suffolk Wildlife Trust, Elveden Estate
Shore Dock *Rumex rupestris*	English Nature	ITE, Plantlife, Rosemary Parslow
Creeping Marshwort *Apium repens*	English Nature	Ashmolean Natural History Society, Oxford University Botanic Gardens
Meadow Clary *Salvia pratensis*	English Nature	Plantlife, Cotswold Rare Plants Group
Stalked Sea-purslane *Atriplex pedunculata*	English Nature	Kew
Thistle Broomrape *Orobanche reticulata*	English Nature	Leeds City Council, Bradford University, Yorkshire Naturalists Union
An Alchemilla *Alchemilla minima*	English Nature	English Nature
Tower Mustard *Arabis glabra*	English Nature	Plantlife
Wild Asparagus *Asparagus officinalis* ssp. *prostratus*	Countryside Council for Wales	The National Trust
Interrupted Brome *Bromus interruptus*	English Nature	Kew
Scottish Small-reed *Calamagrostis scotica*	Scottish Natural Heritage	Royal Botanic Garden Edinburgh
Prickly Sedge *Carex muricata* ssp. *muricata*	English Nature	English Nature
True Fox Sedge *Carex vulpina*	English Nature	English Nature

SPECIES	'CONTACT POINT'	'LEAD PARTNER'
Cornflower *Centaurea cyanus*	Ministry of Agriculture Fisheries and Food	Plantlife
Shetland Mouse-ear *Cerastium nigrescens*	Scottish Natural Heritage	Scottish Natural Heritage
Deptford Pink *Dianthus armeria*	English Nature/Countryside Council for Wales	Plantlife
Red-tipped Cudweed *Filago lutescens*	Ministry of Agriculture Fisheries and Food	Plantlife/English Nature
Broad-leaved Cudweed *Filago pyramidata*	Ministry of Agriculture Fisheries and Food	Plantlife/English Nature
Purple Ramping-fumitory *Fumaria purpurea*	Ministry of Agriculture Fisheries and Food	English Nature
Red Hemp-nettle *Galeopsis angustifolia*	Ministry of Agriculture Fisheries and Food	Plantlife
Corn Cleavers *Galium tricornutum*	Ministry of Agriculture Fisheries and Food	Plantlife
Dune Gentian *Gentianella uliginosa*	Countryside Council for Wales	Countryside Council for Wales
Hawkweeds (Shetland spp. only) *Hieracium* Sect. *Alpestria*	Scottish Natural Heritage	Shetland Amenity Trust
Pygmy Rush *Juncus pygmaeus*	English Nature	English Nature
Cut-grass *Leersia oryzoides*	English Nature	Environment Agency
Sea Lavender *Limonium* (endemic taxa)	Countryside Council for Wales/English Nature	The National Trust/Botanical Society of the British Isles
Marsh Clubmoss *Lycopodiella inundata*	Countryside Council for Wales/English Nature	Plantlife
Pennyroyal *Mentha pulegium*	English Nature	English Nature
Pillwort *Pilularia globulifera*	Scottish Natural Heritage	Countryside Council for Wales/Plantlife

SPECIES	'CONTACT POINT'	'LEAD PARTNER'
Grass-wrack Pondweed *Potamogeton compressus*	English Nature	British Waterways Board
Woolly Willow *Salix lanata*	Scottish Natural Heritage	National Trust for Scotland
Shepherd's Needle *Scandix pecten-veneris*	Ministry of Agriculture Fisheries and Food	Plantlife
Triangular Club-rush *Schoenoplectus triqueter*	English Nature	Environment Agency
Perennial Knawel *Scleranthus perennis* ssp. *prostratus*	English Nature	Wildlife Trusts
Small-flowered Catchfly *Silene gallica*	Ministry of Agriculture Fisheries and Food	Plantlife
Greater Water-parsnip *Sium latifolium*	English Nature	Environment Agency
Ley's Whitebeam *Sorbus leyana*	Countryside Council for Wales	National Botanic Garden of Wales
Cotswold Pennycress *Thlaspi perfoliatum*	English Nature	Plantlife
Spreading Hedge-parsley *Torilis arvensis*	Ministry of Agriculture Fisheries and Food	Plantlife
Broad-fruited Corn-salad *Valerianella rimosa*	Ministry of Agriculture Fisheries and Food	Plantlife
Oblong Woodsia *Woodsia ilvensis*	Scottish Natural Heritage	Royal Botanic Garden Edinburgh

References

1 Akeroyd, J.R. & Beckett, G. (1995) *Petrorhagia prolifera*, an overlooked native species. *Watsonia, 20*, 405–7.

2 Allen, D.E. (1969) *The Victorian Fern Craze*. Hutchinson, London.

3 Allen, D.E. (1986) *The Botanists. A history of the Botanical Society of the British Isles through a hundred and fifty years*. St Pauls Bibliographies, Winchester.

4 Allen, D.E. (1986) The Discoveries of Druce. In: *The Long Tradition*, ed. H.J. Noltie. BSBI Conference Report No. 20, 175–90.

5 Alton, S. (1998) The Millennium Seed Bank Project. *British Wildlife*, 9(5), 273–7.

6 Archer-Briggs, T.R.A. (1880) *The Flora of Plymouth*. Van Voorst, London.

7 Barter, C. (1856) Rare plants of Aberdeenshire, Kincardine etc. *Phytologist, 1*.

8 Bates, Selina & Spurgin, Keith (1994) *Stars in the Grass. The story of Cornish naturalist, Frederick Hamilton Davey 1868–1915*. Dyllansow Truran, Redruth.

9 Beckett, A. & Beckett, G. (1980) *Leucojum vernum*. *BSBI News*, April 1980.

10 Beirne, Bryan P. (1952) *British Pyralid and Plume Moths*. Frederick Warne & Co, London.

11 Biodiversity Challenge (1994) *An agenda for conservation action in the UK*. RSPB, Sandy.

12 Birkinshaw, C.R. (1990) The biology and conservation of Starfruit, *Damasonium alisma*. Unpublished NCC report, Peterborough.

13 Birkinshaw, C.R. (1994) Aspects of the ecology and conservation of *Damasonium alisma* in Western Europe. *Watsonia, 20*, 33–9.

14 Birkinshaw, C.R. & Sanford, M. N. (1996) *Pulmonaria obscura* in Suffolk. *Watsonia, 21*, 169–78.

15 Bowen, H.J.M. (1968) *The Flora of Berkshire*. Privately published, Reading.

16 Bradshaw, Margaret E. (1981) Monitoring grassland plants in Upper Teesdale, England. In: *The Biological Aspects of Rare Plant Conservation*, ed. H. Synge, 241–52. Wiley, Chichester.

17 Brewis, A., Bowman, P. & Rose, F. (1996) *The Flora of Hampshire*. Harley Books, Colchester.

18 Brightman, F.H. (1986) Thomas Johnson's *Iter Plantarum Investigationis ... in Agrum Cantianum*. In: *The Long Tradition, the Botanical Exploration of the British Isles*. The Scottish Naturalist, Kilbarchan.

19 Brookes, Brian S. (1981) The discovery, extermination, translocation and eventual survival of *Schoenus ferrugineus* in Britain. In: *The Biological Aspects of Rare Plant Conservation*, ed. H. Synge, 421–8.

20 Burton, R.M. (1983) *Flora of the London Area*. London Natural History Society.

21 Byfield, A.J. (1986) The Lizard Flora: A history of discovery. In: *The Long Tradition*, ed. H.J. Noltie. BSBI Conference Report No. 20, 135–46.

22 Byfield, A.J. (1991) Classic British wildlife sites – The Lizard Peninsula. *British Wildlife, 3*, 92–105.

23 Byfield, A.J. & Pearman, D. (1996) *Dorset's Disappearing Heathland Flora. Changes in the distribution of Dorset's rarer heathland species 1931 to 1993*. Plantlife and RSPB, London and Sandy.

24 Callaghan, D.A. (1996) The conservation status of *Lythrum hyssopifolia* in the British Isles. *Watsonia, 21*, 179–86.

25 Cerovsky, J. (1995) *Endangered Plants. A comprehensive illustrated guide to the conservation of flowers, trees and shrubs*. Sunburst Books, London.

26 Chapman, S.B. & Rose, R.J. (1994) Changes in the distribution of *Erica ciliaris* and *E.* x *watsonii* in Dorset 1963–1987. *Watsonia, 20*, 89–95.

27 Clapham, A.R. (1969) *Flora of Derbyshire*.

28 Clapham, A.R. (1971) William Harold Pearsall. *Biographical Memoirs of Fellows of the Royal*

Society, *17*, 511–40.
29 Clapham, A.R. (ed. 1978) *Upper Teesdale, the area and its natural history*. Collins, London.
30 Clapham, A.R., Tutin, T.G. & Warburg, E.F. (2nd edn 1962) *Flora of the British Isles*. Cambridge University Press, Cambridge.
31 Clarke, W.A. (1900) *First Records of British Flowering Plants*. West, Newman & Co, London.
32 Clement, E.J. & Foster, M.C. (1995) *Alien Plants of the British Isles*. BSBI, London.
33 Condry, William (1998) *Wildflower Safari. The life of Mary Richards*. Gomer, Ceredigion.
34 Corner, R.W.M. (1998) Northern limits attained by native British plants in North Peary, Greenland. *Watsonia*, *22*, 109–10.
35 Crompton, G. (1981) Surveying rare plants in Eastern England. In: *The Biological Aspects of Rare Plant Conservation*, ed. H. Synge, 117–24. Wiley, Chichester.
36 Curtis, William (1777–98) *Flora Londinensis*, London.
37 Daniels, R.E., McDonnell, E.J. & Raybould, A.F. (1998) The current status of *Rumex rupestris* in England and Wales, and threats to its survival and genetic diversity. *Watonia*, *22*, 33–9.
38 Davey, F. Hamilton (1909) *Flora of Cornwall*. Chegwidden, Penrhyn.
39 David, R.W. (1978) The distribution of *Carex elongata* L. in Britain. *Watsonia*, *12*, 158–60.
40 Davis, Ruth (1997) Back from the brink and into the future. *Plantlife magazine*, Autumn 1997, 6–7.
41 Dony, J.G., Jury, S.L. & Perring, F.H. (2nd edn 1986) *English Names of Wild Flowers*. BSBI, London.
42 Dowlen, C.M. & Ho, T.N. (1995) *Gentianella ciliata* in Wiltshire. *Watsonia*, *20*, 279–81.
43 Druce, G. C. (1897) *The Flora of Berkshire*. Clarendon Press, Oxford.
44 Druce, G.C. (1919) Plant extinctions since 1597. *Rep. Bot. Soc. & Exch. Club., 1919*.
45 Druce, G.C. (1920) Dubious plants of Britain. *Rep. Bot. Soc. & Exch. Club*, 1919.
46 Druce, G.C. (1928) *Flora of Buckinghamshire*. Buncle, Arbroath.
47 Druce, G.C. (1932) *Comital Flora of the British Isles*. Buncle, Arbroath.
48 Duncan, U.K. (1980) *Flora of East Ross-shire*. Buncle, Arbroath.
49 Dunn, A.J. (1997) Biological Flora: *Stachys germanica*. *J. Ecol*, *85*, 531–9.
50 Dunn, S.T. (1905) *Alien Flora of Britain*. West, Newman & Co., London.
51 Ellis, R.G. (1983) *Flowering Plants of Wales*. National Museum of Wales, Cardiff.
52 Everett, Sue (1993) *Cirsium tuberosum*. Tuberous Thistle. In: *The Wiltshire Flora*, ed. Beatrice Gillam, Pisces Publications, Newbury.
53 Ewan, A.H. & Prime, C. T. (1975) Translation of John Ray's *Catalogus Plantarum circa Cantabrigia nascientum*.
54 Field, M.H. (1994) The status of *Bupleurum falcatum* L. (Apiaceae) in the British Isles. *Watsonia*, *20*, 115–7.
55 Firbank, L.G. (1988) Biological Flora: *Agrostemma githago*. *J. Ecol.*, *76*, 1232–46.
56 Fisher, John (1991) *A colour guide to Rare Wild Flowers*. Constable, London.
57 FitzGerald, R. (1989) Lost and Found – *Alopecurus bulbosus* in S. E. England. *Watsonia*, *17*, 425–8.
58 FitzGerald, R. (1990) Wildlife Reports: Higher Plants. *British Wildlife*, *1*, 173–4.
59 FitzGerald, Rosemary & Jermy, Clive (1987) *Equisetum ramosissimum* in Somerset. *Pteridologist*, *1*, 178–81.
60 FitzGerald, Rosemary, Holyoak, D. & Stewart, N. (1997) Survey of *Schoenoplectus* species on the River Tamar. Unpublished report, English Nature, Peterborough.
61 Foley, M.J.Y. (1987) The current distribution and abundance of *Orchis ustulata* in northern England. *Watsonia*, *16*, 409–15.
62 Foley, M.J.Y. (1993) *Orobanche reticulata* populations in Yorkshire. *Watsonia*, *19*, 247–57.
63 Frost, L.C. (1981) The study of *Ranunculus ophioglossifolius* and its successful conservation at the Badgeworth Nature Reserve, Gloucestershire. In: *The Biological Aspects of Rare Plant Conservation*, ed. H. Synge, 481–90.
64 Frost, L.C., Houston, L., Lovatt, C.M. & Beckett, A. (1991) *Allium sphaerocephalon* and introduced *A. carinatum*, *A. roseum* and *Nectaroscordum siculum* on St Vincent's Rocks, Avon Gorge, Bristol. *Watsonia*, *18*, 381–5.
65 Fryer, J. & Hylmo, B. (1994) The native British *Cotoneaster* – Great Orme Berry renamed. *Watsonia*, *20*, 61–63.
66 Gaston, Kevin J. (1994) *Rarity*. Population and community biology series *13*. Chapman & Hall, London.
67 Gerard, J. (Revised and ed. Johnson, T., 1633) *The herball or generall historie of plantes*. Facsimile edition, 1975. Dover Publications, New York.
68 Gillam, Beatrice (ed. 1993) *The Wiltshire Flora*. Pisces Publications, Newbury.
69 Gilmour, John & Walters, Max (1955) *Wild Flowers. Botanising in Britain*. Collins New Naturalist Library, London.
70 Godwin, H. (2nd edn 1975) *The History of the British Flora*. Cambridge University Press, Cambridge.
71 Gornall, R.J. (1987) Notes on a hybrid spearwort, *Ranunculus flammula* x *R. reptans*. *Watsonia*, *16*, 383–8.
72 Graham, G.G. (1988) *The Flora & Vegetation of County Durham*. Durham County Conservation Trust.
73 Grassly, N.C., Harris, S.A. & Cronk, Q.C.B. (1996) British *Apium repens* status assessed using random amplified polymorphic DNA (RAPB). *Watsonia*, *21*, 103–11.
74 Green, P.R., Green, I.P. & Crouch, G.A. (1997) *The Atlas Flora of Somerset*. Privately published, Crewkerne and Yeovil.
75 Grigson, Geoffrey (1955) *The Englishman's Flora*. J.M. Dent & Sons, London.

76 Grose, Donald (1957) *The Flora of Wiltshire*. Wiltshire Archaeological and Natural History Society, Devizes.
77 Halliday, G. (1990) *Crepis praemorsa*, new to western Europe. *Watsonia, 18*, 85–7.
78 Halliday, Geoffrey (1997) *A Flora of Cumbria*. Centre for North–west Regional Studies, University of Lancaster.
79 Hampton, Mark (1996) *Sorbus domestica* – comparative morphology and habitats. *BSBI News*, 32–7.
80 Hampton, M. & Kay, Q.O.N. (1995) *Sorbus domestica* new to Wales and the British Isles. *Watsonia, 20*, 379–84.
81 Hanbury, F.J. & Marshall, E.S. (1899) *Flora of Kent*. Privately published, London.
82 Harper, John L. (1981) The meanings of rarity. In: *The Biological Aspects of Rare Plant Conservation*, ed. H. Synge, 189–204. Wiley, Chichester.
83 Henderson, D.M. & Dickson, J.H. (eds. 1994) *A Naturalist in the Highlands. James Robertson: His life and travels in Scotland 1767–1771*. Scottish Academic Press, Edinburgh.
84 Hind, W.M. (1889) *The Flora of Suffolk*. Gurney & Jackson, London.
85 Howitt, M. (1963) *Flora of Nottinghamshire*. Privately published, Farndon, Notts.
86 Hubbard, C.E. (3rd edn 1984) *Grasses. A guide to their Structure, Identification, Uses and Distribution in the British Isles*. Penguin Books, London.
87 Ingram, R. & Noltie, H.J. (1995) Biological Flora: *Senecio cambrensis*. *J. Ecol, 83*, 537–46.
88 Ingrouille, Martin (1995) *Historical Ecology of the British Flora*. Chapman & Hall, London.
89 Jarman, Derek (1995) *Derek Jarman's Garden*. Thames & Hudson, London.
90 Jermy, A.C. & Crabbe, J.A. (eds. 1978) *The island of Mull: a survey of its flora and environment*. British Museum (Natural History), London.
91 Jermyn, Stanley T. (1974) *Flora of Essex*. Essex Naturalists' Trust, Colchester.
92 Johns, C.A. (1848) *A Week at The Lizard*. Society for Promotion of Christian Knowledge, London.
93 Jones, Dewi (1996) *The Botanists and Guides of Snowdonia*. Carreg Gwalch, Llanrwst.
94 Kay, Q.O.N. & Harrison, J. (1970) Biological Flora: *Draba aizoides*. J. Ecol., *58*, 877–88.
95 Keble Martin, W. (1939) *Flora of Devon*. Buncle, Arbroath.
96 Keble Martin, W. (1965) *The Concise British Flora in Colour*. Ebury Press and Michael Joseph, London.
97 Kent, D.H. (1975) *The Historical Flora of Middlesex*. The Ray Society, London.
98 Key, Roger (1996) *Rare Plants on Lundy*. English Nature, Peterborough.
99 Killick, J., Perry, R. & Woodell, S. (1998) *The Flora of Oxfordshire*. Pisces Publications, Newbury.
100 Knipe, P.R. (1988) *Gentianella ciliata* in Buckinghamshire. *Watsonia, 17*, 94–5.
101 Langeland, K.A. (1996) *Hydrilla verticillata* Royle (Hydrocharitaceae): The perfect aquatic weed. *Castanea, 61*, 292–304.
102 Le Sueur, Frances (1984) *Flora of Jersey*. The Société Jersaise, Jersey.
103 Lees, F. Arnold (1888) *The Flora of West Yorkshire*. Lovell, Reeve & Co., London.
104 Legg, C., Cowie, N. & Sydes, C. (1997) The importance of regeneration studies to the successful management of Scottish rare plants. *Bot. J. Scotl., 49*, 425–32.
105 Lousley, J.E. (1934) *Veronica praecox*. *Rep. Bot. Soc. Exch. Club. 10*, 478–9.
106 Lousley, J.E. (1950) *Wild Flowers of Chalk and Limestone*. Collins New Naturalist Library, London.
107 Lousley, J.E. (1971) *The Flora of the Isles of Scilly*. David & Charles, Newton Abbot.
108 Lousley, J.E. (1976) *Flora of Surrey*. David & Charles, Newton Abbot.
109 Lovatt, C.M. (1983) An early example of rare plant conservation from the Avon Gorge. *BSBI News*, March 1983.
110 Lusby, P. (1996) Practical conservation of *Lychnis viscaria* in Scotland. *Bot. J. Scotl., 48*, 167–75.
111 Lusby, Philip & Wright, Jenny (1996) *Scottish Wild Plants. Their history, ecology and conservation*. Stationery Office, Edinburgh.
112 Maber, R. & Tregoning, A. (1989) *Kilvert's Cornish diary: 19 July – 6 August 1870*. Alison Hodge, Penzance, p. 77.
113 Mabey, Richard (1996) *Flora Britannica*. Sinclair-Stevenson, London.
114 McAllister, H.A. (1996) Plants recently discovered in Scotland. *Bot. J. Scotl. 49*, 267–76.
115 McClintock, David (1966) *Companion to Flowers*. Bell & Sons, London.
116 McClintock, David (1975) *Wild Flowers of Guernsey*. Collins, London.
117 Madge, S. (1994) The status of *Serapias parviflora* in Britain. *Botanical Cornwall, 6*, 51–2.
118 Mansell-Pleydell, J.C. (2nd edn 1895) *Flora of Dorsetshire or a catalogue of plants found in the county of Dorset*. Whittaker, Dorchester.
119 Marren, P.R. (1984) The history of Dickie's Fern in Kincardineshire. *Pteridologist, 1*, 27–32.
120 Marren, P.R. (1988) The past and present distribution of *Stachys germanica* in Britain. *Watsonia, 17*, 59–68.
121 Marren, P.R. (1996) Back from the Brink in Britain. *Plant Talk, 4*, January 1996, 24–5.
122 Marren, P.R., Payne, A.G. & Randall, R.E. (1986) The past and present status of *Cicerbita alpina* in Britain. *Watsonia, 16*, 131–42.
123 Marshall, J.K. (1967) Biological Flora: *Corynephorus canescens*. *J. Ecol., 55*, 207–20.
124 Meredith, T.C. & Grubb, P.J. (1993) Biological Flora: *Peucedanum palustre*. *J. Ecol., 81*, 813–26.
125 Mitchell, J. (1980) Historical notes on *Woodsia ilvensis* in the Moffat Hills, southern Scotland. *Fern Gazette, 12*, 65–8.
126 National Trust for Scotland (1964) *Ben Lawers and its Alpine Flowers*. NTS, Edinburgh.

127 Nelson, E.C. (1994) Historical data from specimens in the Herbarium, National Botanic Gardens, Glasnevin, Dublin, especially on *Cypripedium calceolus*. *BSBI News*, September 1994.

128 Newman, Edward (2nd edn 1854) *History of British Ferns*. Van Voorst, London.

129 Noltie, H.J. (ed. 1986) *The Long Tradition. The Botanical Exploration of the British Isles*. The Scottish Naturalist, Kilbarchan.

130 Page, C. (1988) *Ferns. Their habitats in the British and Irish landscape*. Collins New Naturalist, London.

131 Page, C. (2nd edn 1997) *The Ferns of Britain and Ireland*. Cambridge University Press, Cambridge.

132 Page, S.E. & Risley, J.O. (1985) The ecology and distribution of *Carex chordorrhiza*. *Watsonia, 15*, 253–9.

133 Palmer, J.R. (1982) *Cerastium brachypetalum* Pers. looks native in W. Kent (VC16). *BSBI News, 30*.

134 Palmer, J.R. (1994) Dittander near old hospitals. *BSBI News, 59*.

135 Parker, D.M. (1979) *Saxifraga cespitosa* – in North Wales. *BSBI News*, No. *21*, 22.

136 Parsons, Mark (1986) *Insects Associated Solely with RDB and Notable Plants*. Invertebrate Site Register Rpt No. 63. NCC, Peterborough.

137 Partridge, Frances (1985) *Everything to Lose. Diaries 1945–1960*. Little Brown & Co.

138 Pearman, D. (1997) Presidential Address, 1996: Towards a new definition of rare and scarce plants. *Watsonia, 21*, 225–45.

139 Perring, F.H. (1962) *Bromus interruptus*. A botanical dodo? *Nature in Cambridgeshire, 5*, 28–30.

140 Perring, F.H. (1974) The Last Seventy Years. In: *The Flora of a Changing Britain*, ed. F. H. Perring, 128–35. BSBI, London.

141 Perring, F.H. (1996) A bridge too far – the non-Irish element in the British flora. *Watsonia, 21*, 15–51.

142 Perring, F.H. & Farrell, L. (eds, 1977) *British Red Data Books: 1. Vascular Plants*. Society for the Promotion of Nature Conservation, Lincoln.

143 Perring, F.H. & Sell, P.D. (eds, 1968) *Critical Supplement to the Atlas of the British Flora*. Thomas Nelson & Sons Ltd, London.

144 Perring, F.H. & Walters, S.M. (eds, 1962) *Atlas of the British Flora*. BSBI and Thomas Nelson & Sons Ltd, London.

145 Phillips, Roger (1983) *Wild Food*. Pan Books.

146 Pigott, C.D. (1958) Biological Flora: *Polemonium caeruleum*. *J. Ecol., 46*, 507–25.

147 Pratt, Anne (1855) *The Flowering Plants and Ferns of Great Britain*. Society for the Propagation of Christian Knowledge, London.

148 Preston, C.D. & Croft, J.M. (1997) *Aquatic Plants in Britain and Ireland*. Harley Books, Colchester.

149 Preston, C.D. & Hill, M.O. (1997) The geographical relationships of British and Irish vascular plants. *Bot. J. Linn. Soc., 124*, 1–120.

150 Proctor, J. & Johnston, W.R. (1979) *Lychnis alpina* in Britain. *Watsonia, 11*, 199–204.

151 Proctor, M.C.F. & Groenhof, A.C. (1987) Peroxidase isoenzyme and morphological variation in *Sorbus* L. in South Wales and adjacent areas, with particular reference to *S. porrigentiformis*. *Watsonia, 19*, 21–37.

152 Pyne, Kevin (1997) *Mespilus germanica* in southern Britain. *BSBI News* No. 75, 49–50.

153 Qamaruz-Zaman, F. & Fay, M.E. (1997) Genetic fingerprinting of *Orchis simia* at Hartslock: a preliminary report. Unpublished report, English Nature, Peterborough.

154 Rackham, Oliver (1980) *Ancient Woodland: its history, vegetation and uses in England*. Edward Arnold, London.

155 Randall, R.E. & Thornton, G. (1996) Biological Flora: *Peucedanum officinale*. *J. Ecol., 84*, 475–85.

156 Ratcliffe, D.A. (1959) Biological Flora: *Hornungia patraea*. *J. Ecol, 47*, 241–7.

157 Raven, C.E. (1942) *John Ray, Naturalist – his life and works*. Cambridge University Press, Cambridge.

158 Raven, John & Walters, Max (1956) *Mountain Flowers*. Collins New Naturalist Library, London.

159 Rendell, S. & Rendell, J. (1993) *Steep Holm. The story of a small island*. Alan Sutton, Dover.

160 Rich, T. (1997) Wildlife Reports. Flowering Plants. *British Wildlife, 9*, 124–5.

161 Rich, T. & King, M. (1994) Practical conservation for wild plants. *Plantlife magazine*, Autumn 1994, 6–7.

162 Rich, T.C.G. (1991) *Crucifers of Great Britain and Ireland*. BSBI Handbook No. 6. BSBI, London.

163 Rich, T.C.G. & Jermy, C. (1998) *Plant Crib 1998*. BSBI, London.

164 Rich, T. & others (1996) *Flora of Ashdown Forest*. Privately published, Sussex Botanical Recording Society.

165 Richards, A.J. & Porter, A.F. (1982) On the identity of a Northumberland *Epipactis*. *Watsonia, 14*, 121–8.

166 Rickard, M.H. 1972. The distribution of *Woodsia ilvensis* and *W. alpina* in Britain. *Br. Fern Gaz, 10(5)*, 269–80.

167 Riddelsdell, H.J., Hedley, G.W. & Price, W.R. (1948) *Flora of Gloucestershire*. Chalford House Press, Bristol.

168 Rix, E.M. & Woods, R.G. (1981) *Gagea bohemica* (Zauschner) J.A. & J.H. Schultes in the British Isles, and a general review of the *G. bohemica* species complex. *Watsonia, 13*, 265–70.

169 Rodwell, J.S. (ed. 1991–95) *British plant communities, Volumes 1–4*. Cambridge University Press, Cambridge.

170 Roger, J.G. (1986) George Don 1764–1814. In: *The Long Tradition*, ed. H.J. Noltie. BSBI Conference Report No. 20, 97–108.

171 Rothschild, Miriam & Marren, Peter (1997) *Rothschild's Reserves. Time and fragile nature*. Harley Books, Colchester.

172 Rowell, T.A. (1997) in Laurie Friday (ed.) *Wicken Fen: The making of a wetland nature reserve*. Harley Books, Colchester.

173 Rumsey, F.J., Jermy, A.C. & Sheffield, E. (1998)

The independent gametophytic stage of *Trichomanes speciosum*, the Killarney Fern, and its distribution in the British Isles. *Watsonia*, *22*, 1–19.
174 Salisbury, E. J. (1961) *Weeds and Aliens*. Collins New Naturalist Library, London.
175 Scott, Michael (1996) Rescue mission for rarities. *Plantlife Magazine*, Autumn 1996, 8–9.
176 Scully, Reginald W. (1916) *Flora of County Kerry*. Hodges, Figgis & Co., Dublin.
177 Sell, Peter & Murrell, Gina (1996) *Flora of Great Britain and Ireland. Volume 5. Butomaceae – Orchidaceae*. Cambridge University Press, Cambridge.
178 Sheail, John (1976) *Nature in Trust*. Blackie, Glasgow and London.
179 Simpson, Francis W. (1982) *Simpson's Flora of Suffolk*. Suffolk Naturalists' Society, Ipswich.
180 Stace, Clive (2nd edn 1997) *New Flora of the British Isles*. Cambridge University Press, Cambridge.
181 Stamp, L.D. (1969) *Nature Conservation in Britain*. Collins, London.
182 Stearn, W.T. (1986) John Ray's natural history travels in Britain. In: *The Long Tradition*, ed. H.J. Noltie. BSBI Conference Report No. 20, 43–58.
183 Step, Edward (1908) *Wayside and Woodland Ferns*. Frederick Warne & Co, London.
184 Stewart, A., Pearman, D.A. & Preston, C.D. (1994) *Scarce Plants in Britain*. JNCC, Peterborough.
185 Stokoe, W.J. (1937) *The Observer's Book of Wild Flowers*. Frederick Warne & Co, London.
186 Sturt, N. (1994) Delving into Dittander. *BSBI News*, No. *58*, p. 23.
187 Summerhayes, V.S. (2nd edn 1968) *Wild Orchids of Britain*. Collins New Naturalist Library, London.
188 Sutherland, W. (1858) The Ferns of Aberdeen and Kincardine. *Phytologist*, *2*, 333–7.
189 Taylor, M.B. (1996) *Wildlife Crime. A guide to wildlife law enforcement in the United Kingdom*. HMSO, London.
190 Trist, P.J.O. (Ed. 1979) *An Ecological Flora of Breckland*. EP Publishing Ltd, Wakefield.
191 Trist, P. J. O. (1981) The survival of *Alopecurus bulbosus* Gouan in former sea-flooded marshes in East Suffolk. *Watsonia*, *13*, 313–6.
192 Trist, P.J.O. (1983) The past and present status of *Gastridium ventricosum* as an arable colonist in Britain. *Watsonia*, *14*, 257–1.
193 Trist, P.J.O. (1986) The distribution, ecology, history and status of *Gastridium ventricosum* in the British Isles. *Watsonia*, *16*, 43–54.
194 Turner Ettlinger, D.M. (1997) *Notes on British and Irish Orchids*. Privately published. Dorking.
195 Ubsdell, R.A.E. (1976) Studies on variation and evolution in *Centaurium erythraea* and *C. littorale* in the British Isles. 1. Taxonomy and biometrical studies. *Watsonia*, *11*, 7–31.
196 Verey, David (ed. 1978) *The Diary of a Cotswold Parson by the Rev. F. E. Witts, 1783–1854*. Alan Sutton, Dover.
197 Vickery, Roy (1995) *A Dictionary of Plant-lore*. Oxford University Press, Oxford.
198 Walters, S.M. (1997) Botanical records and floristic studies. In: *Wicken Fen. The making of a wetland nature reserve*, ed. Laurie Friday, 101–22. Harley Books, Colchester.
199 Webb, D.A. (1950) Biological Flora: Mossy Saxifrages (*S. hypnoides*, *S. rosacea*, *S. caespitosa*). *J. Ecol. 38*, 185–213.
200 Webb, D.A. (1986) What are the criteria for presuming native status? *Watsonia*, *16*, 231–6.
201 Wheeler, B.D., Brookes, B.S. & Smith, R.A.H. (1983) An ecological study of *Schoenus ferrugineus* in Scotland. *Watsonia*, *14*, 249–56.
202 White, F. Buchanan (1898) *The Flora of Perthshire*. Perthshire Society of Natural Science, Edinburgh.
203 White, J.W. (1912) *The Bristol Flora*. John Wright & Sons, Bristol.
204 Wild Flower Society (1986) *Wild Flower Magazine* No. 407 (centenary issue), Autumn 1986.
205 Wilson, G.B., Wright, J. and others (1995) Biological Flora: *Lychnis viscaria*. *J. Ecol.*, *83*, 1039–51.
206 Wilson, P.J. (1992) Britain's arable weeds. *British Wildlife*, *3*, 149–61.
207 Wilson, P.J. (1990) The ecology and conservation of rare arable weeds species and communities. Unpublished PhD thesis, University of Southampton.
208 Wilson, P. & Sotherton, N. (1994) *Field Guide to Rare Arable Flowers*. Game Conservancy Ltd, Fordingbridge.
209 Wolley-Dod, A.H. (1937) *The Flora of Sussex*. Saville, Hastings.
210 Woodhead, N. (1951) Biological Flora: *Lloydia serotina*. *J. Ecol. 39*, 198–203.
211 Woods, R. (1993) *Flora of Radnorshire*. National Museum of Wales, Cardiff.
212 Wright, J. (1997) An ecological basis for the conservation management of *Polygonatum verticillatum*. *Bot. J. Scotl. 49*, 489–500.
213 Young, Andrew (1945) *A Prospect of Flowers*. Jonathan Cape, London.
214 Young, Andrew (1950) *A Retrospect of Flowers*. Jonathan Cape, London.

Index

English names are in Roman letters, Latin names in italics. Underlined page numbers mean an illustration.

ISBN 0-85661-114-X
9 780856 611148